Molecular Aspects of Picornavirus Infection and Detection

Molecular Aspects of Picornavirus Infection and Detection

Editors

BERT L. SEMLER

Department of Microbiology and Molecular Genetics
University of California College of Medicine
Irvine, California 92717

ELLIE EHRENFELD

Department of Biochemistry and
Department of Cellular, Viral, and Molecular Biology
University of Utah School of Medicine
Salt Lake City, Utah 84132

American Society for Microbiology
Washington, D.C.

Copyright © 1989 American Society for Microbiology
1913 I Street, N.W.
Washington, DC 20006

Library of Congress Cataloging-in-Publication Data

Molecular aspects of picornavirus infection and
  detection.

  Based on presentations made at the 1988 ICN-UCI
International Conference on Virology, held in Newport
Beach, Calif.
  Includes index.
  1. Picornavirus infections—Pathogenesis—Congresses.
2. Picornavirus infections—Diagnosis—Congresses.
3. Poliomyelitis—Pathogenesis—Congresses.   4. Molecular
microbiology—Congresses.   I. Semler, Bert L.
II. Ehrenfeld, Ellie.   III. ICN-UCI International
Conference on Virology (1988 : Newport Beach, Calif.)
[DNLM: 1. Molecular Biology—congresses.   2. Picor-
naviridae—genetics—congresses.   3. Picornavirus
Infections—diagnosis—congresses.   QW 168.5.P4 M718 1988]
QR201.P45M65   1989   616.9'25   88-34232

ISBN 1-55581-009-8

Cover photo kindly provided by David J. Filman,
Department of Molecular Biology, Research Institute
of Scripps Clinic, La Jolla, Calif.

Dedication

The editors would like to dedicate this volume to
Vadim I. Agol,
whose scientific contributions continue to be
a beacon for picornavirus research

CONTENTS

III. GENETIC DETERMINANTS OF VIRAL DISEASE AND APPLICATIONS TO DIAGNOSIS

CONTRIBUTORS[†]

Gordon Abraham [169] • Department of Virus and Cell Biology, Merck Sharp & Dohme Research Laboratories, West Point, PA 19486

J. W. Almond [307] • Department of Microbiology, Reading University, Reading RG1 5AQ, United Kingdom

Raul Andino [107] • Whitehead Institute for Biomedical Research, Nine Cambridge Center, Cambridge, MA 02142

Mineo Arita [297] • Department of Enteroviruses, National Institute of Health, Mushashimurayama, Tokyo 190-12, Japan

David Baltimore [107] • Whitehead Institute for Biomedical Research, Nine Cambridge Center, Cambridge, MA 02142, and Department of Biology, Massachusetts Institute of Technology, Cambridge, MA 02139

Jonathan Bradley [3] • Department of Microbiology, State University of New York at Stony Brook, Stony Brook, NY 11794

F. Brown [179] • Department of Virology, Wellcome Biotech Ltd., Langley Court, Beckenham, Kent BR3 3BS, United Kingdom

Marie Chow [125] • Department of Biology, Massachusetts Institute of Technology, Cambridge, MA 02142

Jeffrey I. Cohen [27] • Division of Infectious Diseases, Brigham and Women's Hospitals, Boston, MA 02115

Richard J. Colonno [169] • Department of Virus and Cell Biology, Merck Sharp & Dohme Research Laboratories, West Point, PA 19486

Therese Couderc [265] • Unité de Virologie Médicale, Institut Pasteur, 75724 Paris Cedex 15, France

Radu Crainic [265] • Unité de Virologie Médicale, Institut Pasteur, 75724 Paris Cedex 15, France

Patricia Gillis Dewalt [73] • Department of Virus and Cell Biology, Merck Sharp & Dohme Research Laboratories, West Point, PA 19486

G. Dunn [307] • National Institute for Biological Standards and Control, Blanche Lane, South Mimm, Potters Bar, Hertfordshire, EN6 3QG, United Kingdom

Ellie Ehrenfeld [95] • Department of Biochemistry and Department of Cellular, Viral, and Molecular Biology, University of Utah School of Medicine, Salt Lake City, UT 84132

Stephen M. Feinstone [27] • Laboratory of Infectious Diseases, National Institute of Allergy and Infectious Diseases, Bethesda, MD 20892

† Brackets indicate first page number of corresponding chapter.

David J. Filman [125] • Department of Molecular Biology, Research Institute of Scripps Clinic, 10666 North Torrey Pines Road, La Jolla, CA 92037

Marc Girard [265] • Unité de Virologie Moléculaire, UA 545, Centre National de la Recherche Scientifique, Institut Pasteur, 75724 Paris Cedex 15, France

James M. Hogle [125] • Department of Molecular Biology, Research Institute of Scripps Clinic, 10666 North Torrey Pines Road, La Jolla, CA 92037

Richard J. Jackson [51] • Department of Biochemistry, University of Cambridge, Tennis Court Road, Cambridge CB2 1QW, United Kingdom

Robert W. Jansen [27] • Department of Medicine and Department of Microbiology and Immunology, University of North Carolina, Chapel Hill, NC 27514

A. John [307] • National Institute for Biological Standards and Control, Blanche Lane, South Mimm, Potters Bar, Hertfordshire EN6 3QG, United Kingdom

Noriyuki Kawamura [297] • Department of Microbiology, Tokyo Metropolitan Institute of Medical Science, Honkomagome, Bunkyo-ku, Tokyo 113, Japan

Michinori Kohara [297] • Japan Poliomyelitis Research Institute, Higashi-murayama, Tokyo 189, Japan

Richard J. Kuhn [3] • Division of Biology, California Institute of Technology, Pasadena, CA 91125

Nicola La Monica [281] • Department of Microbiology, Columbia University College of Physicians and Surgeons, New York, NY 10032

Donna M. Leippe [155] • Institute for Molecular Virology and Department of Biochemistry, University of Wisconsin, Madison, WI 53706

Stanley M. Lemon [27, 193] • Department of Medicine and Department of Microbiology and Immunology, University of North Carolina, Chapel Hill, NC 27599-7030

Annette Martin [265] • Unité de Virologie Moléculaire, UA 545, Centre National de la Recherche Scientifique, Institut Pasteur, 75724 Paris Cedex 15, France

Philip D. Minor [125, 307] • National Institute for Biological Standards and Control, Blanche Lane, South Mimm, Potters Bar, Hertfordshire EN6 3QG, United Kingdom

Caroline Mirzayan [3] • Department of Microbiology, State University of New York at Stony Brook, Stony Brook, NY 11794

Eric G. Moss [281] • Department of Microbiology, Columbia University College of Physicians and Surgeons, New York, NY 10032

Anne G. Mosser [155] • Institute for Molecular Virology and Department of Biochemistry, University of Wisconsin, Madison, WI 53706

Michael G. Murray [3] • Department of Microbiology, State University of New York at Stony Brook, Stony Brook, NY 11794

Akio Nomoto [297] • Department of Microbiology, Tokyo Metropolitan Institute of Medical Science, Honkomagome, Bunkyo-ku, Tokyo 113, Japan

Robert O'Neill [281] • Department of Microbiology, Columbia University College of Physicians and Surgeons, New York, NY 10032

Ann C. Palmenberg [211] • Institute for Molecular Virology and Department of Veterinary Science, University of Wisconsin, Madison, WI 53706

A. Phillips [307] • National Institute for Biological Standards and Control, Blanche Lane, South Mimm, Potters Bar, Hertfordshire EN6 3QG, United Kingdom

Li-Hua Ping [193] • Division of Infectious Diseases, Department of Medicine, The University of North Carolina at Chapel Hill, Chapel Hill, NC 27599-7030

Robert H. Purcell [27] • Laboratory of Infectious Diseases, National Institute of Allergy and Infectious Diseases, Bethesda, MD 20892

Vincent R. Racaniello [281] • Department of Microbiology, Columbia University College of Physicians and Surgeons, New York, NY 10032

Oliver C. Richards [95] • Department of Biochemistry and Department of Cellular, Viral, and Molecular Biology, University of Utah School of Medicine, Salt Lake City, UT 84132

Michael G. Rossmann [139] • Department of Biological Sciences, Purdue University, West Lafayette, IN 47907

Harley A. Rotbart [243] • Department of Pediatrics and Department of Microbiology/Immunology, University of Colorado School of Medicine, 4200 East 9th Avenue, Box C227, Denver, CO 80262

Roland R. Rueckert [155] • Institute for Molecular Virology and Department of Biochemistry, University of Wisconsin, Madison, WI 53706

Bert L. Semler [73] • Department of Microbiology and Molecular Genetics, California College of Medicine, University of California, Irvine, CA 92717

Rashid Syed [125] • Department of Molecular Biology, Research Institute of Scripps Clinic, 10666 North Torrey Pines Road, La Jolla, CA 92037

Hiroomi Tada [3] • Department of Biochemistry, Jefferson Institute of Molecular Medicine, Thomas Jefferson University, Philadelphia, PA 19107

John Ticehurst [27] • Department of Virus Diseases, Walter Reed Army Institute of Research, Washington, DC 20307-5100

Joanne E. Tomassini [169] • Department of Virus and Cell Biology, Merck Sharp & Dohme Research Laboratories, West Point, PA 19486

Didier Trono [107] • Whitehead Institute for Biomedical Research, Nine Cambridge Center, Cambridge, MA 02142

K. Wareham [307] • Department of Microbiology, Reading University, Reading RG1 5AQ, United Kingdom

G. D. Westrop [307] • Department of Microbiology, Reading University, Reading RG1 5AQ, United Kingdom

Eckard Wimmer [3] • Department of Microbiology, State University of New York at Stony Brook, Stony Brook, NY 11794

Czeslaw Wychowski [265] • Unité de Virologie Moléculaire, UA 545, Centre National de la Recherche Scientifique, Institut Pasteur, 75724 Paris Cedex 15, France

Xiao-Feng Yang [3] • Department of Microbiology, State University of New York at Stony Brook, Stony Brook, NY 11794

PREFACE

In 1984, Gebhard and Friedrich Koch completed a scholarly review of the vast scientific literature, accumulated primarily during the preceding 25 years, that described our knowledge of *The Molecular Biology of Poliovirus* (Springer Verlag, Vienna, 1985). At the time of their writing, a new vision was beginning to appear on the horizon, in which poliovirus (the prototype member of the picornavirus group) could be studied and understood at the level of individual nucleotides and amino acids, rather than just genes and proteins. The vision was based on new methodologies in molecular biology and structural analyses that had begun several years earlier and that showed much promise but, in 1984, few results. Infectious cDNA clones of poliovirus had been constructed, but had not yet been used as a genetic system. Development of theory and methods to resolve the atomic structure of picornavirus particles was well under way, but not yet complete.

By January, 1988, at the convening of the 1988 ICN-UCI International Conference on Virology entitled "Molecular Aspects of Picornavirus Infection and Detection," the new methodologies had truly begun to produce payoffs. The conference included 18 speakers from the United States, Europe, and Japan (whose presentations comprise the chapters of this book) and an enthusiastic audience of 100 scientists. The aim of the conference was to highlight the most recent developments in research on this group of positive-strand RNA viruses that is responsible for a large number of human and animal diseases. Work on poliovirus still predominated, but rapid progress into important areas unique to other picornaviruses, such as hepatitis A virus and human rhinoviruses, was reported.

Many of the experimental approaches described in this volume take advantage of the extensive body of nucleotide sequence information that has been accumulated over the past 8 years for large numbers of picornavirus genomic RNAs. The genetic manipulation of cDNA clones of these viral genomes has led to mutants and recombinant viruses that have revealed functions and specificities that were previously not identified nor understood. Various chapters of this book describe how determinants of virulence have been mapped to 5' noncoding regions of the genome, as well as to specific capsid protein sequences. Type-specific antigenicity can now be readily engineered by insertion or exchange of defined protein sequences among different viruses. The structural requirements of both processing enzymes and precursor substrates for generating functional viral proteins are being elucidated. Molecular approaches are being developed to allow definition of the functions and mechanisms of action of VPg, replicase proteins, and proteinases, as well as the regulation of translational activity and its significance for biological phenotypes. In addition, the use of sophisticated structural determinations, coupled with virion surface binding properties and the development of monoclonal antibody panels, has provided a functional map of viral amino acid sequences involved in attachment to cells, antibodies, and antiviral compounds.

We believe that this volume presents the dynamic and far-reaching status of current research into the molecular, genetic, and biochemical aspects of picornavirus infections.

Bert L. Semler
Department of Microbiology and Molecular Genetics
College of Medicine
University of California, Irvine
Irvine, California 92717

Ellie Ehrenfeld
Department of Cellular, Viral and Molecular Biology
University of Utah Medical School
Salt Lake City, Utah 84132

ACKNOWLEDGMENTS

The editors would like to thank ICN Pharmaceuticals, Inc. for support of the 1988 ICN-UCI Conference on Virology. We are indebted to Hung Fan and Edward Wagner for their guidance in establishing a scientific framework for the virology topics presented in this volume. Finally, we thank Nita Driscoll of the Cancer Research Institute of the University of California, Irvine, for her administrative skills which reduced the conference and this volume to manageable endeavors.

Part I

MOLECULAR BIOLOGY OF VIRAL REPLICATION

Molecular Aspects of Picornavirus Infection and Detection
Edited by Bert L. Semler and Ellie Ehrenfeld
© 1989 American Society for Microbiology, Washington, DC 20006

Chapter 1

Use of Mutagenesis Cartridges in Molecular Genetic Analyses of Poliovirus: Mutations in the Genome-Linked Protein VPg and in Neutralization Antigenic Site I

Jonathan Bradley, Michael G. Murray, Richard J. Kuhn, Hiroomi Tada, Xiao-Feng Yang, Caroline Mirzayan, and Eckard Wimmer

The molecular events occurring during proliferation of picornaviruses are far from being understood. Despite the fact that the picornavirus genome is small and specifies a simple family of proteins, not a single step in the replicative cycle (which includes adsorption, penetration, uncoating, initiation of protein synthesis, protein processing, genome replication, control of host cell macromolecular synthesis, and assembly and release of the virion) has been elucidated. Three decades of research remains to be comprehended in the context of the cascade of events that results in the several thousand-fold multiplication of the infecting virus particle.

The chemical structure of poliovirus has been solved (13, 26, 34, 55,

Jonathan Bradley, Michael G. Murray, Xiao-Feng Yang, Caroline Mirzayan, and Eckard Wimmer • Department of Microbiology, State University of New York at Stony Brook, Stony Brook, New York 11794. Richard J. Kuhn • Division of Biology, California Institute of Technology, Pasadena, California 91125. Hiroomi Tada • Department of Biochemistry, Jefferson Institute of Molecular Medicine, Thomas Jefferson University, Philadelphia, Pennsylvania 19107.

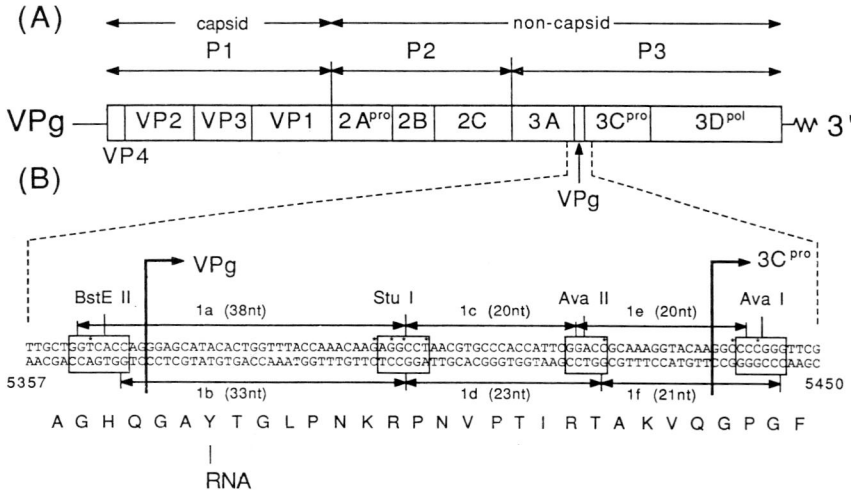

Figure 1. Location and structure of the mutagenesis cartridge for VPg. (A) The genome of poliovirus, shown schematically, is covalently linked to VPg at the 5' end. The 3' end is polyadenylylated. The coding region for the polyprotein is represented by an open box that has been divided into the capsid (P1) and noncapsid (P2 and P3) regions. The divisions do not precisely correspond to relative sizes of the end products of proteolytic processing. Polypeptides 2A^{pro} and 3C^{pro} are proteinases; 3D^{pol} is the RNA-dependent RNA polymerase. 3AB (3B = VPg) is thought to be a precursor for VPg. 3CD, a precursor for 3C^{pro} and 3D^{pol}, has recently been found to be involved in P1 processing in vitro (see text). (B) Structure of the VPg mutagenesis cartridge. The nucleotide sequence shown corresponds to the cDNA encoding nucleotides 5357 through 5450 of the PV-1(M) RNA genome. Nucleotide differences between the wild-type sequence and the mutagenesis cartridge are denoted by asterisks above the appropriate nucleotide. The nucleotides enclosed within a box represent a newly created restriction site. The amino acid sequence is given underneath the nucleotide sequence in the one-letter amino acid code. The 5' end of the RNA is covalently linked to a tyrosine residue as indicated. The oligonucleotides used to construct the cartridge are symbolized by the line with the double arrow along with their appropriate designation (1a, 1b, 1c, 1d, 1e, 1f), and their lengths are given in parentheses. Reproduced from reference 30.

63). All known gene products of poliovirus have been mapped to the polyprotein encoded by the viral genome (13, 24, 26, 34, 51, 61, 63; see Fig. 1A). Thus, we know precisely the linear spatial relationship and amino acid sequence of these component proteins. In addition, we have learned much about the mechanisms by which the viral polypeptides are released from the polyprotein (P. G. Dewalt and B. L. Semler, this volume; see also references 28 and 45), although the functions of many of the gene products are either totally obscure (e.g., 2B, 2C, 3A, VPg) or poorly understood (2A^{pro}, 3C^{pro}, 3D^{pol}).

Crystallographic analyses of the rhinovirion (type 14) (57), the poliovirion (poliovirus type 1 Mahoney [PV-1(M)]) (21), and the mengovirion (36) have resulted in high-resolution models of the capsid proteins (VP1, VP2, VP3, VP4) in the intact viral particle of picornaviruses. What remain to be modeled in the context of the virion are the genomic RNA and its terminally linked, virus-encoded protein, VPg (see Fig. 4). Thus, the relationship of these two to each other and to the capsid proteins in the mature virion is not known. Moreover, many other questions concerning the virion must still be answered. For example, why is the amino terminus of the polyprotein (and its N-terminal cleavage product, VP4) myristoylated (9, 52)? To what extent can the surface protrusions of the virion that elicit and interact with neutralizing antibodies be modified without losing viability of the virion? What is the biological consequence of a small change in the exterior of the capsid with regard to tissue tropism and pathogenicity? What is the nature of the cellular receptor for poliovirus and how does it interact with the virion particle?

Molecular studies of poliovirus have been greatly aided by the engineering of infectious cDNA clones (56, 62) and the construction of transcription vectors from which infectious viral RNA can be synthesized in vitro in virtually unlimited quantities (68). A large number of genetic variants have been generated by site-directed mutagenesis (point mutations, insertions, deletions) and by allele replacements of large segments of the type 1 poliovirus genome with those of other poliovirus strains or types (27) or even of other enterovirus species (64). Their phenotypes are interesting, but have more often than not proved difficult to interpret (see reference 73 for references; see also other chapters in this volume).

We have focused our attention on two small segments of the poliovirus genome that code for amino acid sequences (~20 amino acids) of special biological interest. They are the genome-linked protein, VPg, and the neutralization antigenic site I (N-AgI) of the virion particle. We have developed a strategy (mutagenesis via an oligonucleotide cartridge) which allows us to generate single or multiple amino acid replacements in VPg or N-AgI.

THE MUTAGENESIS CARTRIDGE

Unique restriction sites relatively close to each other (30 to 80 base pairs apart in infectious cDNA of a picornavirus genome) can be conveniently used to excise a DNA segment of interest. The excised cDNA segment can subsequently be replaced by synthetic double-stranded cDNA (a DNA "cartridge") harboring any number of base changes. The cartridge-containing segment can then be engineered back into the full-length cDNA clone. Upon transcription of the modified cDNA clone, mutant viral RNA is produced that can be assayed for new phenotypes in vitro and in vivo. This strategy of mutagenesis is highly specific and approaches absolute efficiency. If possible, cartridge mutagenesis may utilize naturally occurring unique restriction sites fortuitously found in a cDNA segment that codes for a polypeptide of interest. Such is the case in the 5'-terminal half of cDNA corresponding to the coding region for proteinase 3C of PV-1(M). Accordingly, Dewalt and Semler (11) have used this segment to generate mutations in 3C^{pro}.

No suitable restriction sites occur in the cDNA segments surrounding the coding regions of VPg and N-AgI. We therefore created new restriction sites flanking these regions by oligonucleotide mismatch mutagenesis (78) such that the amino acid coding sequence was not disturbed (29, 43; see Fig. 1B and 7B). The cartridge is relatively long (85 base pairs for VPg; Fig. 1B) so that sets of synthetic oligonucleotides could be used that generated new restriction sites within the cartridge on ligation. This facilitated the convenient exchange of complementary pairs of oligonucleotides, as well as restriction enzyme analysis to ascertain that the ligations occurred properly. The oligonucleotides for the VPg mutagenesis cartridge were joined in an order that permitted only one end product in the correct orientation (Fig. 2; R. J. Kuhn, Ph.D. thesis, State University of New York at Stony Brook, Stony Brook, N.Y., 1986). The cartridge for VPg is subsequently ligated into the subclone pNT15 of poliovirus and then engineered into plasmid pT7PV1-5 (Fig. 3), which harbors a phage T7 promoter followed by the full-length poliovirus cDNA (68). A similar strategy was employed to construct a mutagenesis cartridge facilitating the generation of mutants in N-AgI of PV-1(M) (see Fig. 7B; 43).

MUTATIONS GENERATED IN VPg OF PICORNAVIRUSES

The genome-linked protein of poliovirus was discovered in 1977 by Lee et al.; its properties were subsequently described in detail following several investigations (1, 18, 25, 35, 47, 48). VPg is an oligopeptide 22 amino acids in length that maps between polypeptides 3A and 3C (1, 24, 25, 61; Fig. 1A) and is covalently bound to the 5'-terminal phosphate of the viral RNA via a phosphodiester linkage to the O^4-hydroxyl group of a tyrosine residue in position 3 of VPg (2, 58). The structure of the RNA-linked VPg is shown in Fig. 4. The oligopeptide (pI, >10) is rich in basic amino acids and in prolines (1). These positively charged amino acids may facilitate an interaction with the negatively charged phosphates of the genome RNA (for an earlier review, see reference 70). Known

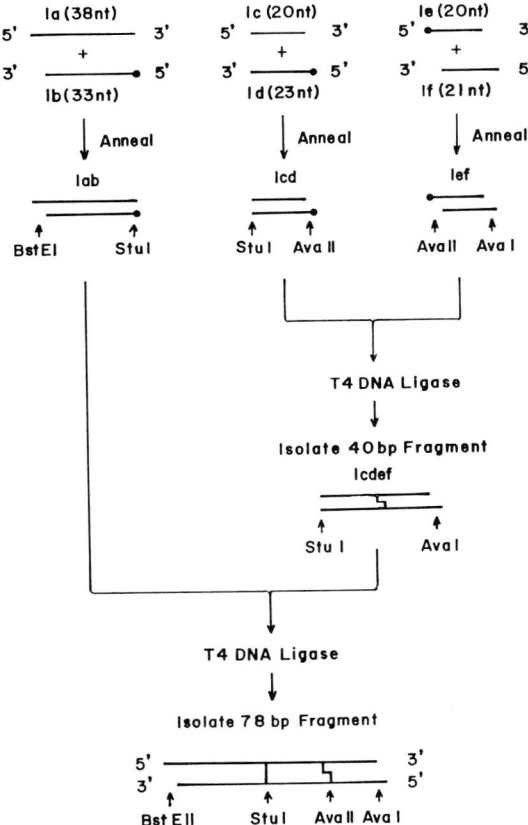

Figure 2. Construction of the synthetic mutagenesis cartridge. Oligonucleotides which were kinased before being annealed to their complementary oligonucleotides are indicated by a solid circle attached to their 5' ends. Kinasing specific oligonucleotides reduced the background of unproductive ligations. Fragments were isolated from acrylamide gels by the crush-and-soak method. The nucleotide and amino acid sequences of the cartridge are shown in Fig. 1. Reproduced from Kuhn (thesis). bp, Base pairs.

amino acid sequences of VPgs of other picornaviruses are shown in Fig. 5. All picornavirus VPgs range between 20 and 24 amino acids in length and are the cleavage products of proteolytic processing at mostly Q-G pairs or E-G pairs. The linkage of the virion RNA to the tyrosine residue at position 3 of all VPgs is invariant. Moreover, all picornavirus VPgs harbor several basic amino acids, although the spacing may differ in VPgs of different genera. Of particular interest is the observation that aphtho-viruses (foot-and-mouth disease viruses) code for three different VPgs

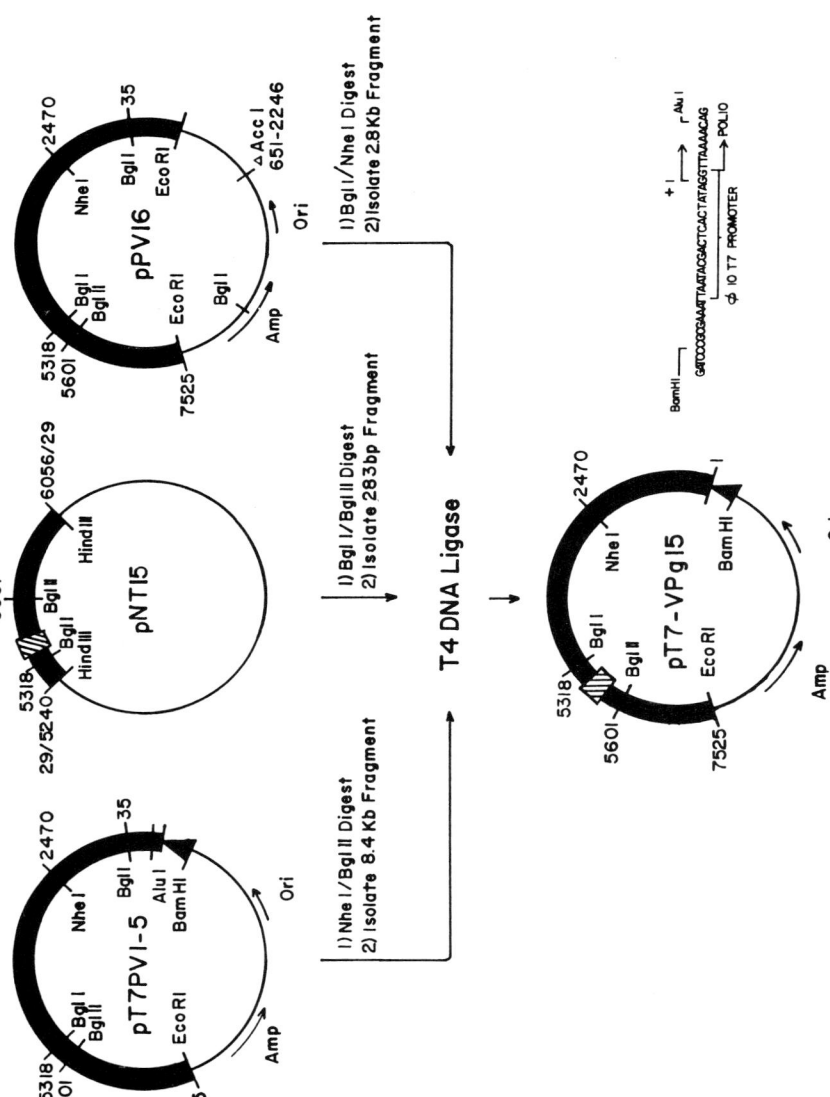

Figure 3. Construction of pT7-VPg15. A *BglI-BglII* fragment of pNT15, containing the synthetic cartridge, was inserted into the full-length infectious clone as outlined here. Subsequent mutants were assembled using the identical strategy. The sequence shown in small type identifies the T7 promoter, the start site for transcription (designated as +1), and the 5' end of the poliovirus sequence. Poliovirus sequences are represented by the heavy line; pBR322 sequences are represented by the thin line; the T7 transcription promoter is indicated by the solid arrow. Reproduced from Kuhn (thesis). Kb, Kilobases; bp, base pairs.

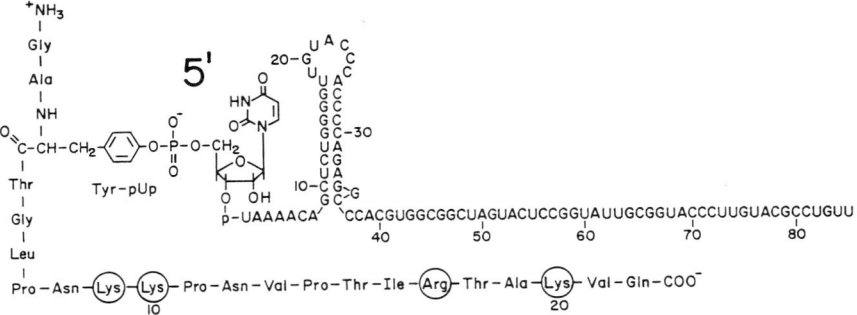

Figure 4. Structure of the 5' terminus of poliovirus RNA. The oligonucleotide ($n = 85$) is a fragment of virion RNA generated by RNase III of *Escherichia coli*. VPg is linked to the RNA via an energy-rich phosphodiester between the O^4-hydroxyl group of tyrosine and a uridylate residue. The position of tyrosine relative to the amino terminus appears to be essential, as determined by site-directed mutagenesis in this region (29). VPg is a cleavage product of the poliovirus polyprotein. Basic amino acids have been circled. Reprinted, with permission, from Kitamura et al. (26).

that map on the genome between 3A and 3C as tandem repeats (20). One copy of any of the three VPgs is found linked to viral RNA in a near-random fashion (23).

The function of VPg is obscure. The protein is almost certainly involved in genome replication since it can be found attached to the nascent strands of the replicative intermediate RNA (18, 47, 48, 54, 70). Two mechanisms have been proposed to explain VPg function: (i) a precursor of VPg (VPg-pU) is uridylylated in a template-dependent reaction with concomitant release of VPg from its membrane anchor, and VPg-pU then serves as a primer for the viral RNA polymerase 3D (47, 60, 65, 66); alternatively, (ii) VPg acts as a nuclease, clipping a hairpin formed during RNA synthesis and thereby linking itself to the RNA (reviewed in reference 19). In addition, VPg could function as a component of the encapsidation signal (47, 48). These topics have recently been extensively reviewed by Kuhn and Wimmer (31).

The mutations generated in VPg via a mutagenesis cartridge are shown in Fig. 6 (29, 30). Initially, we targeted three amino acids for mutagenesis that are conserved in all known human enterovirus VPgs (Fig. 5). These are the tyrosine residue, to which the genomic RNA is covalently bound (Fig. 1A and 4), and the basic amino acids in positions 17 and 20. Other mutations that we made were in the vicinity of the 3C proteinase cleavage sites. In addition, the poliovirus VPg was completely exchanged for that of echovirus 9. The mutant genomic RNAs produced

Bradley et al.

```
Polio 1 Mahoney    G A Y T G • L P N K K P N V P T I R T A K • • • • V Q

Polio 1 Sabin      G A Y T G • L P N K K P N V P T I R T A K • • • • V Q
Polio 2 Lansing    G A Y T G • L P N K R P N V P T I R T A K • • • • V Q
Polio 2 Sabin      G A Y T G • L P N K R P N V P T I R T A K • • • • V Q
Polio 3 Leon       G A Y T G • L P N K R P N V P T I R A A K • • • • V Q
Polio 3 Sabin      G A Y T G • L P N K R P N V P T I R A A K • • • • V Q

Cox B1 b           G A Y T G • M P N Q K P K V P T L R Q A K • • • • V Q
Cox B3             G A Y T G • V P N Q K P R V P T L R Q A K • • • • V Q
Cox B4             G A Y T G • M P N Q K P K V P T L R Q A K • • • • V Q

EV 70              G P Y T G • L P N Q K P K V P T L R T A K • • • • V Q

Echo 9             G A Y T G • M P N K K P K V P T L R Q A K • • • • V Q

Rhino 2            G P Y S G • E P • • K P K T K • I P E R • • R V V T Q
Rhino 9            G P Y S G • E P • • K P K T T R V P E R • • R V V A Q
Rhino 14           G P Y S G N P P H N K L K A P T L R • • • • P V V Q

FMDA 10 a          G P Y S G P L E R Q K P • • L K V R • A K L P • Q Q E
FMDA 12 a          G P Y A G P L E R Q K P • • L K V R • A K L P • Q Q E
FMDO 1 Ka          G P Y A G P L E R Q K P • • L K V R • A K L P • Q Q E
FMDC 10 a          G P Y A G P L E R Q R P • • L K V R • A K L P • Q Q E
FMDA 1 b           G P Y A G P M E R Q K P • • L K V R • V K A P V V K E
FMDA 12 b          G P Y A G P M E R Q K P • • L K V R • A K A P V V K E
FMDO 1 Kb          G P Y A G P M E R Q K P • • L K V R • A K A P V V K E
FMDC 10 b          G P Y A G P M E R Q K P • • L K V R • A R A P V V K E
FMDA 10 c          G P Y E G • • P V K K P V A L K V K • A R N L I V T E
FMDA 12 c          G P Y E G • • P V K K P V A L K V K • A K N L I V T E
FMDO 1 Kc          G P Y E G • • P V K K P V A L K V K • A K N L I V T E

HAV Ch 1           G V Y H G • • • V T K P K Q V I K L D A D • P V E S Q
HAV Hm 175         G V Y H G • • • V T K P K Q V I K L D A D • P V E S Q
HAV MBB            G V Y H G • • • V T K P K Q V I K L D A D • P V E S Q

EMCV               G P Y N E • T A R V K P K T L Q • • • • • • L L D I Q

Mengo.             G P Y N E • T T R I K P K T L Q • • • • • • L L D V Q

TME BeAn           A A Y A G • R A R A Q K Q A L Q • • • • • • V L D I Q
```

Figure 5. Amino acid sequence alignment of the genome-linked protein of picornaviruses. For references, see A. C. Palmenberg, this volume. Abbreviations: Cox, coxsackievirus; EV, enterovirus; FMD, foot-and-mouth disease virus, A, O, and C referring to different serotypes; EMCV, encephalomyocarditis virus; TME, Theiler murine encephalomyelitis virus.

by in vitro transcription were tested for their ability to produce viable virus upon transfection of HeLa cells and to program protein synthesis in a cell-free extract (rabbit reticulocyte lysate). Mutations that yielded viable virus were further characterized for their growth properties at different temperatures and their patterns of protein synthesis in vivo (29, 30).

The mutation of the tyrosine residue in position 3 proved lethal, an observation which supported the importance of this amino acid in VPg function (29). Surprisingly, all mutations of the arginine in position 17 were also lethal, whereas those of lysine in position 20 yielded viable virus, even in the case of K20E (exchange of the lysine 20 to glutamic

```
                                  5        10        15        20          Virus
Polio 1     K L F A G H Q | G A Y T G L P N K K P N V P T I R T A K V Q |  Recovered

pT7-VPg15   - - - - - - -   - - - - - - - - - R - - - - - - - - - - -        +

pT7-VPg21   - - - - - - -   - - - Y - - - - - - - - - - - - - - - - -        -
pT7-VPg16   - - - - - - -   - - T Y - - - - - R - - - - - - - - - - -        -

pT7-VPg22   - - - - - - -   - - - - - M - - - R - - - - - - - - - - -        +
pT7-VPg30   - - - - - - -   - - - - - - - - - R - - M - - - - - - - -        +
pT7-VPg31   - - - - - - -   - - - - - M - - - R - - M - - - - - - - -        +

pT7-VPg28   - - - - - - -   - - - - - - - - - R - K - - - L - Q - - - -      +
pT7-VPg29   - - - - - - -   - - - - - M - - - R - K - - - L - Q - - - -      +

pT7-VPg19   - - - - - - -   - - - - - - - - - R - - - - - - K - - - - -      -
pT7-VPg17   - - - - - - -   - - - - - - - - - R - - - - - - Q - - - - -      -
pT7-VPg18   - - - - - - -   - - - - - - - - - R - - - - - - E - - - - -      -

pT7-VPg27   - - - - - - -   - - - - - - - - - R - - - - - - - - - R - -      +
pT7-VPg23   - - - - - - -   - - - - - - - - - R - - - - - - - - - L - -      +
pT7-VPg24   - - - - - - -   - - - - - - - - - R - - - - - - - - - I - -      +
pT7-VPg25   - - - - - - -   - - - - - - - - - R - - - - - - - - - Q - -      +
pT7-VPg26   - - - - - - -   - - - - - - - - - R - - - - - - - - - E - -      +

pT7-VPg34   - - - - - - -   - - - - - - - - - R - - - - - - - - T - - -      +
pT7-VPg35   - - - - - - -   - - - - - - - - - R - - - - - - - - K A - -      -


pT7-VPg20   - - - - - - -   - - - - - - - - - R - - - - - - - - T A K V Q    +


pT7-VPg37   - - - /\ - - -   - - - - - - - - - R - - - - - - - - - - - -      -
                G H R S T
```

Figure 6. VPg mutants generated by site-directed mutagenesis (VPg21) or by cartridge mutagenesis (all other mutants). For details, see references 29 and 30.

acid; 30). Insertion of a threonine residue in the C-terminal portion yielded virus, whereas an inversion of the A-K residues at positions 19 and 20 to K-A proved lethal. We interpret this to mean that the inversion creates a nonpermissible amino acid (a lysine) in the *P4* position of the Q-G cleavage site (30, 45). Exchanges of amino acids in positions 6, 10, 12, 13, 16, and 18 (see Fig. 6) had little effect on viral growth, and even multiple exchanges were tolerated. Mutants pT7-VPg22, -29, and -30 (with one methionine residue) and pT7-VPg31 (with two methionine residues) are of interest for studies of VPg function as they permit labeling of VPg with [35S]methionine (29). Finally, mutant pT7-VPg29 generated a hybrid virus in which the poliovirus VPg was replaced with that of echovirus 9. This virus was impaired in growth (approximately 1 log less virus relative to wild-type virus in one-step growth kinetics), but it showed no temperature sensitivity phenotype (30).

We also generated three insertion mutations in the carboxy-terminal portion of polypeptide 3A just upstream of VPg (Fig. 1A), one of which, pT7-VPg37, is shown in Fig. 6. In all instances the five-amino-acid insertion proved to be lethal (30). It is possible that these mutations disturb the function of polypeptide 3AB (3B = VPg), a membrane-associated protein thought to be the precursor of VPg (60).

The translation in vitro of each mutant viral RNA generally supported a normal viral phenotype. That is, RNA transcripts that yielded virus also showed wild-type translation and proteolytic processing patterns following sodium dodecyl sulfate-polyacrylamide gel electrophoresis (29, 30). There are notable exceptions, however, where "viable RNA" yielded aberrant protein patterns upon translation in vitro and "nonviable RNA" yielded normal protein patterns. We have cautiously interpreted the latter observation to mean that the mutation in VPg impaired VPg function rather than proteolytic processing. Interestingly, we found that those RNAs that failed to direct the synthesis of 3BCD and 3CD also failed to show proteolytic processing of the capsid precursor P1 (30). We conclude that 3BCD or 3CD is necessary for cleavage of P1 in vitro, an observation confirming previous results by Ypma-Wong and Semler (77). The importance of 3CD in P1 processing has been corroborated recently by results of other studies (44, 76; R. J. Jackson, this volume).

In summary, we have found that VPg allows considerable flexibility in amino acid exchanges without loss of function(s), although some amino acids are highly sensitive to change (tyrosine 3 and arginine 17). We are currently pursuing further allele replacements of VPgs, using amino acid sequences of other human enteroviruses as well as those of viruses of other genera (see Fig. 5). Ultimately, we hope to generate conditional lethal mutants for a direct study of VPg function.

POLIOVIRUS N-AgI

Poliovirus exhibits three independent neutralization antigenic sites (N-AgI, N-AgII, and N-AgIII) that elicit and bind to neutralizing antibodies (38, 71; see also P. D. Minor, G. Dunn, A. John, A. Phillips, G. D. Westrop, K. Wareham, and J. W. Almond, this volume). N-AgI is formed by a continuous sequence of 20 amino acids mapping to residues 89 to 107 of capsid protein VP1 (Fig. 7B and C). Amino acids 95 to 105 of VP1 protrude from the virion surface in the form of a loop (21; see Fig. 7C).

The region encoding N-AgI is very heterologous among the three poliovirus serotypes, a phenomenon first recognized by Toyoda et al. (67). Emini et al. (15, 16) suggested that the intertypic variability in this region indicates a structure specifying a type-specific antigenic site. The first direct evidence that the region of 89–104 of VP1 of poliovirus type 3

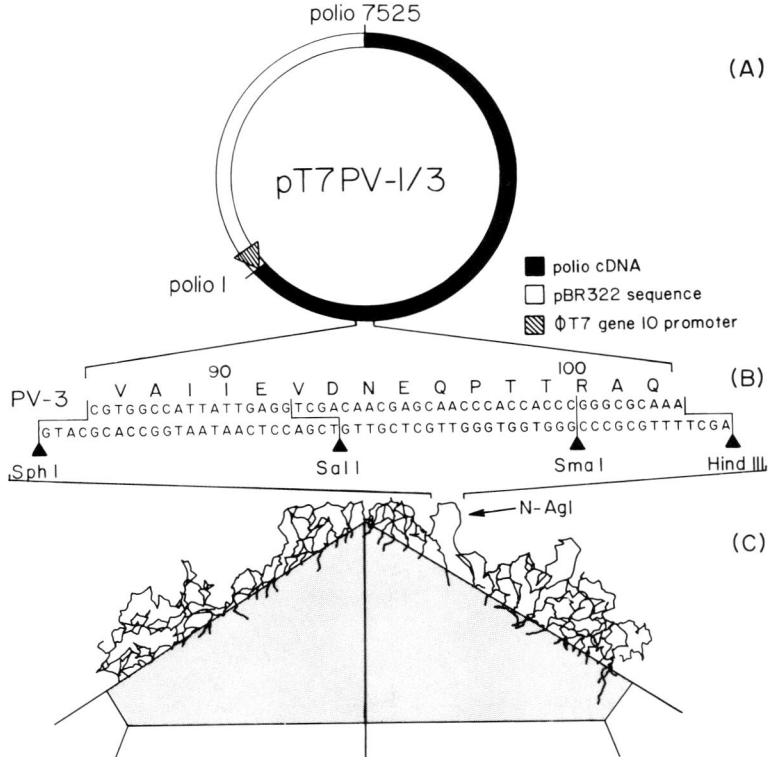

Figure 7. Schematic representation of the strategy to construct a poliovirus type 1/type 3 hybrid virus. (A) Plasmid pT7PV-1/3 is a derivative of transcription vector pT7PV1-5 (68). It contains the phage T7 (φT7) promoter for T7 RNA polymerase and the cDNA sequence of PV-1(M), in which a small DNA segment encoding the N-AgI amino acid sequence was replaced with that of PV-3(L). (B) Sequence corresponding to a region specifying N-AgI of PV-3(L). (C) The loop is located near the apex of the poliovirion (21); redrawn from a figure in reference 21.

(Leon) [PV-3(L)] interacts with neutralizing antibodies was obtained by mapping neutralization escape mutations (17, 39, 40). Moreover, a peptide corresponding to the sequence 93–103 of VP1 of PV-1(M) elicited neutralizing antibodies in rabbits (15, 75). Finally, using a molecular genetic approach, Van der Werf et al. (69) showed that the binding site of a neutralizing monoclonal antibody (N-MAb), C3, to the virion was region 95–105 of VP1.

The extent of amino acid variability in the region 86–108 of VP1 of different polioviruses is shown in Fig. 8. Numerous amino acid exchanges

```
                          90        95        100       105
PV-1 (Mahoney)a   C V T I M T V D N P A S T T N K D K L F A V W
PV-1 (Sabin)      - - A - I - - - - S - - - K - - - - - - T - -
Neutralization  1 - - - - - - - - - - - - - - D - - - - - - - -
escape mutants  2 - - - - - - - - - - - - - - K - - - - - - - -
of PV-1

PV-2 (Lansing)    - - A - I E - - - D - P - K R A S - - - S - -
PV-2 (Sabin)      - - A - I E - - - D - P - K R A S R - - S - -
Neutralization  1 - - - - - - - - - - - - - - - P - - - - - - -
escape mutants  2 - - - - - - - - - - - - - - - - D - - - - - -
of PV-2         3 - - - - - - - - - - - - - - E - - - - - - - -
                4 - - - - - - - - - - - - - - N - - - - - - - -

PV-3 (Leon)       - - A - I E - - - E Q P - - R A Q - - - - M -
PV-3 (Sabin)      - - A - I E - - - E Q P - - R A Q - - - - M -
Neutralization  1 - - - - - - N - - - - - - - - - - - - - - - -
escape mutants  2 - - - - - G - - - - - - - - - - - - - - - - -
of PV-3         3 - - - - - K - - - - - - - - - - - - - - - - -
                4 - - - - - - - - - - - I - - - - - - - - - - -
                5 - - - - - - - - - - - A - - - - - - - - - - -
                6 - - - - - - - - - - - N - - - - - - - - - - -
                7 - - - - - - - - - - - - N - - - - - - - - - -
                8 - - - - - - - - - - - - I - - - - - - - - - -
                9 - - - - - - - - - - - - Q - - - - - - - - - -
               10 - - - - - - - - - - - - W - - - - - - - - - -
               11 - - - - - - - - - - - - - T - - - - - - - - -
               12 - - - - - - - - - - - - - V - - - - - - - - -
               13 - - - - - - - - - - - - - - L - - - - - - - -
               14 - - - - - - - - - - - - - - P - - - - - - - -
               15 - - - - - - - - - - - - - - R - - - - - - - -
               16 - - - - - - - - - - - - - - H - - - - - - - -
```

Figure 8. Amino acid sequences of the region spanning N-AgI of various poliovirus types and strains. Amino acid sequence is given in the single-letter code and is numbered according to the sequence of VP1 of PV-1(M). The numbering may be slightly different in VP1 polypeptides of other serotypes due to deletions or insertions. Data compiled from references 4, 12, 16, 38–40, and 71.

are tolerated without apparent loss of viability. It is likely that the whole region can be exchanged among polioviruses, and even between different picornavirus genera. To this end we constructed a mutagenesis cartridge whose restriction sites flank amino acids 87 to 102 of VP1.

CONSTRUCTION OF A MUTAGENESIS CARTRIDGE FOR N-AgI

Murray et al. (42, 43) used a naturally occurring restriction site (SphI, nucleotide 2732) and a newly created restriction site (HindIII, nucleotide 2786) to generate a cartridge useful in exchanging amino acid residues in the region 87–102 of VP1 (see Fig. 7B and C). As with the cartridge for VPg, synthetic oligonucleotides were chosen such that the region is divided into segments by unique restriction sites. These segments can be easily assembled from short synthetic oligonucleotides and ligated into a subclone. The fragment encoding the amino acid replacements via the cartridge was then engineered back into the full-length poliovirus transcription vector (Fig. 7A). In vitro RNA transcription with phage T7 RNA polymerase produced genomic RNA that was tested for the ability to produce virus upon transfection into HeLa cells.

WHY MODIFY THE N-AgI OF POLIOVIRUS?

Several compelling reasons underlie our objective of generating variants mapping to N-AgI of poliovirus. First, we would like to define precisely those amino acids that determine the serotype of poliovirus at this site. This is particularly interesting in view of the observation that inbred animals, such as BALB/c mice, respond very differently to the antigenic stimulus of poliovirus type 1, type 2, and type 3 (see discussions by Minor et al. [references 49, 71, and 72 and this volume]). Second, we have aimed to generate multivalent antigenic hybrid viruses containing neutralization antigenic sites of two different polioviruses and also of polioviruses and other picornaviruses (see below). Third, this highly exposed region of N-AgI could be used to test the cleavage specificity of the viral proteinases 2A and 3C when specific proteinase cleavage sites are engineered into this loop on the surface of the poliovirus particle.

The second objective is potentially important for the development of novel vaccines. Because of the apparent tolerance to changes seen in the sequence of N-AgI (Fig. 8), it is feasible to replace the N-AgI of the attenuated poliovirus strains with N-Ags of other picornaviruses such as human hepatitis A virus (HAV; also known as enterovirus 72). Simple mixing of antigenic sites among the attenuated polioviruses themselves could possibly lead to improved poliovirus vaccines. The attenuated poliovirus strains selected by Albert Sabin (59) have been used with great success for more than 2 decades to control poliomyelitis in the United States and elsewhere in the world. Although the live poliovirus vaccine can be considered one of the safest ever developed, isolated cases of poliomyelitis are occurring. These cases are likely to be vaccination related (for a discussion and references see references 6 and 49). It had originally been concluded that such cases of vaccine-associated poliomyelitis were due to genetic variation toward neurovirulence of the Sabin type 2 and type 3 strains, but not of the Sabin type 1 strain, during virus proliferation in vaccine recipients (74). In other words, the attenuation phenotype of the Sabin 1 strain appears to be more stable than those of the Sabin 2 and Sabin 3 strains (50); hence, a hybrid virus consisting predominantly of Sabin 1, but carrying the protective N-AgI of either Sabin 2 or Sabin 3, may represent a safer vaccine. Therefore, we (42, 43; M. G. Murray, R. J. Kuhn, and E. Wimmer, *Abstr. VII Int. Congr. Virol.*, Edmonton, Alberta, Canada, abstr. no. R16.40, p. 137, 1987) and others (see Minor et al. and M. Girard, A. Martin, T. Couderc, R. Crainic, and C. Wychowski, this volume) have exchanged the N-AgI of polioviruses.

A recent paper by Nkowane et al. (46), however, has dampened the overly enthusiastic expectation that such hybrids of attenuated polioviruses may eliminate vaccine-related poliomyelitis. Nkowane et al. (46) concluded that the Sabin 1 strain is not entirely safe and may cause poliomyelitis upon vaccination, although this occurs much less frequently than with the Sabin 2 and Sabin 3 vaccine strains. This observation, previously not taken into consideration (5), casts some doubt as to the feasibility of developing hybrid poliovirus vaccine strains that are superior to the existing vaccine strains. This is particularly relevant in light of the fact that an estimated 22.3×10^6 doses of poliovirus vaccine are administered in the United States annually, yet not a single case of poliomyelitis was recorded in the United States during 1987, and in 1986 there were only two cases (7, 8). On the basis of these new findings, can we really improve the existing oral poliovirus vaccines by mixing N-Ags of different poliovirus strains?

CONSTRUCTION OF TYPE 1-TYPE 3 HYBRID VIRUSES

The first report of an antigenic hybrid poliovirus was the description of a hybrid between PV-1(M) and PV-3(L) (Murray et al., *Abstr. VII Int. Congr. Virol.*). The construction of the virus was straightforward once a mutagenesis cartridge for N-AgI of PV-1(M) was available (Fig. 7; 42, 43; Murray et al., *Abstr. VII Int. Congr. Virol.*). The hybrid virus, W1/3-1D-1 (see Fig. 9), was impaired in growth (see Fig. 10), but it was apparently not temperature sensitive (42). W1/3-1D-1 was neutralized by polyclonal anti-type 1 and anti-type 3 antibodies and also by an N-MAb directed against N-AgI of PV-3(L), but not by an N-MAb directed against N-AgI of PV-1(M) or by anti-type 2 polyclonal antibodies as was expected (42, 43). In addition to serological characterization, nucleotide sequence analysis of transfection-derived viral RNA was used to confirm a bona fide hybrid virus. Significantly, when this virus was injected into rabbits or monkeys, it elicited anti-type 1 and anti-type 3 hyperimmune responses, but no significant response against poliovirus type 2 (42, 43). Thus, this hybrid virus not only interacts with neutralizing antibodies, but also elicits the production of high-titered neutralizing antibodies in rodents and primates.

Strain W1/3-1D-1 is a derivative of type 1 (Mahoney) in which nine amino acids have been replaced. The amino acid replacements map not only to the loop region of N-AgI but also to the β sheet B region of VP1 (Fig. 9). Since some neutralization escape mutants map to the β sheet B

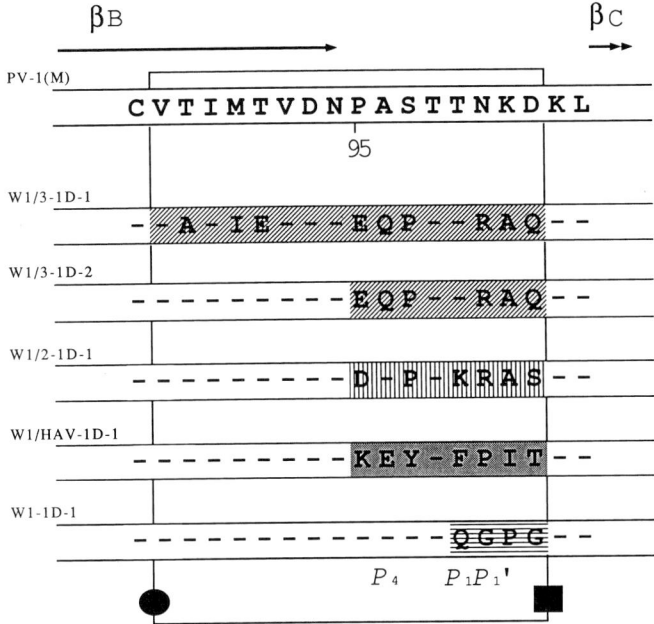

Figure 9. Schematic representation of the hybrid polioviruses showing β sheet B and β sheet C, flanking the N-AgI loop (21). Amino acids of: ●——●, PV-1(M); ▨, PV-3(L); ▥, PV-2(L); and ▩, HAV. The amino acids that are common between the respective serotypes are represented by dashes. ▱, Potential 3C proteinase cleavage site. The mutagenesis cartridge is enclosed in the large box: ●, naturally occurring *Sph*I restriction site; ■, newly generated *Hin*dIII restriction site (42).

(see Fig. 8), this structural element, and not only the loop, must be considered part of N-AgI. For this reason we included the β sheet B in our mutagenesis cartridge. As mentioned above, strain W1/3-1D-1 is impaired in viral growth, a phenomenon that may be related to the three amino acid replacements in the β sheet B region possibly disturbing interactions with other structural elements during morphogenesis of the capsid promoter. We logically constructed the virus W1/3-1D-2 (Fig. 9) in which only the loop region of PV-1(M) is exchanged with that of PV-3(L). One-step growth kinetics experiments revealed that W1/3-1D-2 does not grow to significantly higher titers than W1/3-1D-1. Thus virus yield with either hybrid virus is less than 10% that of PV-1(M).

CONSTRUCTION OF A TYPE 1-TYPE 2 HYBRID VIRUS

Our strategy to generate antigenic type 1-type 3 hybrid viruses was extended to generate a hybrid virus in which the N-AgI of PV-1(M) was replaced with the N-AgI of poliovirus type 2 (Lansing) [PV-2(L)]. First, we investigated whether or not replacement of the region yielded viable virus that expresses type 1 and type 2 N-Ags. Second, we tested whether a PV-1(M)/PV-2(L) hybrid virus exhibited altered host range.

Polioviruses are strictly primate viruses, a host range restriction that is based upon the fact that only primate cells have the capacity to express a suitable receptor (10). Mouse L cells, for example, are entirely resistant to poliovirus infection unless the human gene for the poliovirus receptor has been transferred into the rodent cell (37). PV-2(L), on the other hand, has been adapted to proliferate in mouse brain cells (3), where it causes a poliomyelitislike syndrome (22). Racaniello and his colleagues have mapped the genetic element(s) allowing PV-2(L) to grow in mouse brain cells to the region of VP1 that specifies N-AgI of PV-2(L) (32, 33; see V. R. Racaniello, N. La Monica, E. G. Moss, and R. O'Neill, this volume). It was therefore prudent to test a PV-1(M)/PV-2(L) hybrid for mouse neurovirulence.

Using the strategy of cartridge mutagenesis, synthetic oligonucleotides corresponding to the sequence in VP1 implicated by La Monica and colleagues for PV-2(L) neurovirulence were synthesized. The oligonucleotides were annealed and ligated into the subclone originating from PV-1(M) cDNA, and the full-length polio T7 transcription vector was reconstructed. RNA obtained by in vitro transcription with phage T7 RNA polymerase yielded virus after transfection of the RNA into HeLa cells. The resulting virus, W1/2-1D-1, was found to be neutralized by type 1- and type 2-specific polyclonal antibodies and also by an N-MAb directed against N-AgI of PV-2 (41). In contrast to strains W1/3-1D-1 and W1/3-1D-2, strain W1/2-1D-1 was found to exhibit growth kinetics nearly identical to those of the wild-type strain PV-1(M) (Fig. 10).

Most recently, hybrid virus W1/2-1D-1 was tested by V. R. Racaniello and E. G. Moss for mouse neurovirulence by intracerebral injection into 20-day-old mice. The hybrid virus killed mice at a 50% lethal dose of 6.1 $\log_{10}$ PFU (41). A similar result has been obtained by Girard et al. (this volume).

CONSTRUCTION OF A POLIOVIRUS-HAV HYBRID VIRUS

Infectious hepatitis caused by HAV is a widespread disease in the United States (over 24,000 cases in 1987; 8) and around the world, yet there is currently no hepatitis A vaccine commercially available. The sequence of the HAV genome is known (reference 53 and references therein), but the precise location of the N-Ags of HAV has not been determined. Nevertheless, as outlined by Lemon and Ping (S. M. Lemon

Figure 10. One-step growth curve comparing the replication of viruses W1/3-1D-1 and W1/2-1D-1 with that of PV-1(M). Confluent HeLa cell monolayers growing in 60-mm tissue culture plates were infected at a multiplicity of infection of 20 in a volume of 0.3 ml. Virus was adsorbed for 30 min at room temperature; then prewarmed medium was added, and the cells were placed at 37°C. At the indicated time points, the medium was aspirated, and the cells were washed twice with phosphate-buffered saline, scraped from the plate, and collected in a 1.5-ml tube. Any remaining cells were collected by washing the plate with 1.0 ml of phosphate-buffered saline and combining this wash with the cells. Cells were pelleted, suspended in 100 μl of buffer (0.1 M NaCl, 0.1 M Tris, 0.0015 M $MgCl_2$, pH 7.4), and lysed by three cycles of freeze-thawing. Nuclei were pelleted by centrifugation, and the virus-containing supernatant was collected and titrated by plaque assay. Symbols: ○, W1/3-1D-1; □, W1/2-1D-1; ▲, PV-1(M).

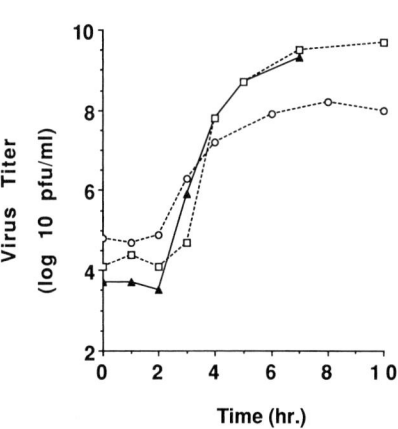

and L.-H. Ping, this volume), two N-Ags appear to exist, of which one maps to amino acids of VP3 and the other to a region in VP1 that corresponds to N-AgI of poliovirus. The latter antigen correlates well to an antigenic region on the surface of HAV predicted by Emini et al. (14). We have constructed a viable virus, W1/HAV-1D-1, in which N-AgI of PV-1(M) is replaced by an HAV-specific sequence that is likely to represent an HAV N-Ag (Fig. 9). Serological characterization of W1/HAV-1D-1 is in progress, and nucleotide sequence analysis of transfection-derived W1/HAV-1D-1 has confirmed a bona fide PV-1/HAV hybrid virus.

PROTEINASE 3C CLEAVAGE SITE IN N-AgI

The poliovirus polyprotein is cleaved by two virus-encoded proteinases, 2A and 3C, that map in the P2 and P3 regions of the polyprotein (references 28 and 45 and references therein) (see Fig. 1A). As has been pointed out above, 3C[pro] cleaves at certain Q-G sites, but the *P4* amino acid, that is, the residue in position −4 relative to the cleavage site, may also participate in enzyme-substrate recognition (45). In addition, we consider it probable that the folding of the polypeptide around the Q-G pair may play a role.

We have constructed a mutant PV-1(M) virus whose N-AgI has been modified such that its loop region contains a potential 3C^{pro} cleavage site (Fig. 9). Interestingly, the virus carrying this modification was found to be viable. Moreover, its capsid polypeptide VP1 remained uncleaved during the various steps of biosynthesis and morphogenesis. We are currently testing whether this putative cleavage site can be cleaved with purified proteinase 3C (44), either in the context of the virion or in denatured VP1.

CONCLUDING REMARKS

The strategy of poliovirus genome mutagenesis via a cartridge has yielded numerous genetic VPg and N-AgI variants of poliovirus that are of great interest. We have defined amino acids in VPg that appear to be absolutely essential, as well as others that can be replaced by virtually any amino acid. We are confident that saturation mutagenesis using mixed oligonucleotides will enable us to isolate conditional lethal mutations mapping to VPg that will allow us to determine the function(s) of this fascinating picornavirus gene product.

Modification of the N-AgI of PV-1(M) has yielded, for the first time, a true antigenic hybrid virus (Murray et al., *Abstr. VII Int. Congr. Virol.*), a result meanwhile reported also by others (see reference 5 and Girard et al., this volume). If hybrid viruses are to replace the currently existing Sabin vaccine strains, they must grow well and must show low lesion scores and low spread values in the central nervous system of monkeys. The PV-1/2 hybrid virus grows well. The PV-1/3 hybrids, in contrast, are impaired in growth, but it may be possible to isolate variants that replicate efficiently and yet retain their antigenic hybrid phenotype. These experiments are currently in progress. The PV-1/2 hybrid has acquired an extended host range phenotype, in that this hybrid virus by virtue of a six-amino-acid substitution is neurovirulent in mice (41). This intriguing result remains to be explained since we know neither the function of the human receptor for poliovirus, nor the route of entry that PV-2(L) utilizes for mouse brain cells.

Finally, the N-AgI of poliovirus can be used as a carrier for other antigenic sites or even be modified to contain recognition sites for enzymes. It will be fascinating to see whether our PV-1(M)/HAV hybrid is active in that it can be neutralized by anti-HAV antisera and induce a protective immune response in primates.

Acknowledgments. We thank our colleagues for many stimulating discussions; Quentin Reuer, Christopher Hellen, and James Harber for critically reading and editing this manuscript; and Atsuko Kameda for excellent technical assistance.

This work was supported, in part, by Public Health Service grants AI 15122 and CA 28146 from the National Institutes of Health.

Addendum in Proof

We have recently found that viral RNAs of nonviable VPg mutants, when transfected into HeLa cells, can replicate. This observation suggests that these mutants are blocked in morphogenesis (Q. Reuer and E. Wimmer, unpublished data).

Literature Cited

1. **Adler, C. J., M. Elzinga, and E. Wimmer.** 1983. The genome-linked protein of picornavirus. VIII. Complete amino acid sequence of poliovirus and carboxy-terminal analysis of its precursor, P3-9. *J. Virol.* **46:**349–355.
2. **Ambros, V., and D. Baltimore.** 1978. Protein is linked to the 5′ end of poliovirus RNA by a phosphodiester linkage to tyrosine. *J. Biol. Chem.* **253:**5263–5266.
3. **Armstrong, C.** 1939. Successful transfer of the Lansing strain of poliomyelitis virus from the cotton rat to the white mouse. *Publ. Health Rep.* **54:**2302–2305.
4. **Blondel, B., R. Crainic, O. Fichot, G. Dufraisse, A. Candrea, D. Diamond, M. Girard, and F. Horand.** 1986. Mutations conferring resistance to neutralization with monoclonal antibodies in type 1 poliovirus can be located outside or inside the antibody-binding site. *J. Virol.* **57:**81–90.
5. **Burke, K. L., G. Dunn, M. Ferguson, P. D. Minor, and J. W. Almond.** 1988. Antigen chimaeras of poliovirus as potential new vaccines. *Nature* (London) **332:**81–82.
6. **Centers for Disease Control.** 1986. Poliomyelitis—United States 1975–1984. *Morbid. Mortal. Weekly Rep.* **35:**180–182.
7. **Centers for Disease Control.** 1987. Notifiable diseases of low frequency. *Morbid. Mortal. Weekly Rep.* **35:**810.
8. **Centers for Disease Control.** 1988. Notifiable diseases of low frequency. *Morbid. Mortal. Weekly Rep.* **36:**840.
9. **Chow, M., J. F. E. Newman, D. Filman, J. M. Hogle, D. J. Rowlands, and F. Brown.** 1987. Myristylation of picornavirus capsid protein VP4 and its structural significance. *Nature* (London) **327:**482–486.
10. **Crowell, R. L., and B. J. Landau.** 1983. Receptors in the initiation of picornavirus infections, p. 1–42. *In* H. Fraenkel-Conrat and R. R. Wagner (ed.), *Comprehensive Virology*. Plenum Publishing Corp., New York.
11. **Dewalt, P. G., and B. L. Semler.** 1987. Site-directed mutagenesis of proteinase 3C results in a poliovirus deficient in synthesis of viral RNA polymerase. *J. Virol.* **61:**2162–2170.
12. **Diamond, D. C., B. A. Jameson, J. Bonin, M. Kohara, S. Abe, H. Itoh, T. Komatsu, M. Arita, S. Kuge, A. D. M. E. Osterhaus, R. Crainic, A. Nomoto, and E. Wimmer.** 1985. Antigenic variation and resistance to neutralization in poliovirus type 1. *Science* **229:**1090–1093.
13. **Emini, E. A., M. Elzinga, and E. Wimmer.** 1982. Carboxy-terminal analysis of poliovirus proteins: termination of poliovirus RNA translation and location of unique poliovirus polyprotein cleavage sites. *J. Virol.* **42:**194–199.
14. **Emini, E. A., J. V. Hughes, D. S. Perlow, and J. Boger.** 1985. Induction of hepatitis A virus-neutralizing antibody by a virus-specific synthetic peptide. *J. Virol.* **55:**836–839.
15. **Emini, E. A., B. A. Jameson, and E. Wimmer.** 1983. Priming for and induction of anti-poliovirus neutralizing antibodies by synthetic peptides. *Nature* (London) **304:**699–703.

16. **Emini, E. A., S. Y. Kao, A. J. Lewis, R. Crainic, and E. Wimmer.** 1983. Functional basis of poliovirus neutralization determined with monospecific neutralizing antibodies. *J. Virol.* **46:**466–474.

17. **Evans, D. M. A., P. D. Minor, G. C. Schild, and J. W. Almond.** 1983. Critical role of an eight amino acid sequence of VP1 in neutralization of poliovirus type 3. *Nature* (London) **304:**459–462.

18. **Flanegan, J. B., R. F. Pettersson, V. Ambros, M. J. Hewlett, and D. Baltimore.** 1977. Covalent linkage of a protein to a defined nucleotide sequence at the 5'-terminus of virion and replicative intermediate RNAs of poliovirus. *Proc. Natl. Acad. Sci. USA* **74:** 961–965.

19. **Flanegan, J. B., D. C. Young, G. J. Tobin, M. M. Stokes, C. D. Murphy, and S. M. Oberste.** 1987. Mechanism of RNA replication by the poliovirus RNA polymerase, HeLa cell host factor, and VPg, p. 273–284. *In* M. A. Brinton and R. R. Rueckert (ed.), *Positive Strand RNA Viruses.* Alan R. Liss, Inc., New York.

20. **Forss, S., and H. Schaller.** 1982. A tandem repeat gene in a picornavirus. *Nucleic Acids Res.* **10:**6441–6450.

21. **Hogle, J. M., M. Chow, and D. J. Filman.** 1985. Three-dimensional structure of poliovirus at 2.9 angstrom resolution. *Science* **229:**1358–1363.

22. **Jubelt, B., B. Gallez-Hawkis, O. Narayan, and R. T. Johnson.** 1980. Pathogenesis of human poliovirus infection in mice. I. Clinical and pathological studies. *J. Neuropathol. Exp. Neurol.* **39:**138–148.

23. **King, A. M. Q., D. V. Sangar, T. J. R. Harris, and F. Brown.** 1980. Heterogeneity of the genome-linked protein of foot-and-mouth disease virus. *J. Virol.* **34:**627–634.

24. **Kitamura, N., C. Adler, and E. Wimmer.** 1980. Structure and expression of the picornavirus genome. Ann. N.Y. Acad. Sci. **354:**183–201.

25. **Kitamura, N., C. J. Adler, P. G. Rothberg, J. Martinko, S. G. Nathenson, and E. Wimmer.** 1980. The genome-linked protein of picornaviruses. VIII. Genetic mapping of poliovirus VPg by protein and RNA sequence studies. *Cell* **21:**295–302.

26. **Kitamura, N., B. L. Semler, P. G. Rothberg, G. R. Larsen, C. J. Adler, A. J. Dorner, E. A. Emini, R. Hanecak, J. J. Lee, S. van der Werf, C. W. Anderson, and E. Wimmer.** 1981. Primary structure, gene organization and polypeptide expression of poliovirus RNA. *Nature* (London) **219:**547–553.

27. **Kohara, M., S. Abe, T. Komatsu, K. Tago, M. Arita, and A. Nomoto.** 1988. A recombinant virus between Sabin 1 and Sabin 3 vaccine strains of poliovirus as a possible candidate for a new type 3 poliovirus live vaccine strain. *J. Virol.* **62:**2828–2835.

28. **Kräusslich, H. G., and E. Wimmer.** 1988. Viral proteinases. Annu. Rev. Biochem. **57:**701–754.

29. **Kuhn, R. J., H. Tada, M. F. Ypma-Wong, J. J. Dunn, B. L. Semler, and E. Wimmer.** 1988. Construction of a mutagenesis cartridge for poliovirus VPg: isolation and characterization of viable and nonviable mutants. *Proc. Natl. Acad. Sci. USA* **85:**519–523.

30. **Kuhn, R. J., H. Tada, M. F. Ypma-Wong, B. L. Semler, and E. Wimmer.** 1988. Mutational analysis of the genome-linked protein VPg of poliovirus. *J. Virol.* **62:**4207–4215.

31. **Kuhn, R. J., and E. Wimmer.** 1987. The replication of picornaviruses, p. 17–51. *In* D. J. Rowlands, B. W. J. Mahy, and M. A. Mayo (ed.), *The Molecular Biology of Positive Strand RNA Viruses.* Academic Press, Inc. (London), Ltd., London.

32. **La Monica, N., W. Kupsky, and V. R. Racaniello.** 1987. Reduced mouse neurovirulence

of poliovirus type 2 Lansing antigenic variants selected with monoclonal antibodies. *Virology* **161**:429–437.

33. **La Monica, N., C. Meriam, and V. R. Racaniello.** 1986. Mapping of sequences required for mouse neurovirulence of poliovirus type 2 Lansing. *J. Virol.* **57**:515–525.

34. **Larsen, G. R., C. W. Anderson, A. J. Dorner, B. L. Semler, and E. Wimmer.** 1982. Cleavage sites within the poliovirus capsid protein precursors. *J. Virol.* **41**:340–344.

35. **Lee, Y. F., A. Nomoto, B. M. Detjen, and E. Wimmer.** 1977. A protein covalently linked to poliovirus genome RNA. *Proc. Natl. Acad. Sci. USA* **74**:59–63.

36. **Luo, M., G. Vriend, G. Kamer, I. Minor, E. Arnold, M. G. Rossmann, U. Boege, D. G. Scraba, G. M. Duke, and A. C. Palmenberg.** 1987. The atomic structure of mengo virus at 3.0 angstrom resolution. *Science* **235**:182–191.

37. **Mendelsohn, C., B. Johnson, K. A. Lionetti, P. Nobis, E. Wimmer, and V. R. Racaniello.** 1986. Transformation of a human poliovirus receptor gene into mouse cells. *Proc. Natl. Acad. Sci. USA* **83**:7845–7849.

38. **Minor, P. D., D. M. A. Evans, M. Ferguson, G. C. Shild, J. W. Almond, and G. Stanway.** 1987. Molecular basis of antigenicity of poliovirus, p. 539–553. *In* M. A. Brinton and R. R. Rueckert (ed.), *UCLA Symposium on Positive-Strand RNA Viruses: Keystone Colorado.* Alan R. Liss, Inc., New York.

39. **Minor, P. D., D. M. A. Evans, M. Ferguson, G. C. Schild, G. Westrop, and J. W. Almond.** 1985. Principal and subsidiary antigenic sites of VP1 involved in the neutralization of poliovirus type 3. *J. Gen. Virol.* **65**:1159–1165.

40. **Minor, P. D., G. C. Schild, Y. Bootman, D. M. A. Evans, M. Ferguson, P. Reeve, M. Spitz, G. Stanway, A. Y. Cahn, R. Hauptmann, L. D. Clarke, R. C. Mountford, and J. W. Almond.** 1983. Location and primary structure of the antigenic site for poliovirus neutralization. *Nature* (London) **301**:674–679.

41. **Murray, M. G., J. Bradley, X.-F. Yang, E. Wimmer, E. G. Moss, and V. R. Racaniello.** 1988. Poliovirus host range is determined by a short amino acid sequence in neutralization antigenic site I. *Science* **241**:213–215.

42. **Murray, M. G., R. J. Kuhn, M. Arita, N. Kawamura, A. Nomoto, and E. Wimmer.** 1988. Poliovirus type 1/type 3 antigenic hybrid virus constructed in vitro elicits type 1 and type 3 neutralizing antibodies in rabbits and monkeys. *Proc. Natl. Acad. Sci. USA* **85**:3203–3207.

43. **Murray, M. G., R. J. Kuhn, and E. Wimmer.** 1988. Poliovirus type 1/type 3 antigenic hybrid virus constructed in vitro elicits type 1/type 3 neutralizing antibodies in rabbits, p. 197–204. *In* R. A. Lerner, R. M. Chanock, F. Brown, and H. Ginsberg (ed.), *Vaccines '88.* Cold Spring Harbor Laboratory, Cold Spring Harbor, N.Y.

44. **Nicklin, M. J. H., K. S. Harris, P. V. Pallai, and E. Wimmer.** 1988. Poliovirus proteinase 3C: large-scale expression, purification, and specific cleavage activity on natural and synthetic substrates in vitro. *J. Virol.* **62**:4586–4593.

45. **Nicklin, M. J. H., H. Toyoda, M. G. Murray, and E. Wimmer.** 1986. Proteolytic processing in the replication of polio and related viruses. *Bio/Technology* **4**:33–42.

46. **Nkowane, B. M., S. G. F. Wassilak, W. A. Orenstein, K. J. Bart, L. B. Schonberger, A. R. Hinman, and O. M. Kew.** 1987. Vaccine-associated paralytic poliomyelitis, United States: 1973 through 1984. *J. Am. Med. Assoc.* **257**:1335–1340.

47. **Nomoto, A., B. M. Detjen, R. Pozzatti, and E. Wimmer.** 1977. The location of the polio genome protein in viral RNAs and its implication for RNA synthesis. *Nature* (London) **268**:208–213.

48. **Nomoto, A., N. Kitamura, F. Golini, and E. Wimmer.** 1977. The 5'-terminal structures of poliovirion RNA and poliovirus mRNA differ only in the genome-linked protein VPg. *Proc. Natl. Acad. Sci. USA* **74**:5345–5349.

49. **Nomoto, A., and E. Wimmer.** 1987. Genetic studies of the antigenicity and the attenuation phenotype of poliovirus. *Symp. Soc. Gen. Microbiol.* **40:**107–134.

50. **Omata, T., M. Kohara, S. Kuge, T. Komatsu, S. Abe, B. L. Semler, A. Kameda, H. Itoh, M. Arita, E. Wimmer, and A. Nomoto.** 1986. Genetic analysis of the attenuation phenotype of poliovirus type 1. *J. Virol.* **58:**348–358.

51. **Pallansch, M., O. M. Kew, B. L. Semler, D. R. Omilianowski, C. W. Anderson, E. Wimmer, and R. R. Rueckert.** 1984. Protein processing map of poliovirus. *J. Virol.* **49:**873–880.

52. **Paul, A. V., A. Shultz, S. E. Pincus, S. Oroszlan, and E. Wimmer.** 1987. Capsid protein VP4 of poliovirus is N-myristylated. *Proc. Natl. Acad. Sci. USA* **84:**7827–7831.

53. **Paul, A. V., H. Tada, K. von der Helm, T. Wissel, R. Kiehn, E. Wimmer, and F. Deinhardt.** 1987. The entire nucleotide sequence of the genome of human hepatitis A virus (isolate MBB). *Virus Res.* **8:**153–171.

54. **Pettersson, R. F., V. Ambros, and D. Baltimore.** 1978. Identification of a protein linked to nascent poliovirus RNA and to the polyuridylic acid of negative-strand RNA. *J. Virol.* **27:**357–365.

55. **Racaniello, V. R., and D. Baltimore.** 1981. Molecular cloning of poliovirus cDNA and determination of the complete nucleotide sequence of the viral genome. *Proc. Natl. Acad. Sci. USA* **78:**4887–4891.

56. **Racaniello, V. R., and D. Baltimore.** 1981. Cloned poliovirus complementary DNA is infectious in mammalian cells. *Science* **214:**916–919.

57. **Rossmann, M. G., E. Arnold, J. W. Erickson, E. A. Frankenberger, J. P. Griffith, H. J. Hecht, J. Johnson, G. Kamer, M. Luo, A. G. Mosser, R. R. Rueckert, B. Sherry, and G. Vriend.** 1985. Structure of a human common cold virus and functional relationship to other picornaviruses. *Nature* (London) **317:**145–153.

58. **Rothberg, P. G., T. J. R. Harris, A. Nomoto, and E. Wimmer.** 1978. The genome-linked protein of picornaviruses. O^4-(5'-uridylyl) tyrosine is the bond between the genome-linked protein and the RNA of poliovirus. *Proc. Natl. Acad. Sci. USA* **75:**4868–4872.

59. **Sabin, A. B., and C. R. Boulger.** 1973. History of Sabin attenuated poliovirus oral live vaccine strains. *J. Biol. Stand.* **1:**115–118.

60. **Semler, B. L., C. W. Anderson, R. Hanecak, L. F. Dorner, and E. Wimmer.** 1982. A membrane-associated precursor to poliovirus VPg identified by immunoprecipitation with antibodies directed against a synthetic heptapeptide. *Cell* **28:**405–412.

61. **Semler, B. L., C. W. Anderson, N. Kitamura, P. G. Rothberg, W. L. Wishart, and E. Wimmer.** 1981. Poliovirus replication proteins: RNA sequence encoding P3-1b and the sites of proteolytic processing. *Proc. Natl. Acad. Sci. USA* **78:**3464–3468.

62. **Semler, B. L., A. J. Dorner, and E. Wimmer.** 1984. Production of infectious poliovirus from cloned cDNA is dramatically increased by SV40 transcription and replication signals. *Nucleic Acids Res.* **12:**5123–5141.

63. **Semler, B. L., R. Hanecak, C. W. Anderson, and E. Wimmer.** 1981. Cleavage sites in the polypeptide precursors of poliovirus protein P2-X. *Virology* **114:**589–594.

64. **Semler, B. L., V. H. Johnson, and S. Tracy.** 1986. A chimeric plasmid from cDNA clones of poliovirus and coxsackievirus produces a recombinant virus that is temperature sensitive. *Proc. Natl. Acad. Sci. USA* **83:**1777–1781.

65. **Takeda, N., C.-F. Yang, R. J. Kuhn, and E. Wimmer.** 1987. Uridylylation of the genome-linked protein of poliovirus in vitro is dependent upon an endogenous RNA template. *Virus Res.* **8:**193–204.

66. **Takegami, T., R. J. Kuhn, C. W. Anderson, and E. Wimmer.** 1983. Membrane-dependent uridylylation of the genome-linked protein VPg of poliovirus. *Proc. Natl. Acad. Sci. USA* **80:**7447–7451.

67. **Toyoda, H., M. Kohara, Y. Kataoka, T. Suganuma, T. Omata, N. Imura, and A. Nomoto.** 1984. Complete nucleotide sequences of all three poliovirus serotype genomes: implication for genetic relationship, gene function and antigenic determinants. *J. Mol. Biol.* **174:**561–585.

68. **van der Werf, S., J. Bradley, E. Wimmer, F. W. Studier, and J. J. Dunn.** 1986. Synthesis of infectious poliovirus RNA by purified T7 RNA polymerase. *Proc. Natl. Acad. Sci. USA* **83:**2330–2334.

69. **van der Werf, S., C. Wychowski, P. Bruneau, B. Blondel, R. Crainic, F. Horodniceanu, and M. Girard.** 1983. Localization of a poliovirus type 1 neutralization epitope in viral capsid polypeptide VP1. *Proc. Natl. Acad. Sci. USA* **80:**5080–5084.

70. **Wimmer, E.** 1979. The genome-linked protein of picornaviruses: discovery, properties and possible functions, p. 175–190. *In* R. Perez-Bercoff (ed.), *The Molecular Biology of Picornaviruses.* Plenum Publishing Corp., New York.

71. **Wimmer, E., E. A. Emini, and D. C. Diamond.** 1986. Mapping neutralization domains of viruses, p. 159–173. *In* A. L. Notkins and M. B. A. Oldstone (ed.), *Concepts in Viral Pathogenesis II.* Springer-Verlag, New York.

72. **Wimmer, E., B. A. Jameson, and E. A. Emini.** 1984. Poliovirus antigenic sites and vaccines. *Nature* (London) **308:**19.

73. **Wimmer, E., R. J. Kuhn, S. Pincus, C.-F. Yang, H. Toyoda, M. J. H. Nicklin, and N. Takeda.** 1987. Molecular events leading to picornavirus genome replication. *J. Cell Sci.* Suppl. 7, p. 251–276.

74. **World Health Organization Consultative Group.** 1982. The relation between acute persisting spinal paralysis and poliomyelitis vaccine: results of a ten-year inquiry. Bull. W.H.O. **60:**231–242.

75. **Wychowski, C., S. Van der Werf, O. Siffert, R. Crainic, P. Bruneau, and M. Girard.** 1983. A poliovirus type 1 neutralization epitope is located within amino acid residues 93 to 104 of viral capsid polypeptide VP1. *EMBO J.* **2:**2019–2024.

76. **Ypma-Wong, M. F., P. G. Dewalt, V. H. Johnson, J. G. Lamb, and B. L. Semler.** 1988. Protein 3CD is the major poliovirus proteinase responsible for cleavage of the P1 capsid precursor. *Virology* **166:**265–270.

77. **Ypma-Wong, M. F., and B. L. Semler.** 1987. In vitro molecular genetics as a tool for determining the differential cleavage specificities of the poliovirus 3C proteinase. *Nucleic Acids Res.* **15:**2069–2088.

78. **Zoller, M. J., and M. Smith.** 1984. Laboratory methods. Oligonucleotide-directed mutagenesis: a simple method using two oligonucleotide primers and a single-stranded DNA template. *DNA* **3:**479–488.

Molecular Aspects of Picornavirus Infection and Detection
Edited by Bert L. Semler and Ellie Ehrenfeld
© 1989 American Society for Microbiology, Washington, DC 20006

Chapter 2

Replication of Hepatitis A Virus: New Ideas from Studies with Cloned cDNA

John Ticehurst, Jeffrey I. Cohen, Stephen M. Feinstone,
Robert H. Purcell, Robert W. Jansen,
and Stanley M. Lemon

INTRODUCTION

Although data have accumulated rapidly since hepatitis A virus (HAV) was identified in 1973 (32), many features of this important human pathogen remain enigmatic. Pieces of the HAV puzzle are sometimes fitted together by drawing on knowledge of other picornaviruses. However, HAV has many distinctive characteristics (for reviews, see references 21, 22, 40, 57, 98, and 111 and J. I. Cohen, *Hepatology*, in press) and probably is best classified as the only member of a new genus (72, 110) rather than as enterovirus type 72 (66). In particular, in vitro replication of HAV typically has little effect on the host cell, requires days for completion, and yields relatively small amounts of progeny virus. Because of these characteristics, it has been difficult to study the HAV replication cycle.

Few phenotypes have been analyzed in detail. When inoculated into

John Ticehurst • Department of Virus Diseases, Walter Reed Army Institute of Research, Washington, D.C. 20307-5100. Jeffrey I. Cohen • Division of Infectious Diseases, Brigham and Women's Hospitals, Boston, Massachusetts 02115. Stephen M. Feinstone and Robert H. Purcell • Laboratory of Infectious Diseases, National Institute of Allergy and Infectious Diseases, Bethesda, Maryland 20892. Robert W. Jansen and Stanley M. Lemon • Department of Medicine and Department of Microbiology and Immunology, University of North Carolina, Chapel Hill, North Carolina 27514.

susceptible primates by the oral or intravenous route, wild-type HAV undergoes replication in the liver, followed by hepatitis. Although such wild-type virus grows poorly, if at all, in cell culture, variants have been selected for in vitro multiplication (25, 82). Optimal growth rates have been obtained with different HAV strains by using particular combinations of cells and temperatures (13, 99), but markers for host range and temperature sensitivity have not been identified. Persistently infected cell lines are readily established and usually cannot be distinguished from uninfected cells (13, 24, 26, 84, 98, 99, 102, 111) unless indirect methods are used to detect the virus (58, 99). During multiple passages in cell culture, HAV progressively becomes attenuated and may even lose the ability to infect primates (51, 80, 81). Similarly, passaging of HAV in marmosets selected for a host-range alteration that made the virus more virulent for marmosets but attenuated for chimpanzees (14). Cytolytic variants (2, 3, 24, 69, 116, 124) and mutants resistant to neutralization with monoclonal antibody (108) have been recently selected in cell culture. The HM175 strain of HAV (41) is the best characterized of all HAV isolates and has the widest range of available phenotypes (84; Table 1).

Cloned HAV cDNA (113, 117) has proven to be a practical tool for studying the relationship between the HAV genome and viral replication. As cited in succeeding sections of this paper, quantitative hybridization analysis (46, 48, 112) has been useful for detecting HAV RNA in infected cells or in virions when standard techniques, such as radiolabeling of HAV RNA molecules or isolation of wild-type virus, were impractical.

In addition, cloned cDNAs have been used for determining the nucleotide sequences of several HAV strains and variants (Table 2). These viruses are only distantly related to other picornaviruses at both the nucleotide and amino acid levels (72, 110; Table 3; J. Ticehurst, J. Cohen, T. Chestnut, and R. Purcell, manuscript in preparation; see A. C. Palmenberg, this volume). There is, however, extensive similarity among the sequences of HAV strains isolated from widely separated geographic regions (Tables 2, 4, and 5; Fig. 1), a finding in agreement with results of oligonucleotide mapping of HAV strains (119). As determined by hybridization with strain HM175 probes, the genome of a virus strain (PA21) recovered from owl monkeys has distinct nucleotide sequences in its P1 and P2 regions, but the neutralization epitopes of strain PA21 appear to be conserved with those of other HAV strains (59).

A more detailed analysis of the few genomic changes that accompanied adaptation of HAV strain HM175 to growth in cell culture (18, 47; Tables 4 and 5; Fig. 1), and possibly other genetic markers, can now be performed by using infectious HAV cDNA (17, 19; Cohen, in press). As

Table 1. Variants of HAV HM175

Designation	Passage history[a]	Virulence[b]	Other phenotypes	References
HM175/wt	Marmoset × 3	Wild type	cDNA not infectious	17, 20, 113
HM175/P16	Marmoset × 6 AGMK × 10 BS-C-1 × 6	Probably wild type[c]		13, 47
HM175/P35[d]	AGMK × 35	Attenuated	cDNA infectious	17–19, 51
HM175/P59	Marmoset × 6 AGMK × 17 BS-C-1 × 42	Not known		90
pHM175[e]	Marmoset × 6 AGMK × 10 BS-C-1 × 5 Subculture in BS-C-1 × 20	Not known	Persistent infection in subcultured BS-C-1; cytolytic on acute passage in FRhK-4	24
HM175A [f]	HM175/pi[e] → BS-C-1 × 8	Not known	Cytolytic	2, 3
HM175/S18	Marmoset × 6 AGMK × 10 BS-C-1 × 12 with selection[g]	Wild-type revertant[c]	Neutralization resistant	60, 108

[a] In vivo passages made by intravenous inoculation of homogenate of liver.
[b] Determined by intravenous inoculation of chimpanzees, marmosets, or owl monkeys.
[c] Virulence of P16 has not been determined, but virulent, neutralizable HAV was selected in vivo from neutralization-resistant mutant HM175/S18 (60).
[d] Alternative designation for HM175/7 MK-5 (18).
[e] Other, independent, persistently infecting variants named HM175/pi (Mihalik and Feinstone, unpublished data) or not named (102). The latter variant also persistently infects AGMK or BGMK cells.
[f] Unnamed, independent, cytolytic variant obtained by acute passage of a persistently infecting variant of HM175 in FRhK-4 cells (69, 102).
[g] Selection for growth in the presence of a monoclonal antibody to HAV during passages 6 through 10 in BS-C-1 cells.

we will discuss, following a more detailed review of HAV replication, such approaches have yielded new insights concerning the distinctive characteristics of HAV growth.

REPLICATION

A Primary Extrahepatic Site of Replication, If It Exists, Has Been Difficult to Demonstrate

Natural infection with HAV usually follows ingestion of virus-contaminated material. Virus is subsequently shed in the feces, and it is

Table 2. Sequences Determined from Cloned HAV cDNAs

Strain	Origin	Nucleotides[a]	Region[b]	References
HM175	Australia	wt: 1–7478	Complete	9, 20, 41, 113
		P16: 0–7476[c]	Complete	47
		P35: 1–7473	Complete	18
		P59: 29–3107	5′ through 1D	1, 90
LA	California	1–7478	Complete	68[d]
MBB	Africa	1–7470	Complete	75
CR326	Costa Rica	236–3290	5′ into 2A	61
HAS15	Arizona	845–4217	1B into 2C	14, 71
MS1	New York	2208–3164	1D into 2A	14, 87
Unnamed	Italy	12–199	5′	116

[a] Positions numbered relative to the 5′ terminus of the HM175/wt sequence.
[b] Nomenclature per reference 92.
[c] Base 0 is a 5′-terminal addition of U.
[d] Sequence recently revised (R. Ralston, personal communication).

clear that this HAV is largely derived from infected hepatocytes and gains access to the intestines via the biliary system. However, HAV has been detected in human saliva (84), suggesting that the virus may replicate in the oropharynx or salivary glands in a manner similar to poliovirus (30). Despite several attempts (55, 64, 65) to demonstrate HAV, HAV RNA, or viral antigen in the gastrointestinal tract of experimentally infected marmosets, HAV RNA and antigen have been detected in the small intestine in only one set of studies (49, 50). In another recent study, oropharyngeal inoculation of a chimpanzee was followed by isolation of HAV from the serum, saliva, and throat washings, all prior to detection of viral antigen in the liver. However, the latter result may reflect tissue sampling errors and the insensitivity of immunofluorescence analysis compared with isolation of virus (J. Cohen, S. Feinstone, and R. Purcell, manuscript in preparation). The sequence of such events and their role in the pathogenesis of hepatitis A have not been well defined in natural or other experimental infections.

HAV May Cause Hepatitis without Virus-Induced Cytolysis

Viral antigen is first detected in the hepatocytes of experimentally infected primates as early as 1 to 2 weeks after intravenous inoculation and may persist there throughout the acute phase of disease (40, 51, 64, 112; Fig. 2). Cytopathology is minimal during the early phase of maximal viral replication, although a direct cytopathic effect may result from inocula containing large doses of HAV (40). Within hepatocytes and Kupffer cells, virus is found in vesicular structures (97), but the mechanism by which virus is released to the biliary tract is not known.

Table 3. Comparisons of Amino Acid Sequences of HAV and Other Picornaviruses[a]

Protein	Function in PV[b]	Length (amino acids)					% Identical[c] HAV versus:			
		HAV[d]	PV	RV2	EMCV	FMDV	PV	RV2	EMCV	FMDV
P1										
1A	VP4	23[e]	69	69	69	82	NA	11	15	6
1B	VP2 (1AB protease)	222	272	261	256	218	6	23	24	28
1C	VP3	246	238	237	231	221	24	25	20	18
1D	VP1	300	302	289	287	213	15	18	19	19
P2										
2A	1D/2A protease, host shutoff	189	149	136	147	16	17	21	17	NA
2B	Transcription	107	97	95	136	154	18	19	21	13
2C	Gua$^{s/r}$, possible GTP binding (transcr.)	335	329	322	325	318	26	27	30	29
P3										
3A	Pre-VPg (transcr.)	74	87	77	88	153	12	14	15	18
3B	VPg (transcr.)	23	22	21	20	23/23/24	26	41	23	33
3C	Major protease at Gln/Gly	219	183	183	205	213	18	27	25	21
3D	RNA polymerase	489	461	460	460	470	32	29	28	28

[a] Sequence references: HAV, HAV HM175/wt (20); PV, poliovirus type 1 Mahoney (53, 85); RV2, rhinovirus 2 (103); EMCV, encephalomyocarditis virus (73; G. Duke and A. Palmenberg, personal communication); FMDV, foot-and-mouth disease virus type A12 (89).

[b] Reviewed in references 91, 106, and 123. Other references as follows: capsid protein structure (43); possible 1AB protease activity (5); protein 2A (10, 11, 114); undefined roles of 2B, 2C, 3A, and 3B in viral RNA transcription (transcr.) (8, 10, 23, 28, 34, 39, 56, 95, 109); guanidine sensitivity marker (Gua$^{s/r}$) of 2C (52, 78), 3C (42), and 3D (33, 63).

[c] Calculated from alignments made with computer programs (38, 111; Ticehurst et al., in preparation). NA, Not aligned.

[d] Lengths predicted from amino acid sequence determinations (36, 61) and analysis with computer programs (20, 111); alternative genome map predicts similar lengths (29). Boldface numerals indicate that the predicted length of HAV protein is shorter than that of most or all other analogous proteins; underlined numerals indicate the HAV protein is longer.

[e] May be cleaved at Gln/Gly (residues 6–7) and myristylated as a 17-mer (16; see Palmenberg, this volume).

Table 4. Nucleotide Differences between HAV Strains and Variants

Region (length in nucleotides)	MBB versus LA[a]	HM175/wt versus:				
		LA	MBB	P16	P35	P59
		No. of nucleotide differences (% of sequence total)				
Whole genome[b] (7,470–7,478)	637 (8.5)	618 (8.3)	370 (5.0)	20 (0.3)	24 (0.3)	
Deletions	8	2	9	2	5	
Insertions	2	2	1	1	0	
5' Noncoding[c] (726–734)	39 (5.3)	28 (3.8)	23 (3.2)	6 (0.8)	7 (1.0)	9 (1.2)
Nt 99 to 207	17	9	14	3	7	2
Deletions	8	2	9	2	5	1
Insertions	1	1	1	0	0	0
Nt 636 to 725	0	0	0	1	0	3
3' Noncoding (63–64)	13 (21)	7 (11)	6 (10)	1 (1.6)	1 (1.6)	
Insertions	1	1	0	0	0	

[a] Calculations made with revised LA sequence (68; Ralston, personal communication).
[b] Values for the P1, P2, and P3 regions are similar to those for the whole genome.
[c] LA and CR236 sequences have 0.8% difference in nucleotides 236 to 734 and are identical over the last 280 bases (nucleotides 455 to 734). Nucleotide (Nt) positions refer to HM175/wt, numbered from the 5' terminus.

Table 5. Amino Acid Differences between HAV Strains and Variants[a]

Strain or variant	No. of amino acids different/total amino acids (%)			
	HM175/wt	LA	MBB	CR326
LA[b]	20/2,227 (0.90)			
MBB	19/2,227 (0.85)[c]	31/2,227 (1.4)		
CR326[d]	9/852 (1.1)	7/852 (0.82)	9/852 (1.1)	
HAS15[e]	16/1,127 (1.4)	10/1,127 (0.89)	20/1,127 (1.8)	12/812 (1.5)
HM175/P16	8/2,227 (0.36)			
HM175/P35	12/2,227 (0.54)			
HM175/P59 [f]	2/791 (0.25)			

[a] Partial sequence of HAV MS1 (1D: amino acids 492–791, relative to HM175/wt sequence) is also highly homologous with other HAV 1D sequences (87).
[b] Calculations made with revised LA sequence (68; Ralston, personal communication).
[c] P1 sequences of MBB and HM175 are 100% identical (see Fig. 1).
[d] Amino acids 1 to 852.
[e] Amino acids 38 to 1167.
[f] Amino acids 1 to 791; calculation made by using those differences present in a majority of the cDNA clones analyzed (1).

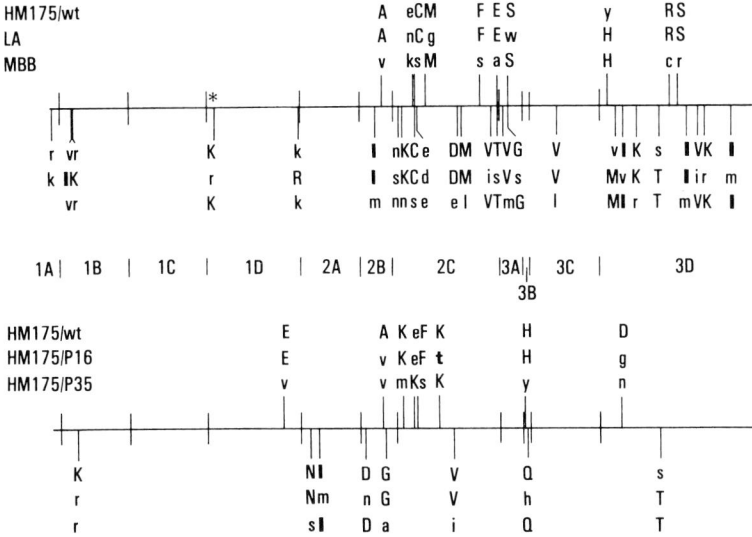

Figure 1. Amino acid differences between HAVs. Predicted amino acid sequences from HAV strains HM175, LA, and MBB (upper portion of figure) and from variants of HAV HM175 (lower) were compared, and changes were plotted on linear maps of the HAV polyprotein (LA sequence revised from reference 68; Ralston, personal communication). Differences are indicated by single-letter code; uppercase letters indicate consensus residues among the five complete sequences and the partial sequences of strains CR326 and HAS15 (Table 2). Lines that extend upward from the horizontal line of the polyprotein map indicate changes that produced scores of ≤0 from the mutation data matrix (94). This matrix compares amino acid sequences from pairs of proteins on phylogenetic trees; a negative score means that the pair would be expected to occur less frequently in related sequences than random chance would permit. Downward lines indicate positive scores from the mutation data matrix. Other vertical lines, between designations for proteins (center) and crossing the horizontal line of the polyprotein map, mark proposed sites of cleavage during protein maturation (20, 111). Asterisk marks deletion of six amino acids (residues 26 to 31 of 1D) in the HAS15 sequence that correlates with the altered electrophoretic mobility of VP1 of HAS15 (71, 88). Alignments made and kindly provided by A. Palmenberg (personal communication; this volume) were helpful in preparing this figure.

Hepatocellular injury usually develops concurrently with the appearance of serum anti-HAV and may involve cellular immune mechanisms (for reviews, see references 40 and 98). Thus, infection of liver cells with HAV may result in noncytolytic replication of virus. Replication of attenuated HAV, as measured by immunofluorescence analysis of liver tissue and hybridization with extracts of feces, is reduced in quantity and duration compared with wild-type virus (51).

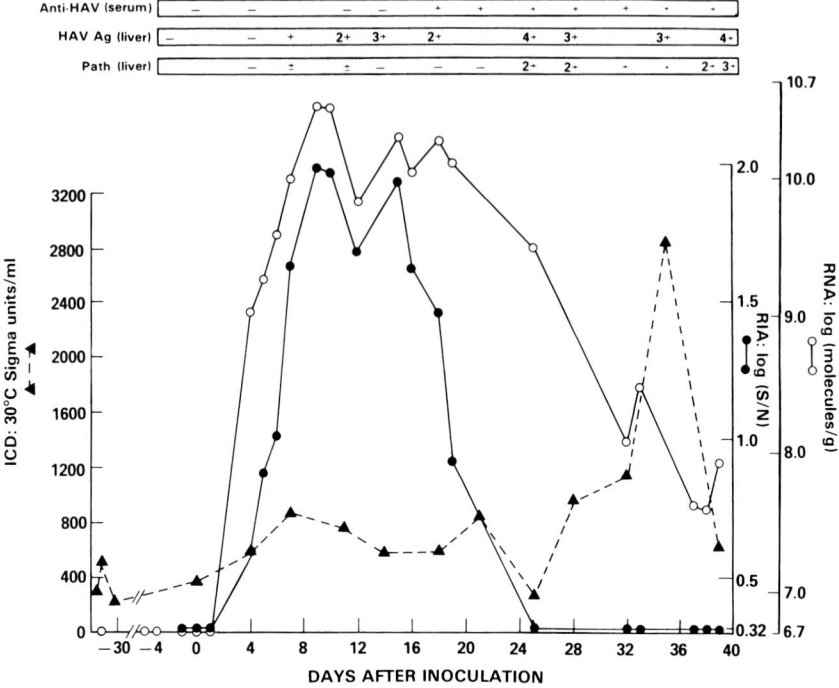

Figure 2. Course of hepatitis A in an experimentally infected marmoset. A marmoset was inoculated intravenously with a human fecal suspension containing 2×10^6 50% chimpanzee infective doses of HAV HM175. ICD, Serum isocitric dehydrogenase (a marker for biochemical evidence of hepatitis). RIA, Results from radioimmunoassay for HAV antigen in 20% (wt/vol) fecal suspensions. RNA, Molecules of HAV RNA per gram of stool, calculated from densitometric measurements of autoradiograms from hybridization of cloned HAV cDNA probes with extracts of 5% (wt/vol) fecal suspensions. RIA and RNA values are plotted as $\log_{10}$. Anti-HAV, Results of radioimmunoassay of sera for antibody to HAV. HAV Ag (liver), Results of immunofluorescence analysis of liver tissue, scored $-$ to $4+$. Path (liver), Histopathologic changes in liver tissue, scored $-$ to $4+$. Figure is reproduced from reference 112.

Revertant Selected In Vivo from Neutralization-Resistant Inoculum

A neutralization escape mutant of HAV, HM175/S18, was selected for its ability to grow in the presence of a monoclonal anti-HAV antibody (108; Table 1) and was determined to have a single-amino-acid alteration in its capsid proteins (Asp to His at residue 70 of VP3; see S. M. Lemon and L.-H. Ping, this volume). This virus had stable growth characteristics in cell culture. When it was intravenously inoculated into six owl monkeys, the animals developed viremia and hepatitis (60). The virus recovered from the feces of two monkeys studied in detail was no longer resistant to neutralization by the cognate monoclonal antibody and had the nucleotide sequence of wild-type strain HM175 (HM175/wt) in the coding region for VP3 (L.-H. Ping and S. Lemon, unpublished results). These data suggest that the neutralization-resistant virus was unable to replicate efficiently in owl monkeys and that there are strong biologic constraints to antigenic variation among HAV strains in vivo.

In Vitro Replication Is Slow and Usually Not Cytolytic

In contrast with most other picornaviruses, HAV was difficult to adapt to growth in cell culture (25, 82). Once adapted, HAV usually replicates slowly, reaching maximal titers (usually lower than 10^9 tissue culture infective doses per ml) after days or weeks, without inhibition of host cell metabolism (13, 21, 22, 26, 40, 57, 111; Cohen, in press). The virus gradually accumulates in the cytoplasm and spreads to adjacent cells by an unknown mechanism. Ultrastructural changes in such HAV-infected cells are similar to those in hepatocytes from experimentally infected primates, with virus particles clustered in vesicular structures (6).

After multiple passages of several HAV strains in cell culture, genomic defects were found in portions of the virus populations (70). Internal deletions spanned nucleotides 930–4380, 1140–3820, or 1370–3240 in the P1 region, and various 3' truncations included large portions of the P3 region. It is not known whether such defective particles have an effect on the replication cycle of HAV.

The replicative events in cells that became persistently infected with HAV have been studied under one-step growth conditions (26). Three phases after inoculation were identified: (i) maximum virion replication, days 2 to 8, when minus- and plus-strand viral RNAs and the rate of infectious HAV synthesis reached their highest levels (day 8); (ii) maximum viral antigen production, days 9 to 14, when plus-strand HAV RNA and infectious HAV levels remained high, but minus-strand HAV RNA became undetectable; and (iii) minimal viral metabolic activity, after day 14. Nearly all of the virus remained cell associated, a characteristic of many HAV-cell systems in vitro. In a similar study, nearly 100% of the HAV inoculum was absorbed onto cells within 20 min, but a large portion of the inoculum was recovered as intact virus (121). During the period of maximal virus production, virus was released into the culture medium, but an average of 80% of the HAV remained cell associated.

Several laboratories have identified variants of HAV that cause cytolysis after 2 to 7 days of growth (2, 3, 24, 69, 116, 124; Table 1). The development of cytopathic effect depended on the interaction between particular HAV strains and specific cell lines at certain passage levels. In most cases, cytolytic growth was selected by passaging HAV from persistently infected cells at intervals of 3 to 7 days. Other isolates of HAV were cytopathic for FRp/3 rhesus kidney (116) and A549 human lung carcinoma (124) cell lines within several passages after direct isolation from human feces.

Anderson et al. studied the replicative events of cytolytic variant HM175A (Table 1) in BS-C-1 continuous green monkey kidney cells (3). It appeared that the inoculum rapidly penetrated cells and was efficiently uncoated, but a large proportion of the input virus was recoverable. Increased HAV RNA production was detected 6 h after inoculation, reaching maximal levels at 12 to 24 h, a period that coincided with the logarithmic phase of HAV production. Cytopathic effect occurred after 2 to 3 days, yielding 20 infectious doses of HAV per cell. There has been no evidence for direct suppression of host macromolecular synthesis in cells infected with HM175A or other cytolytic variants of HAV. While cytolytic variants of HAV will be extremely useful for studying replication, they are atypical for HAV. Other variants of HAV attain maximal accumulation in vitro within 3 days but are not cytolytic (7, 54, 62, 84).

HAV Is Unusually Stable and Is Not Affected by Several Drugs That Inhibit Poliovirus

The sensitivity of HAV and enteroviruses to physical effects and chemical agents is generally similar (21, 40, 57, 111). However, HAV has greater resistance to heat and is the only picornavirus that is stable at 60°C (74, 101). Cytolytic variant HM175A (Table 1) appeared to have increased susceptibility to heat (2). The heat stability of neutralization-resistant mutants (108) has not been tested. In addition, HAV appears to be more chlorine resistant than other picornaviruses (77). HAV also survived longer than poliovirus in a variety of environmental samples (105).

Several agents that inhibit the growth of poliovirus and some other picornaviruses do not affect HAV at subtoxic concentrations, including arildone, disoxaril (formerly known as WIN 51711), 3′-methylquercetin, 2,4-dichloropyrimidine, guanidine, and 2-(α-hydroxybenzyl)benzimidazole (27, 57, 100, 104; A. Widell, doctoral thesis, University of Lund, Malmö, Sweden, 1988; S. Lemon, unpublished observations). The first two drugs represent a group of antiviral agents that, by filling the hydrophobic interior of VP1, stabilize the capsid and prevent uncoating of enteroviruses and rhinoviruses (104; see M. Rossmann, this volume). The latter agents interfere with intracellular replicative events: 3′-methylquercetin inhibits synthesis of poliovirus RNA, and 2,4-dichloropyrimidine inhibits the morphogenesis of poliovirus (27). Markers for guanidine sensitivity have been located within protein 2C of poliovirus (52, 78). HAV replication was enhanced by 5,6-dichloro-1-β-D-ribofuranosylbenzimidazole (122). The latter primarily inhibits transcription by cellular RNA polymerase II and did not enhance growth of poliovirus or echovirus type 6.

INSIGHTS AND APPROACHES

RNA Noncoding Regions and Several Proteins of HAV Are Particularly Distinct and May Reflect Constraints for Replication in Hepatocytes

Although the genomic organization of HAV and enteroviruses is similar (20, 111), the nucleotide sequence of HAV is not homologous with those of other picornaviruses (44, 79, 110; Ticehurst et al., in preparation; see Palmenberg, this volume). In addition, the low G+C content (38%) in HAV is approximated only by that of rhinoviruses (39 to 40%) (15, 20, 68, 75, 103, 107). HAV and rhinoviruses have a preference for A or U in the third base position of codons (72).

The high proportion of pyrimidines near the 5' terminus of the HAV genome (95% of nucleotides 99 to 138 and 68% of nucleotides 31 to 249 in HM175/wt [20]) is unique. Nucleotides 99 to 207 seem particularly susceptible to mutation, often occurring as deletions, in different HAV strains and during adaptation of strain HM175 to growth in cell culture (47; Table 4). The segment from nucleotides 636 to 725 is highly conserved between different HAVs (Table 4) and contains a short polypyrimidine tract similar to those found near the initiation codon of other picornaviruses; it thus may have an essential function. The 3' noncoding region has the highest percentage of differences among HAV strains and variants (Table 4).

There is substantial amino acid sequence similarity among HAV and other picornaviruses in proposed functional segments of proteins 1AB (cleavage site and active Ser for autoproteolysis of VP0 [5, 9]), 2C (domains for possible binding of GTP [28, 39] and Gly correlating with guanidine resistance [20, 78]), 3B (Tyr for linking VPg to viral RNA [118, 123]), 3C (closely spaced Cys and His in active site [4, 20, 42]), and 3D (Tyr-Gly-Asp-Asp in active site [4, 20]). However, overall similarity is low (Table 3; Ticehurst et al., in preparation). HAV proteins 1A, 1D, 2A, 2B, and 3A are particularly distinct. The predicted lengths of HAV proteins 1A, 1B, and 3A are shorter than those of most or all such proteins of other picornaviruses, while the lengths of 1C, 1D, 2A, 2C, 3C, and 3D are longer. Seven different dipeptides in the HAV polyprotein are thought to be cleaved by the 3C protease; many of these sites and the adjacent residues are different from those in other picornaviral polyproteins (20,

29, 36, 61, 120). Thus, it cannot be concluded from these data that HAV proteins have the same functions (1A, 2A, 2B, and 3A, in particular) or enzymatic specificity (3C) as the analogous proteins in other picornaviruses. In contrast to the apparently distant relationship between proteins of HAV and other picornaviruses, amino acid sequences representing several HAV strains are ≥98% identical and particularly similar in the P1 region (Table 5, Fig. 1).

The most important in vivo site of HAV replication, in terms of pathogenesis and the quantity of virus produced, is the liver. No other picornavirus has such affinity for hepatocytes. The many distinctive features of the HAV genome may reflect requirements for successful replication in the liver and may account for the usual in vitro phenotype of slow, nonlytic replication. A more stable virion could be one of the secondary effects of tissue-specific constraints on the HAV genome.

HAV May Be Deficient in Functions for Efficient Viral Translation

Once HAV RNA levels begin to rise in infected cells in vitro, the accumulation of RNA and then virus is rapid (3, 26; K. Mihalik, J. Ticehurst, and S. Feinstone, unpublished results). If early events are particularly slow, what inhibits them?

One can speculate from considerations of poliovirus mutants, some of which have replicative characteristics that resemble those of HAV (see D. Trono, R. Andino, and D. Baltimore, this volume). First, a mutant with altered protein 2A has normal kinetics of viral RNA synthesis but produces only low levels of viral proteins and does not inhibit host protein synthesis (10, 11). Protein 2A of wild-type poliovirus autocatalytically cleaves its junction with protein 1D (114) and is also involved, perhaps indirectly, in the cleavage of p220 of the cap-binding protein complex and the subsequent shutoff of host cell translation (for a review, see reference 106). In mutant-infected cells, p220 is not cleaved.

Second, poliovirus mutants with changes in the 5′ noncoding region (nucleotides 224, 270, or 392) have delayed synthesis of viral RNA, are deficient in viral protein synthesis and inhibition of host cell translation, and are not complemented by the 2A mutant (115). These mutants have a wild-type phenotype when inoculated at a high multiplicity of infection. A 5′ noncoding segment of several hundred nucleotides that includes these mutations is thus thought to stimulate viral protein synthesis by an unknown cis-active mechanism. Similar mutants have also been recently described in other laboratories (76; S. Dildine and B. Semler, submitted for publication).

Third, the importance of secondary RNA structure within the 5′ noncoding regions of polioviruses has been suggested by analysis of a small-plaque mutant and its revertant that apparently destabilized and restored, respectively, a predicted 5'-terminal stem-and-loop figure (86). In addition, a nucleotide substitution at base 472 of the poliovirus 3 genome has been associated with a decrease in neurovirulence and a change in predicted secondary structure (31).

Fourth, an insertion in the 3' noncoding region has been shown to result in a temperature-sensitive poliovirus (93). It appeared that small amounts of viral proteins were synthesized at the nonpermissive temperature because the 2A mutant, described above, was complemented by this mutant.

The size and amino acid sequence of HAV 2A are quite different from those of poliovirus 2A (Table 3); HAV 2A also lacks the closely spaced Cys and His residues found in the proposed active site of poliovirus 2A (114). The 2A functions of enteroviruses and rhinoviruses may therefore be absent in HAV. It is not known whether the HAV polyprotein has a primary cleavage at 1D/2A, as in enteroviruses and rhinoviruses, or at 2A/2B, as in cardioviruses and aphthoviruses (5).

Mutations among HAV strains and variants (Table 4) in the hypervariable segment (nucleotides 99 to 207) and in the 3' noncoding region may result from selection for better growth in particular cells. The only sequence reported thus far from a cytolytic strain of HAV (116; Table 2) contains this hypervariable segment: it resembles other HAV sequences for the first 159 nucleotides, but there is no detectable similarity for the remaining 29 bases. Thus, the replication of wild-type HAV, as it is studied in vitro, could be analogous to that of polioviruses with mutations in the noncoding regions. Furthermore, it is difficult to detect viral proteins in HAV-infected cells (62, 118), and in vitro translation of HAV RNA appears to be relatively inefficient (37). HAV may have a 5' noncoding region that is deficient for stimulating viral translation in cell culture.

The 5' noncoding regions of wild-type and cell culture-adapted variants of HAV strain HM175 were compared to determine whether certain nucleotide differences contribute to alterations in predicted RNA structures (Fig. 3). Similarities had previously been found in the predicted secondary structures of 5' noncoding region RNA from different HAV strains, including the presence of stem-and-loop figures involving the first 40 nucleotides that resemble those of 5'-terminal sequences from other picornaviruses (18, 20, 75, 86a; V. Rivera, B.A. thesis, Princeton University, Princeton, N.J., 1986; J. Ticehurst, unpublished data). Several stem-and-loop figures are shared by the three predicted structures shown in Fig. 3, but the structures of HM175/wt and HM175/P16 are the most similar. Most of the changes between HM175/wt and P16 result

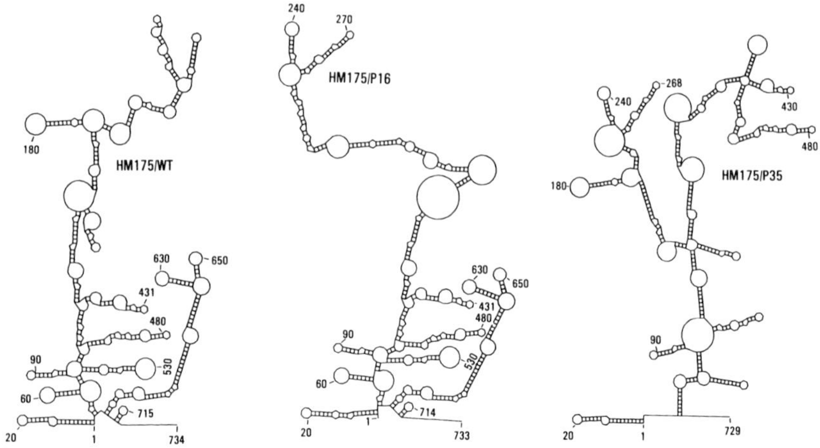

Figure 3. Predicted secondary structures in 5' noncoding regions of HM175 variants. Sequences from HM175/wt (left), P16 (middle), and P35 (right) were analyzed for possible secondary structures by using computer programs (45, 96). Predicted structures were unfolded to emphasize stem-and-loop figures in common between variants. All three sequences have similar structures around nucleotide positions 20, 90, 430, and 480. HM175/wt and P16 sequences also share structures at positions 60, 530, 630, and 650, HM175/wt and P35 at position 180, and HM175/P16 and P35 at positions 240 and 270.

from an A-to-G mutation at nucleotide 152. The structure of the 5' noncoding region of HM175/P59, assuming that its first 28 nucleotides are identical to those of HM175/wt, is very similar to the structures of HM175/wt and P16 (not shown). The base changes in the highly conserved segment (Table 4, nucleotides 636 to 725) have little effect on predicted structures. Most of the changes in the HM175/P35 structure result from a U-to-C change at nucleotide 124 (18). Some of the differences in these predicted RNA structures reflect the different passage histories of the HM175 variants (Table 1), but it remains uncertain whether these structural changes relate to cell culture adaptation or changes in virulence.

Mutations Necessary for Efficient Growth in Cell Culture Are in Nonstructural Genes

Although adaptation and quantitative improvement of HAV replication in cell culture may involve noncoding region mutations, none of those between HM175/wt and P35 are required for growth because cDNA hybrids containing the HM175/wt noncoding regions are infectious for cultured cells (17). The progeny virus of such hybrid cDNAs was attenuated for marmosets (as measured by levels of serum isocitric dehydrogenase), suggesting that changes in the noncoding regions are not necessary for the attenuated phenotype (J. Cohen, B. Rosenblum, S. Feinstone, J. Ticehurst, and R. Purcell, manuscript in preparation). Chimeras containing the wild-type P1 region are also infectious in vitro, so necessary mutations for efficient growth in cell culture are located in the P2 and P3 regions. When the nonstructural proteins of the HM175/P16 and P35 cell culture-adapted variants are compared (Fig. 1), the only alterations in common from HM175/wt are in 2B (Ala to Val at residue 72) and in 3D (Asp to Gly in HM175/P16 and to Asn in HM175/P35 at residue 67, Ser to Thr at residue 192).

The HAV Capsid Is Probably Different from Those of Other Picornaviruses

Several characteristics of the HAV capsid proteins may be related to the stability of the virus and its resistance to arildone and disoxaril. Although the three major HAV capsid proteins are thought to have β-barrel structures as determined for poliovirus 1, rhinovirus 14, and mengovirus, alignment of amino acid sequences was difficult because of minimal similarity (67, 72; see Palmenberg, this volume). Furthermore, 1A of HAV is considerably shorter in length than other 1A proteins (Table 3); the predicted size of VP4 is only 1.7 kilodaltons if 1A (2.5 kilodaltons) is cleaved to expose a site for myristylation (16; see Palmenberg, this volume). VP4 molecules of ≤2.5 kilodaltons have not been detected in mature HAV (K. Mihalik and S. Feinstone, unpublished data), and earlier reports which tentatively identified VP4 molecules of 7 to 14 kilodaltons (22) have not been confirmed.

The distant relationship between the capsid proteins of HAV and other picornaviruses, as well as the resistance of HAV to incubation at 60°C and to arildone and disoxaril, infers significant structural differences. An in vivo advantage for the naturally occurring HAV capsid is suggested by selection in monkeys of neutralization-sensitive virus from a neutralization-resistant inoculum (60), by the high level of antigenic conservation among HAV strains (59, 108; see Lemon and Ping, this volume), and by nearly identical amino acid sequences of P1 regions from different HAV strains (Fig. 1).

Efforts to Prevent Hepatitis A Will Also Benefit from Molecular Approaches to Understanding HAV Replication

Vaccine development is a practical reason for understanding HAV replication. Improved growth of HAV in cell culture would make prospective killed vaccines (12, 35, 83) more cost-effective. HM175/P35 is one of several candidates for a live virus vaccine (51, 80), and further elucidation of its molecular biology may help to identify critical determinants of cell culture adaptation, virulence, and vaccine stability.

Future studies of HAV replication will use sequence analysis and manipulation of infectious cDNA for more precise phenotypic characterization. For example, these approaches will be valuable for localizing sequences responsible for cytolytic, neutralization-resistant (see Lemon and Ping, this volume), and temperature-sensitive (Z.-Q. Wang and S. Lemon, unpublished results) phenotypes. Other phenotypes might be selected by environmental pressure, mutagenesis, or careful analysis of existing variants. HAV mutants that preferentially replicate in tissues other than liver might be advantageous for discovering a suspected extrahepatic primary site of replication in humans and possibly for developing a live attenuated vaccine. Methods such as in situ hybridization may help define extrahepatic and other replication events. It may also be possible to complement HAV by providing a *trans*-acting function or by making a chimera with another picornavirus, but such experiments should be carefully considered for their biohazard potential. The next few years should yield many more interesting findings about the replication of this unusual picornavirus.

Acknowledgments. Among many people who have had significant roles in the work cited here, we especially thank David Baltimore, William Bancroft, Bahige Baroudy, Leonard Binn, Robert Chanock, Taylor Chestnut, Richard Daemer, Ian Gust, Ruth Karron, Jacob Maizel, Kathleen Mihalik, Thomas Miele, John Newbold, Ann Palmenberg, Vincent Racaniello, Betsy Rosenblum, Max Shapiro, Jack Stapleton, Manfred Weitz, and Doris Wong. We also appreciate data and manuscripts, prior to their publication, from Gregory Duke, Ian Gust, Timo Hyypiä, Robert Ralston, Victor Rivera, Howard Thomas, Didier Trono, and Anders Widell. In addition, J.T. and S.M.L. thank the organizers of the 1988 ICN-UCI International Conference on Virology.

This work was supported in part by contract DAMD17-85C-5272 from the U.S. Army Medical Research and Development Command and by Public Health Service grant AI22279 from the National Institutes of Health.

Literature Cited

1. **Anderson, B. N., B. C. Ross, and I. D. Gust.** 1988. Sequence changes in capsid proteins of high-passage hepatitis A virus HM175, p. 55–58. *In* A. J. Zuckerman (ed.), *Viral Hepatitis and Liver Disease.* Alan R. Liss, Inc., New York.
2. **Anderson, D. A.** 1987. Cytopathology, plaque assay, and heat inactivation of hepatitis A virus strain HM-175. *J. Med. Virol.* 22:35–44.

3. **Anderson, D. A., S. A. Locarnini, B. C. Ross, A. G. Coulepis, B. N. Anderson, and I. D. Gust.** 1987. Single-cycle growth kinetics of hepatitis A virus in BSC-1 cells, p. 497–507. *In* M. A. Brinton and R. R. Rueckert (ed.), *Positive Strand RNA Viruses.* Alan R. Liss, Inc., New York.

4. **Argos, P., G. Kamer, M. J. H. Nicklin, and E. Wimmer.** 1984. Similarity in gene organization and homology between proteins of animal picornaviruses and a plant comovirus suggest common ancestry of these virus families. *Nucleic Acids Res.* **12:**7251–7257.

5. **Arnold, E., M. Luo, G. Vriend, M. G. Rossmann, A. C. Palmenberg, G. D. Parks, M. J. H. Nicklin, and E. Wimmer.** 1987. Implications of the picornavirus capsid structure for polyprotein processing. *Proc. Natl. Acad. Sci. USA* **84:**21–25.

6. **Asher, L. V. S., L. N. Binn, and R. H. Marchwicki.** 1987. Demonstration of hepatitis A virus in cell culture by electron microscopy with immunoperoxidase staining. *J. Virol. Methods* **15:**323–328.

7. **Balayan, M. S., A. G. Andzhaparidze, E. A. Tol'skaia, and M. S. Kolesnikova.** 1979. Possibility of reproducing hepatitis A virus infection in cell systems. *Vopr. Virusol.* **6:**675–676.

8. **Baron, M. H., and D. Baltimore.** 1982. Antibodies against the chemically synthesized genome-linked protein of poliovirus react with native virus-specified proteins. *Cell* **28:**395–404.

9. **Baroudy, B. M., J. R. Ticehurst, T. A. Miele, J. V. Maizel, Jr., R. H. Purcell, and S. M. Feinstone.** 1985. Sequence analysis of hepatitis A virus cDNA coding for capsid proteins and RNA polymerase. *Proc. Natl. Acad. Sci. USA* **82:**2143–2147.

10. **Bernstein, H. D., P. Sarnow, and D. Baltimore.** 1986. Genetic complementation among poliovirus mutants derived from an infectious cDNA clone. *J. Virol.* **60:**1040–1049.

11. **Bernstein, H. D., N. Sonenberg, and D. Baltimore.** 1985. Poliovirus mutant that does not selectively inhibit host cell protein synthesis. *Mol. Cell. Biol.* **5:**2913–2923.

12. **Binn, L. N., W. H. Bancroft, K. H. Eckels, R. H. Marchwicki, D. R. Dubois, L. V. S. Asher, J. W. LeDuc, C. J. Trahan, and D. S. Burke.** 1988. Inactivated hepatitis A virus vaccine produced in human diploid MRC-5 cells, p. 91–93. *In* A. J. Zuckerman (ed.), *Viral Hepatitis and Liver Disease.* Alan R. Liss, Inc., New York.

13. **Binn, L. N., S. M. Lemon, R. H. Marchwicki, R. R. Redfield, N. L. Gates, and W. H. Bancroft.** 1984. Primary isolation and serial passage of hepatitis A virus strains in primate cell cultures. *J. Clin. Microbiol.* **20:**28–33.

14. **Bradley, D. W., C. A. Schable, K. A. McCaustland, E. H. Cook, B. L. Murphy, H. A. Fields, J. W. Ebert, C. Wheeler, and J. E. Maynard.** 1984. Hepatitis A virus: growth characteristics of in vivo and in vitro propagated wild and attenuated strains. *J. Med. Virol.* **14:**373–386.

15. **Callahan, P. L., S. Mizutani, and R. J. Colonno.** 1985. Molecular cloning and complete sequence determination of RNA genome of human rhinovirus type 14. *Proc. Natl. Acad. Sci. USA* **82:**732–736.

16. **Chow, M., J. F. E. Newman, D. Filman, J. M. Hogle, D. J. Rowlands, and F. Brown.** 1987. Myristylation of picornavirus capsid protein VP4 and its structural significance. *Nature* (London) **327:**482–486.

17. **Cohen, J. I., B. Rosenblum, S. Feinstone, J. Ticehurst, and R. H. Purcell.** 1988. Use of infectious hepatitis A virus cDNA to study viral attenuation, p. 133–137. *In* R. M. Chanock, R. A. Lerner, F. Brown, and H. Ginsburg (ed.), *Modern Approaches to New Vaccines Including Prevention of AIDS.* Cold Spring Harbor Laboratory, Cold Spring Harbor, N.Y.

18. **Cohen, J. I., B. Rosenblum, J. R. Ticehurst, R. J. Daemer, S. M. Feinstone, and R. H.**

Purcell. 1987. Complete nucleotide sequence of an attenuated hepatitis A virus: comparison with wild-type virus. *Proc. Natl. Acad. Sci. USA* **84:**2497–2501.

19. Cohen, J. I., J. R. Ticehurst, S. M. Feinstone, B. Rosenblum, and R. H. Purcell. 1987. Hepatitis A virus cDNA and its RNA transcripts are infectious in cell culture. *J. Virol.* **61:**3035–3039.

20. Cohen, J. I., J. R. Ticehurst, R. H. Purcell, A. Buckler-White, and B. M. Baroudy. 1987. Complete nucleotide sequence of wild-type hepatitis A virus: comparison with different strains of hepatitis A virus and other picornaviruses. *J. Virol.* **61:**50–59.

21. Coulepis, A. G., B. N. Anderson, and I. D. Gust. 1987. Hepatitis A. *Adv. Virus Res.* **32:**129–169.

22. Coulepis, A. G., S. A. Locarnini, E. G. Westaway, G. A. Tannock, and I. D. Gust. 1982. Biophysical and biochemical characterization of hepatitis A virus. *Intervirology* **18:**107–127.

23. Crawford, N. M., and D. Baltimore. 1983. Genome-linked protein VPg of poliovirus is present as free VPg and VPg-pUpU in poliovirus-infected cells. *Proc. Natl. Acad. Sci. USA* **80:**7452–7455.

24. Cromeans, T., M. D. Sobsey, and H. A. Fields. 1987. Development of a plaque assay for a cytopathic, rapidly replicating isolate of hepatitis A virus. *J. Med. Virol.* **22:**45–56.

25. Daemer, R. J., S. M. Feinstone, I. D. Gust, and R. H. Purcell. 1981. Propagation of human hepatitis A virus in African green monkey kidney cell culture: primary isolation and serial passage. *Infect. Immun.* **32:**388–393.

26. de Chastonay, J., and G. Siegl. 1987. Replicative events in hepatitis A virus-infected cells. *Virology* **157:**68–75.

27. de Chastonay, J., and G. Siegl. 1988. Effect of three known antiviral substances on hepatitis A virus replication, p. 956–960. *In* A. J. Zuckerman (ed.), *Viral Hepatitis and Liver Disease*. Alan R. Liss, Inc., New York.

28. Dever, T. E., M. J. Glynias, and W. C. Merrick. 1987. GTP-binding domain: three consensus sequence elements with distinct spacing. *Proc. Natl. Acad. Sci. USA* **84:**1814–1818.

29. Diamond, D. C., E. Wimmer, K. von der Helm, and F. Deinhardt. 1986. The genomic map of hepatitis A virus: an alternate analysis. *Microbiol. Pathogen.* **1:**217–219.

30. Dulbecco, R., and H. Ginsburg. 1980. *Virology*, p. 1096–1117. Harper and Row, Publishers, Inc., Philadelphia.

31. Evans, D. M. A., G. Dunn, P. D. Minor, G. C. Schild, A. J. Cann, G. Stanway, J. W. Almond, K. Currey, and J. V. Maizel, Jr. 1985. Increased neurovirulence associated with a single nucleotide change in a noncoding region of the Sabin type 3 poliovaccine genome. *Nature* (London) **314:**548–550.

32. Feinstone, S. M., A. Z. Kapikian, and R. H. Purcell. 1973. Hepatitis A: detection by immune electron microscopy of a virus-like antigen associated with acute illness. *Science* **182:**1026–1028.

33. Flanegan, J. B., and D. Baltimore. 1977. Poliovirus-specific primer-dependent RNA polymerase able to copy poly(A). *Proc. Natl. Acad. Sci. USA* **74:**2677–2680.

34. Flanegan, J. B., R. F. Petterson, V. Ambros, M. J. Hewlett, and D. Baltimore. 1977. Covalent linkage of a protein to a defined nucleotide sequence at the 5' terminus of the virion and replicative intermediate RNAs of poliovirus. *Proc. Natl. Acad. Sci. USA* **74:**961–965.

35. Flehmig, B., A. Haage, U. Heinracy, and M. Pfisterer. 1988. Studies with an inactivated hepatitis A vaccine, p. 100–105. *In* A. J. Zuckerman (ed.), *Viral Hepatitis and Liver Disease*. Alan R. Liss, Inc., New York.

36. Gauss-Müller, V., E. Lottspeich, and F. Deinhardt. 1986. Characterization of hepatitis A virus structural proteins. *Virology* 155:732–736.

37. Gauss-Müller, V., K. von der Helm, and F. Deinhardt. 1984. Translation *in vitro* of hepatitis A virus RNA. *Virology* 137:182–184.

38. Goad, W. B., and M. I. Kanehisa. 1982. Pattern recognition in nucleic acid sequences. I. A general method for finding local homologies and symmetries. *Nucleic Acids Res.* 10:247–263.

39. Gorbalenya, A. E., V. M. Blinov, and E. V. Koonin. 1985. Prediction of nucleotide-binding properties of virus-specific proteins from their primary structure. *Mol. Genet. Mikrobiol. Virusol.* 11:30–36.

40. Gust, I. D., and S. M. Feinstone. 1988. *Hepatitis A.* CRC Press, Boca Raton, Fla.

41. Gust, I. D., N. I. Lehmann, S. Crowe, M. McCrorie, S. A. Locarnini, and C. R. Lucas. 1984. The origin of the HM175 strain of hepatitis A virus. *J. Infect. Dis.* 151:365–366.

42. Hanecak, R., B. L. Semler, C. W. Anderson, and E. Wimmer. 1982. Proteolytic processing of poliovirus polypeptides: antibodies to polypeptide P3-7c inhibit cleavage at glutamine-glycine pairs. *Proc. Natl. Acad. Sci. USA* 79:3973–3977.

43. Hogle, J. M., M. Chow, and D. J. Filman. 1985. Three-dimensional structure of poliovirus at 2.9 Å resolution. *Science* 229:1358–1367.

44. Hyypiä, T., M. Maaronen, P. Auvinen, P. Stålhandske, U. Pettersson, G. Stanway, P. Hughes, J. Ryan, J. Almond, M. Stenvik, and T. Hovi. 1987. Nucleic acid sequence relationships between enterovirus serotypes. *Mol. Cell. Probes* 1:169–176.

45. Jacobson, A. B., L. Good, J. Sinometti, and M. Zuker. 1984. Some simple computational methods to improve the folding of large RNAs. *Nucleic Acids Res.* 12:45–52.

46. Jansen, R. W., J. E. Newbold, and S. M. Lemon. 1985. Combined immunoaffinity cDNA-RNA hybridization assay for detection of hepatitis A virus in clinical specimens. *J. Clin. Microbiol.* 22:984–989.

47. Jansen, R. W., J. E. Newbold, and S. M. Lemon. 1988. Complete nucleotide sequence of a cell culture-adapted variant of hepatitis A virus: comparison with wild-type virus with restricted capacity for *in vitro* replication. *Virology* 163:299–307.

48. Jiang, X., M. K. Estes, and T. G. Metcalf. 1987. Detection of hepatitis A virus by hybridization with single-stranded RNA probes. *Appl. Environ. Microbiol.* 53:2487–2495.

49. Karayiannis, P., T. Jowett, M. Enticott, D. Moore, M. Pignatelli, F. Brenes, P. J. Scheuer, and H. C. Thomas. 1986. Hepatitis A virus replication in tamarins and host immune response in relation to pathogenesis of liver cell damage. *J. Med. Virol.* 18:261–276.

50. Karayiannis, P., M. J. McGarvey, M. A. Fry, and H. C. Thomas. 1988. Detection of hepatitis A virus RNA in tissues and feces of experimentally infected tamarins by cDNA-RNA hybridization, p. 117–120. *In* A. J. Zuckerman (ed.), *Viral Hepatitis and Liver Disease.* Alan R. Liss, Inc., New York.

51. Karron, R. A., R. Daemer, J. Ticehurst, E. D'Hondt, H. Popper, K. Mihalik, J. Phillips, S. Feinstone, and R. H. Purcell. 1988. Studies of prototype live hepatitis A virus vaccines in primate models. *J. Infect. Dis.* 157:338–345.

52. Kirkegaard, K., and D. Baltimore. 1986. The mechanism of RNA recombination in poliovirus. *Cell* 47:433–443.

53. Kitamura, N., B. Semler, P. G. Rothberg, G. R. Larsen, C. J. Adler, A. J. Dorner, E. A. Emini, R. Hanecak, J. J. Lee, S. van der Werf, C. W. Anderson, and E. Wimmer. 1981. Primary structure, gene organization and polypeptide expression of poliovirus RNA. *Nature* (London) 291:547–553.

54. Kojima, H., T. Shibayama, A. Sato, S. Suzuki, F. Ichida, and C. Hamada. 1981.

Propagation of human hepatitis A virus in conventional cell lines. *J. Med. Virol.* 7:273–286.

55. **Krawczynski, K. K., D. W. Bradley, B. L. Murphy, J. W. Ebert, T. E. Anderson, I. L. Doto, A. Nowoslawski, W. Duermeyer, and J. E. Maynard.** 1981. Pathogenetic aspects of hepatitis A virus infection in enterally inoculated marmosets. *Am. J. Clin. Pathol.* 76:698–706.

56. **Lee, Y. F., A. Nomoto, B. M. Detjen, and E. Wimmer.** 1977. A protein covalently linked to poliovirus genome RNA. *Proc. Natl. Acad. Sci. USA* 74:59–63.

57. **Lemon, S. M.** 1985. Type A viral hepatitis: new developments in an old disease. *N. Engl. J. Med.* 313:1059–1067.

58. **Lemon, S. M., L. N. Binn, and R. H. Marchwicki.** 1983. Radioimmunofocus assay for quantitation of hepatitis A virus in cell cultures. *J. Clin. Microbiol.* 17:834–839.

59. **Lemon, S. M., S.-F. Chao, R. W. Jansen, L. N. Binn, and J. W. LeDuc.** 1987. Genomic heterogeneity among human and non-human strains of hepatitis A virus. *J. Virol.* 61:735–742.

60. **Lemon, S. M., J. T. Stapleton, J. W. LeDuc, D. Taylor, R. Marchwicki, and L. N. Binn.** 1988. Cell-culture-adapted variant of hepatitis A virus selected for resistance to neutralizing monoclonal antibody retains virulence in owl monkeys, p. 70–73. *In* A. Zuckerman (ed.), *Viral Hepatitis and Liver Disease.* Alan R. Liss, Inc., New York.

61. **Linemeyer, D. L., J. G. Menke, A. Martin-Gallardo, J. V. Hughes, A. Young, and S. W. Mitra.** 1985. Molecular cloning and partial sequencing of hepatitis A viral cDNA. *J. Virol.* 54:247–255.

62. **Locarnini, S. A., A. C. Coulepis, E. G. Westaway, and I. D. Gust.** 1981. Restricted replication of human hepatitis A virus in cell culture: intracellular biochemical studies. *J. Virol.* 37:216–225.

63. **Lundquist, R. E., E. Ehrenfeld, and J. V. Maizel.** 1974. Isolation of a viral polypeptide associated with the poliovirus replication complex. *Proc. Natl. Acad. Sci. USA* 71:4774–4777.

64. **Mathiesen, L. R., J. Drucker, D. Lorenz, J. A. Wagner, R. J. Gerety, and R. H. Purcell.** 1978. Localization of hepatitis A antigen in marmoset organs during acute infection with hepatitis A virus. *J. Infect. Dis.* 138:369–377.

65. **Mathiesen, L. R., A. M. Moller, R. H. Purcell, W. T. London, and S. M. Feinstone.** 1980. Hepatitis A virus in the liver and intestine of marmosets after oral inoculation. *Infect. Immun.* 28:45–48.

66. **Melnick, J. L.** 1982. Classification of hepatitis A virus as enterovirus type 72 and of hepatitis B virus as hepadnavirus type 1. *Intervirology* 18:105–106.

67. **Mesyanzhinov, V. V., E. N. Peletskaya, V. M. Zhdanov, A. V. Efimov, and A. V. Finkelstein.** 1987. Prediction of secondary structure, spatial organization and distribution of antigenic determinants for hepatitis A virus proteins. *J. Biomol. Struct. & Dyn.* 5:447–458.

68. **Najarian, R., D. Caput, W. Gee, S. J. Potter, A. Renard, J. Merryweather, G. Van Nest, and D. Dina.** 1985. Primary structure and gene organization of human hepatitis A virus. *Proc. Natl. Acad. Sci. USA* 82:2627–2631.

69. **Nasser, A. M., and T. G. Metcalf.** 1987. Production of cytopathology in FRhK-4 cells by BS-C-1-passaged hepatitis A virus. *Appl. Environ. Microbiol.* 53:2967–2971.

70. **Nüesch, J., S. Krech, and G. Siegl.** 1988. Detection and characterization of subgenomic RNAs in hepatitis A virus particles. *Virology* 165:419–427.

71. **Ovchinnikov, Y. A., E. D. Sverdlov, S. A. Tsarev, S. G. Arsenian, T. O. Rokhlina, V. E. Chizhikov, N. A. Petrov, G. G. Prikhod'ko, V. M. Blinov, S. K. Vasilenko, L. S. Sandakhchiev, Y. Y. Kusov, V. I. Grabko, G. P. Fleer, M. S. Balayan, and S. G.**

Molecular Aspects of Picornavirus Infection and Detection
Edited by Bert L. Semler and Ellie Ehrenfeld
© 1989 American Society for Microbiology, Washington, DC 20006

Chapter 3

Comparison of Encephalomyocarditis Virus and Poliovirus with Respect to Translation Initiation and Processing In Vitro

Richard J. Jackson

Picornaviruses have always held a particular fascination for those of us whose primary interests lie in the mechanism and control of translation of eucaryotic mRNA. One reason is that in vitro translation proves to be a powerful method for studying the pathway of proteolytic processing of viral products; although such studies were originally limited to the use of virion RNA (13, 27–29), the range of questions they can answer has been greatly extended in recent years by cloning in transcription vectors (23, 28, 37, 39).

The other reason is that the specificity of initiation at the correct site on picornavirus RNAs poses a challenge to our understanding of how eucaryotic ribosomes select initiation sites; while it may be premature to say that picornaviral RNAs break the rules of the generally accepted scanning ribosome model (18), it seems unquestionable that they "bend" these rules to an extreme not found with other classes of mRNA or viral RNA. Here again, the exploitation of DNA cloning in transcription vectors offers a good chance that the problems will soon be solved. A further possible approach, not yet exploited to any great extent, is the use

Richard J. Jackson • Department of Biochemistry, University of Cambridge, Tennis Court Road, Cambridge CB2 1QW, United Kingdom.

of oligodeoxynucleotides and RNase H to achieve truncation of viral RNA at specific sites (20).

This article discusses these two facets of cell-free translation of picornaviral RNAs and places particular emphasis on the differences between the various picornavirus groups, especially between cardio- and enteroviruses.

PROCESSING IN CELL-FREE SYSTEMS

Most of the available information on picornaviral processing pathways is derived from studies using either poliovirus or encephalomyocarditis virus (EMC). It is common practice to try to amalgamate the results from these two systems into a single common model, and while there are undoubted similarities, one of the conclusions from our results described below is that there are distinct differences which argue against too much uncritical extrapolation between the two systems. Table 1 summarizes what appear to me to be the main outstanding questions concerning the differences between the two systems and also includes some comments on those processing steps of foot-and-mouth disease virus (FMDV) products which seem to differ from both the entero- and cardiovirus systems. The subdivision of the table into the historic categories of primary and secondary cleavages is done for convenience and is not intended to imply the previous assumption that primary cleavages are carried out by host cell proteases. Possibly the only exception, the only cleavage for which no virus-coded protease has been yet discovered, is the cleavage between P1-2A and -2B in the EMC and FMDV systems. This processing step remains a major mystery, for on the one hand its extreme rapidity and resistance to inhibition (14) suggest an intramolecular mechanism (i.e., involving a virus-coded protease), but on the other hand the failure to identify any virus-coded protease leaves open the possibility that it might indeed be a host-cell or reticulocyte lysate protease. However, supporters of the host-cell protease hypothesis should appreciate that even though lysates undoubtedly do contain proteases, such proteases have never been previously observed to effect such a rapid cleavage as that observed at the EMC P1-2A/2B junction.

For most other processing steps, the proteases involved have now been firmly identified, and the questions of current interest are those concerning the structural requirements for cleavage at each site (40) and whether the cleavage occurs exclusively by an intramolecular event (cleavage in *cis*) or in *trans*.

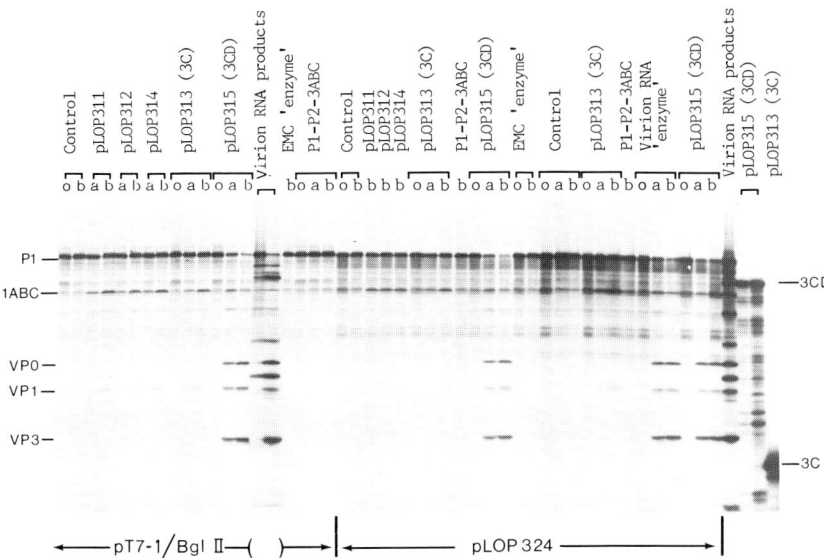

Figure 2. Poliovirus capsid precursors are cleaved by 3CD rather than 3C. RNA was synthesized in vitro by T7 RNA polymerase from plasmid pLOP324 (see text) cut with *Sma*I, or from pT7-1 (39) cut with *Bgl*II in the middle of the 3C coding region. To generate labeled P1, these RNAs were translated in vitro for 75 min at a final concentration of 80 mM added KCl, using RNA concentrations determined empirically to optimize the ratio of P1 synthesis to aberrant product synthesis. The assays were then supplemented with 10 μg of RNase per ml and 0.5 mM cycloheximide before mixing with an equal volume of unlabeled translation assays which had been incubated under the same conditions with either no RNA (control), poliovirion RNA (type 1 Mahoney strain), EMC RNA, or RNA generated from the following plasmids: pLOP311, pLOP312, pLOP313, pLOP314, and pLOP315 (defined in the text), or a plasmid coding for P1-P2-3ABC, constructed by fusing the large *Eco*RI-*Bgl*II fragment of pT7-1 to the *Bgl*II-*Sst*I fragment of pLOP313. After mixing, samples were taken for gel electrophoresis either immediately (o) or after 20 (a) or 80 (b) min of incubation. The tracks at the right display the labeled translation products after 180 min of incubation of virion RNA (repeated in the center of the gel together with the products after 45 min of incubation), the products of translation of two concentrations of pLOP315 RNA, and pLOP313 RNA. Translations of poliovirion RNA and of pT7-1/*Bgl*II RNA were supplemented with 1.05 mg of HeLa A fraction per ml, as defined in Fig. 5 and the text, to optimize synthesis of authentic products and minimize aberrant products.

INITIATION OF TRANSLATION OF PICORNAVIRUS RNAs

Translation of picornaviral RNAs is of particular interest by virtue of the question as to how these uncapped RNAs are efficiently translated from a single initiation site (or two sites in the case of FMDV) which is situated over 700 residues into the interior of the RNA and which is not the 5'-proximal AUG codon (4, 8, 17, 25, 32). Here again, there are striking differences between the cell-free translation characteristics of RNAs from different picornavirus groups, particularly in the rabbit reticulocyte lysate system. As uncapped RNAs they would be expected to be translated relatively inefficiently and with a low optimum salt concentration (especially if KCl is used), as is the case with cowpea mosaic virus M RNA, an uncapped RNA with a less complex 5' untranslated region than is found in picornaviral RNAs. In our hands poliovirus RNA partially conforms to these expectations in that the KCl optimum is about 75 mM and translation efficiency is low. It differs in that it is significantly less efficient than cowpea mosaic virus M RNA and gives a much higher proportion of incorrect products resulting from initiation at numerous internal (incorrect) sites (6, 31), especially at low salt or at high RNA concentrations (see below) or if potassium acetate is used in place of KCl. In contrast, EMC RNA is very efficient as a message, being among the most efficient mRNAs that we have encountered (FMDV RNA may be translated even more efficiently, though it is given to few of us to experience this personally!), and has a high salt optimum of about 120 mM KCl. Clearly EMC RNA is anomalous (for an uncapped RNA) for its high translational efficiency and high salt optimum, while poliovirus is anomalous for its very low efficiency and unusually high susceptibility to incorrect initiation site selection.

However, in extracts from HeLa cells or L cells, poliovirus RNA is almost as efficient as EMC RNA and shows little synthesis of aberrant products (6) even when potassium acetate is used. Our experience suggests that the difference between the relative efficiencies of poliovirus and EMC RNAs in the two types of cell-free system can almost entirely be explained by an increase in the fidelity and efficiency of poliovirus RNA translation in HeLa or L-cell extracts relative to reticulocyte lysates. While it is true that EMC translation appears to be less efficient in HeLa or L-cell extracts than in reticulocyte lysates, so too is the translation of other RNAs such as tobacco mosaic virus RNA or cowpea mosaic virus RNA, and we believe this merely reflects a lower general efficiency of these extracts from nucleated cells and a considerably larger pool of unlabeled methionine (some 10-fold larger in our hands). Our concern should therefore be focused on why poliovirus RNA is a much better message (and not on why EMC RNA appears to be a poorer message) in HeLa or L-cell extracts than in reticuloycte lysates.

Unique Features of Initiation on EMC RNA

The current version of the scanning ribosome model for initiation site selection on eucaryotic mRNAs (18) asks us to imagine that 40S ribosomal subunits scan FMDV or EMC RNA from the 5' end, through the poly(C) tract, past several putatively nonfunctional AUG codons, until they reach the correct AUG initiation site (or the two correct sites, in the case of FMDV). Yet the extraordinarily complex secondary structure of the 5' region of FMDV RNA (22) would seem to eliminate the possibility of such a scanning mechanism, according to the results presented by Kozak (19). Indeed, our experience with constructs of capped influenza virus NS mRNA suggests that hairpin loops having a calculated stability of −30 kCal/mol are sufficient to inhibit in vitro translation by over 80% (M. C. Dasso and R. J. Jackson, unpublished observations). While it can be argued that this secondary structure in the 5' segment of FMDV RNA may be a computer-generated fantasy that has not been proved to exist in reality, the fact that different FMDV serotypes have compensatory base changes such as to maintain this secondary structure is a strong argument that it really does exist (22). EMC RNA also has the potential for extensive secondary structure at the 5' end (38), albeit less spectacular than in FMDV, but nevertheless sufficient to impede scanning according to current ideas (18).

One of the puzzles of initiation on these RNAs is that if they are translated so efficiently, and if the ribosomes do indeed scan from the 5' end over a distance in excess of 800 nucleotide residues, one would expect to see a queue of 40S subunits building up over this long 5' untranslated region. Yet in our hands, EMC RNA binds only one ribosome in a standard ribosome binding assay; in fact, we always used this as a control for cases where we had evidence for the simultaneous binding of two (80S) ribosomes, as in cowpea mosaic virus M RNA (30), tobacco mosaic virus RNA, or papaya mosaic virus RNA (13). (We have considered the possibility that our sucrose gradient analysis conditions may be such as to disrupt any complexes of scanning 40S subunits on mRNA, but the fact that we do detect two, or sometimes three, 40S subunits on globin mRNA in the presence of edeine makes this an unlikely explanation of our failure to see a queue of 40S subunits on EMC RNA.)

Another peculiarity of initiation on EMC RNA is that it is much less ATP dependent than when capped mRNAs such as tobacco mosaic virus RNA are examined (13). According to current ideas, the ATP is required in initiation not for the scanning process per se, but acts first at the recognition of the cap (on capped mRNAs) by eIF-4F, and more especially in the action of eIF-4A as an RNA-unwinding protein (21). Taken at

a b c d e f   g h i j k l m

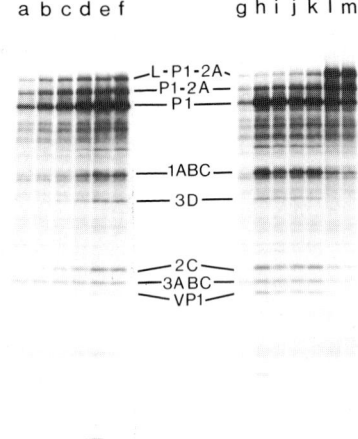

Figure 3. Effect of potassium salts on the pattern of products of EMC RNA translation. EMC RNA was translated at 36 µg/ml with 0.2 mCi of [^{35}S]methionine per ml under conditions described previously (14) except that the following concentrations of KCl or KSCN were added, and sufficient potassium acetate was also added to give a final concentration of 100 mM added K^+ in each assay: (a) 0, (b) 20 mM, (c) 40 mM, (d) 60 mM, (e) 80 mM, (f) 100 mM KCl; (g) 0, (h) 5 mM, (i) 10 mM, (j) 15 mM, (k) 20 mM, (l) 25 mM, (m) 30 mM KSCN. Incubation was for 90 min.

face value, our results suggest that initiation on EMC RNA is less dependent on such unwinding than is the case with most other RNAs.

A further peculiarity of EMC RNA translation was revealed when we were comparing the efficiency of translation in HeLa or L-cell extracts (which we routinely supplement with potassium acetate) versus reticulocyte lysates (in which we normally use KCl). This comparison necessitated controls with potassium acetate in the reticulocyte lysate system, and we were surprised to find that overall translational efficiency was much lower than with KCl and the product pattern was less specific in a way that was suggestive of initiation at internal sites (Fig. 3). In case this was a peculiarity of one batch of lysate, we examined four other batches; all showed the lower efficiency and less specific product pattern, although the precise degree of inferiority of potassium acetate was batch dependent. This response of EMC RNA to the two potassium salts was precisely the reverse of what was observed with other RNAs (tobacco mosaic virus RNA, poliovirus RNA, cowpea mosaic virus RNA, bacteriophage MS2 RNA), which were translated more efficiently in the presence of potassium acetate than with KCl (although with the uncapped RNAs it should be added that potassium acetate promoted, and KCl restricted, the appearance of aberrant products resulting from internal initiation).

The anomalous response of EMC RNA translation to these two potassium salts prompted the testing of a wider range of salts, which revealed that the addition of low concentrations of KSCN (5 to 10 mM) to translations supplemented with 90 to 100 mM potassium acetate strongly stimulated overall translational efficiency and improved the quality of the product pattern (Fig. 3). Translation with this combination of KSCN and potassium acetate was at least as efficient as, if not slightly better than, when 100 to 120 mM KCl was used. No other potassium salt tested (glutamate, bromide, phosphate, or sulfate) had this specific activating effect, and no other defined mRNA (not even poliovirus RNA) showed the same response to traces of KSCN, although when a mixed mRNA population (from starfish oocytes) was translated in the presence of potassium acetate, the addition of KSCN did enhance the synthesis of a very few proteins.

It is tempting to speculate that the low ATP requirement and the specific KSCN (or KCl) activation of initiation at the authentic site are manifestations of an initiation mechanism that is peculiar to EMC RNA. That this might be a mechanism which bypasses the normal scanning process was suggested by the observation that oligodeoxynucleotides complementary to any region within the 5'-proximal 450 residues did not inhibit translation (33), as would be expected according to a scanning ribosome model (18, 19). More compelling evidence for internal initiation has recently been provided by studies of a dicistronic construct in which the downstream cistron was translated with higher efficiency and more rapidly than the upstream, provided the intercistronic region included a critical segment, from about nucleotide 250, of the EMC 5' untranslated sequences (16). If ribosomes can gain access to the authentic initiation site of EMC RNA without scanning from the 5' end of the RNA, the question arises as to how the correct site is identified. Three possibilities deserve consideration: (i) initial binding at an internal site, probably situated 3' to the poly(C) tract, followed by conventional scanning until the correct AUG initiation codon is reached; (ii) binding to an internal site as in (i), followed by transfer of the 40S subunit to the correct initiation site without linear scanning of the intervening RNA sequence; (iii) direct binding to the correct initiation site. The results of Shih et al. make (iii) seem the least likely possibility (33) but do not definitively eliminate it. Further work is required to define the minimal EMC sequence necessary to activate translation of the downstream cistron of dicistronic mRNAs.

Initiation of Poliovirus RNA Translation: Activation by HeLa Cell Factors

As mentioned above, the translation of poliovirus RNA in the reticulocyte lysate system is anomalous for its very low efficiency and strong tendency for internal initiation and synthesis of incorrect products (6, 31), especially at high RNA concentrations. The dose response is most bizarre: maximal synthesis of P1 and other correct P1-derived products is

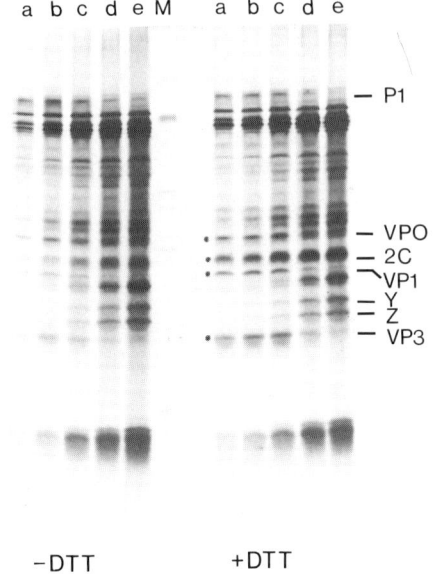

a b c d e M   a b c d e

— P1

— VPO
— 2C
— VP1
— Y
— Z
— VP3

−DTT   +DTT

Figure 4. Dose response of poliovirion RNA in the reticulocyte lysate. Poliovirus RNA (type 1 Mahoney) was translated for 180 min under standard conditions (15), except that the final concentration of added KCl was 70 mM and dithiothreitol (DTT) was present at 2.5 mM where indicated. The final concentration of viral RNA in each assay was: (a) 4, (b) 8, (c) 16, (d) 32, and (e) 64 μg/ml. Lane M was loaded with radioactive marker proteins diluted in unlabeled translation assay mix. Y and Z are two products originating from strong internal initiation sites as described previously (6).

observed at low RNA concentrations (about 5 μg/ml); higher concentrations result in **reduced** synthesis of P1 (or P1-derived polypeptides) and more synthesis of incorrect products. This is observed both with virion RNA (Fig. 4) and, to an even greater extent, with RNA produced by in vitro transcription (Fig. 5). (Qualitatively, there is little difference between the two types of RNA preparation, and the aberrant translation products from one are very similar to the other, but it is our experience that virion RNA is translated more efficiently for reasons that are not yet clear.) On the other hand, HeLa or L-cell extracts give almost exclusively the correct products (6) over a wide RNA concentration range and show a more typical dose response to RNA concentration. The addition of relatively small amounts (20% by volume) of HeLa or L-cell extract to the reticulocyte lysate system enhances the synthesis of authentic products and suppresses internal initiation (6, 31). These observations suggest that efficient and faithful translation of poliovirus RNA requires one or more factors that are either missing from reticulocyte lysates or, more likely, are present in very low concentrations relative to HeLa cell extracts.

In collaboration with S. Milburn and J. W. B. Hershey (University of California, Davis), my colleagues and I have been looking for such factors in HeLa cell extracts. The presence of such factors in the HeLa ribosomal salt wash was presaged by the observations of Brown and Ehrenfeld (1)

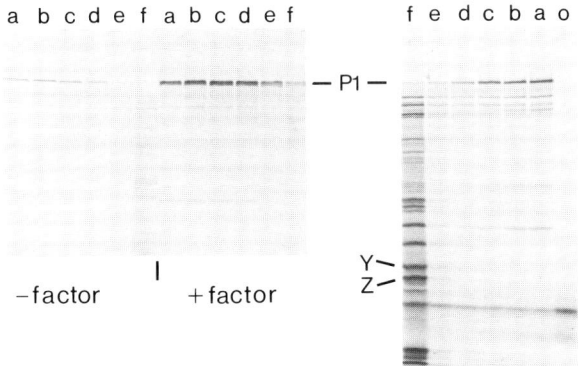

Figure 5. Effect of HeLa A fraction on dose response of poliovirus RNA generated by in vitro transcription. RNA was generated by transcription of SmaI-cut pT7-1 DNA (39) by T7 RNA polymerase and translated for 90 min in the reticulocyte lysate system with a final concentration of 70 mM added KCl and 5 mM dithiothreitol. Where indicated, the assays shown in the left panel were supplemented with HeLa cell A-cut factor preparation (the 0 to 40% saturated ammonium sulfate precipitate of the 0.5 M KCl ribosomal wash fraction) at a final concentration of 1.05 mg of protein per ml. Final RNA concentrations were: (a) 4, (b) 6, (c) 9, (d) 14, (e) 22, and (f) 33 μg/ml. The same concentrations of a different RNA preparation were used for the right panel, where the gel was run so as to display the full range of products including Y and Z as defined in Fig. 4. The labeled products visible in track o, the no-RNA control, are peptidyl-tRNA, as the samples were not treated with RNase before gel electrophoresis (15).

that the addition of HeLa ribosomal salt wash to reticulocyte lysates elicited a more authentic product pattern, whereas reticulocyte ribosomal salt wash seemed to enhance the synthesis of all products, both authentic and incorrect (although in our hands reticulocyte ribosomal salt wash, or ammonium sulfate fractions derived from it, has very little effect). It was therefore not very surprising to find that the addition of HeLa ribosomal salt wash to reticulocyte lysates enhanced the synthesis of correct poliovirus products and inhibited the synthesis of incorrect ones (Fig. 5). The effect is seen not only with in vitro-synthesized RNA, as in Fig. 5, but also with type I virion RNA as well as virion RNAs from attenuated and neurovirulent type III strains (kindly provided by J. W. Almond and P. Minor). The activity was routinely found in the 0 to 40% saturated ammonium sulfate precipitate (known as the A fraction) derived from this salt wash fraction, and in some, but not all, preparations there was activity also in the 40 to 70% ammonium sulfate cut. No activity was found in the HeLa cell postribosomal supernatant fraction or in subfractions derived from it.

One could envisage this HeLa cell activity operating in one of two ways. It might act primarily to suppress the internal initiation that gives rise to aberrant products, in which case the enhancement of authentic product synthesis could be regarded as the passive consequence of inhibition of this competing internal initiation. Alternatively, the HeLa factor might act primarily by enhancing initiation at the correct site (nucleotide 743), and the suppression of internal initiation would then be the passive consequence of more efficient competing initiation at the correct site. A third possibility is that the HeLa fraction contains both types of activity, one enhancing initiation at the correct site and the other suppressing internal initiation. To guard against the latter possibility, assays have been done under two conditions: (i) at 4 to 5 μg of poliovirus RNA (synthesized in vitro) per ml, when synthesis is predominantly of authentic poliovirus products (albeit in low yield) and comparatively little internal initiation is observed; (ii) at 30 μg/ml, when synthesis is predominantly of incorrect products and authentic P1 is a very minor product (Fig. 5). The former assays for specific enhancement of correct initiation; the latter determines suppression of internal initiation. Using these assays, we obtained evidence that the HeLa A fraction does contain both these types of activity, as they could be resolved by gel filtration. The activity enhancing correct initiation behaved with an apparent size of about 50 kilodaltons, while the activity suppressing internal initiation was about 120 kilodaltons. They could also be separated by ion-exchange chromatography on cation exchangers (Pharmacia FPLC Mono-S); the activity enhancing correct initiation did not stick to this column at 75 to 100 mM salt, but the internal initiation-suppressing activity bound very tightly.

None of the recognized initiation factors could substitute for the HeLa cell activity. In particular, eIF-4A, which Thach's group have suggested might be the factor limiting overall efficiency of poliovirus RNA translation (5), was found to have no effect on efficiency or product pattern at either low or high RNA concentrations, even when very high concentrations of eIF-4A (250 μg/ml) were tested. This is perhaps not surprising, since we are looking for a factor which is more abundant or qualitatively different in HeLa cells as opposed to reticulocyte lysates, and yet eIF-4A is one of the most abundant initiation factors in the reticulocyte lysate system. Indeed, none of the recognized initiation factors is either strikingly more abundant or qualitatively very different in HeLa extracts than in reticulocyte lysates (7). It is therefore far from improbable that we are pursuing a hitherto unrecognized "factor" or, alternatively, a hitherto unrecognized activity that modifies one of the recognized initiation factors.

As an alternative test of whether the HeLa activity acts primarily to enhance initiation at the correct site or to suppress initiation at incorrect sites, I have examined, in collaboration with B. L. Semler and M.-F. Ypma-Wong, the effect of truncating the RNA on (i) the efficiency of translation in the reticulocyte lysate, (ii) the dose response to RNA concentration, and (iii) the response to HeLa A fraction. As our previous studies identified most of the strong internal initiation sites as lying in the P3-coding region (6), the use of runoff transcripts synthesized from a cloned poliovirus genome cut with restriction enzymes at the end of the P1-coding region allowed us to test whether elimination of the strongest competing internal initiation sites automatically enhanced synthesis from the authentic site and reduced the requirement for HeLa factors. The outcome was that translation of this truncated RNA was only marginally more efficient than that of full-length RNA and showed a similar enhancement by HeLa factor. The unusual dose-response characteristics were still observed, and at high doses aberrant products were formed, albeit different products of internal initiation than are observed with full-length RNA. My colleagues and I have made similar observations with RNA transcribed from a clone constructed by J. Jore and B. Enger-Valk comprising the complete 5′ untranslated region and the complete P1-coding region, ending with a translation termination codon. Wimmer's group have also noted the low translational efficiency of RNA transcribed from a similar construct, but in their case the use of high potassium acetate levels probably exaggerated the poor translatability (23).

On the other hand, RNA transcribed from a clone lacking the 5′-proximal 670 residues of the poliovirus genome was translated significantly more efficiently than full-length RNA and showed a more typical dose response to increasing RNA concentration, with the synthesis of aberrant products at high RNA concentration much reduced, though not totally eliminated. The activation of translation by HeLa cell factor or crude HeLa (or L-cell) postmitochondrial supernatant was also largely abolished. In terms of translational efficiency and dose response, this RNA with the 5′ deletion typically gives a result in the **absence** of HeLa factor that is very similar to that of full-length RNA in the **presence** of HeLa A fraction illustrated in Fig. 5. Again, we have made similar observations of enhanced efficiency, more typical dose responses, reduced synthesis of aberrant products, and reduced response to HeLa factors by using RNA transcribed from clone pLOP324 (constructed by J. Jore and B. Enger-Valk) and comprising just the P1-coding region inserted between appropriate translation initiation and termination codons. A similar construct made in Wimmer's laboratory also showed greatly enhanced translational efficiency as compared with one which included all the 5' untranslated region (23).

Taken together, these results imply that the 5' untranslated region of poliovirus RNA is responsible for the poor translation of the viral RNA in the reticulocyte lysate system. They suggest that initiation at the correct site requires the interaction of one or more factors with this region of the RNA, and that such factors are very much more abundant in HeLa or L cells than in reticulocyte lysates (or are qualitatively different so that the HeLa or L-cell factors have higher specific activity). My co-workers and I argue that initiation at the authentic site requires a critical density of factors bound to this region: either several molecules of one type of factor, or different species of factor. We further suggest that such binding must be noncooperative, to explain the bizarre dose response whereby the addition of **more** RNA to the reticulocyte lysate results in **less** initiation at the authentic site. We propose that synthesis of aberrant products from internal initiation sites occurs primarily by default, whenever the factor/RNA ratio falls below the critical level to direct ribosomes to the authentic site. (However, our results do not exclude, and in fact favor, the proposition that HeLa cells also contain a factor that suppresses internal initiation by a more active mechanism than merely directing ribosomes to the competing authentic site.) According to these views, the critical HeLa factor that enhances initiation at the correct site should be identifiable not only by its property of enhancing the synthesis of authentic poliovirus products, but also by its specific binding to the 5' untranslated region of poliovirus RNA.

Acknowledgments. I would like to thank my colleagues in Cambridge for their indispensable help with this work: Ann Kaminski for all the in vitro transcription work and subsequent translation of the transcripts, Mary Dasso for advice on these procedures, and Rebecca Lane for HeLa and L-cell extracts. It should be apparent that my own role in this work is merely that of a simple-minded cell-free translator, and I am therefore deeply indebted to my many colleagues who have supplied initiation factors, virion RNAs, and various clones: Jeff Almond, Betty Enger-Valk, John Hershey, Jan Jore, Sue Milburn, Phil Minor, Bert Semler, Eckard Wimmer, and Mary Frances Ypma-Wong.

Part of this work was supported by a grant from the Medical Research Council, and the transatlantic collaboration was supported by grants first from the Royal Society and subsequently from NATO Scientific Affairs Division.

Literature Cited

1. **Brown, B. A., and E. Ehrenfeld.** 1979. Translation of poliovirus RNA *in vitro*: changes in cleavage pattern and initiation sites by ribosomal salt wash. *Virology* **97**:396–405.

2. **Butterworth, B. E., and B. D. Korant.** 1974. Characterization of the large picornaviral polypeptides produced in the presence of zinc ions. *J. Virol.* **14**:282–291.

3. **Campbell, E. A., and R. J. Jackson.** 1983. Processing of the encephalomyocarditis virus capsid precursor protein studied in rabbit reticulocyte lysates incubated with N-formyl-[^{35}S]methionine-tRNA$_f^{Met}$. *J. Virol.* **45**:439–441.

4. **Clarke, B. E., D. V. Sangar, J. N. Burroughs, S. E. Newton, A. R. Carroll, and D. J. Rowlands.** 1985. Two initiation sites for foot-and-mouth disease virus polyprotein *in vivo*. *J. Gen. Virol.* **66**:2615–2626.

5. **Daniels-McQueen, S., B. M. Detjen, J. A. Grifo, W. C. Merrick, and R. E. Thach.** 1983. Unusual requirements for optimum translation of polio viral RNA *in vitro*. *J. Biol. Chem.* **258**:7195–7199.

6. **Dorner, A. J., B. L. Semler, R. J. Jackson, R. Hanecak, E. Duprey, and E. Wimmer.** 1984. In vitro translation of poliovirus RNA: utilization of internal initiation sites in reticulocyte lysate. *J. Virol.* **50**:507–514.

7. **Duncan, R., and J. W. B. Hershey.** 1983. Identification and quantitation of levels of protein synthesis initiation factors in crude HeLa cell lysates by two-dimensional polyacrylamide gel electrophoresis. *J. Biol. Chem.* **258**:7228–7235.

8. **Forss, S., K. Strebel, E. Beck, and H. Schaller.** 1984. Nucleotide sequence and genome organisation of foot-and-mouth disease virus. *Nucleic Acids Res.* **12**:6587–6601.

9. **Gorbalenya, A. E., Y. V. Svitkin, and V. I. Agol.** 1981. Proteolytic activity of the nonstructural polypeptide p22 of encephalomyocarditis virus. *Biochem. Biophys. Res. Commun.* **98**:952–960.

10. **Gorbalenya, A. E., Y. V. Svitkin, Y. A. Kazachkov, and V. I. Agol.** 1979. Encephalomyocarditis virus-specific polypeptide p22 is involved in the processing of the viral precursor polypeptides. *FEBS Lett.* **108**:1–5.

11. **Hanecak, R., B. L. Semler, C. W. Anderson, and E. Wimmer.** 1982. Proteolytic processing of poliovirus polypeptides: antibodies to polypeptide P3-7c inhibit cleavage at glutamine-glycine pairs. *Proc. Natl. Acad. Sci. USA* **79**:3973–3977.

12. **Hanecak, R., B. L. Semler, H. Ariga, C. W. Anderson, and E. Wimmer.** 1984. Expression of a cloned gene segment of poliovirus in E. coli: evidence for autocatalytic production of the viral proteinase. *Cell* **37**:1063–1073.

13. **Jackson, R. J.** 1982. The control of initiation of protein synthesis in reticulocyte lysates, p. 362–418. *In* R. Perez-Bercoff (ed.), *Protein Biosynthesis in Eukaryotes*. Plenum Publishing Corp., New York.

14. **Jackson, R. J.** 1986. A detailed kinetic analysis of the *in vitro* synthesis and processing of encephalomyocarditis virus products. *Virology* **149**:114–127.

15. **Jackson, R. J., and T. Hunt.** 1983. Protein synthesis in rabbit reticulocyte lysates. *Methods Enzymol.* **96**:50–74.

16. **Jang, S. K., H.-G. Kräusslich, M. J. H. Nicklin, G. M. Duke, A. C. Palmenberg, and E. Wimmer.** 1988. A segment of the 5' nontranslated region of encephalomyocarditis virus RNA directs internal entry of ribosomes during in vitro translation. *J. Virol.* **62**:2636–2643.

16a. **Jore, J., B. de Geus, R. J. Jackson, P. H. Pouwels, and B. E. Enger-Valk.** 1988. Poliovirus protein 3CD is the active protease for processing of the precursor protein P1 in vitro. *J. Gen. Virol.* **69**:1627–1636.

17. **Kitamura, N., B. L. Semler, P. G. Rothberg, G. R. Larsen, C. J. Adler, A. J. Dorner,**

E. A. Emini, R. Hanecak, J. J. Lee, S. van der Werf, C. W. Anderson, and E. Wimmer. 1981. Primary structure, gene organisation, and polypeptide expression of poliovirus RNA. *Nature* (London) **291**:547–553.

18. Kozak, M. 1983. Comparison of initiation of protein synthesis in procaryotes, eucaryotes, and organelles. *Microbiol. Rev.* **47**:1–45.

19. Kozak, M. 1986. Influence of mRNA secondary structure on initiation by eukaryotic ribosomes. *Proc. Natl. Acad. Sci. USA* **83**:2850–2854.

20. Minshull, J., and T. Hunt. 1986. The use of single-stranded DNA and RNase H to promote quantitative "hybrid arrest of translation" of mRNA/DNA hybrids in reticulocyte lysate cell-free translations. *Nucleic Acids Res.* **214**:6433–6451.

21. Moldave, K. 1985. Eukaryotic protein synthesis. *Annu. Rev. Biochem.* **54**:1109–1149.

22. Newton, S. E., A. R. Carroll, R. O. Campbell, B. E. Clarke, and D. J. Rowlands. 1985. The sequence of foot-and-mouth disease virus RNA to the 5′ side of the poly(c) tract. *Gene* **40**:331–336.

23. Nicklin, M. J. H., H. G. Krausslich, H. Toyoda, J. J. Dunn, and E. Wimmer. 1987. Poliovirus polypeptide precursors: expression *in vitro* and processing by exogenous 3C and 2A. *Proc. Natl. Acad. Sci. USA* **84**:4002–4006.

24. Palmenberg, A. C. 1987. Genome organisation, translation and processing in picornaviruses, p. 1–15. *In* D. J. Rowlands, B. W. J. Mahy, and M. Mayo (ed.), *The Molecular Biology of the Positive Strand RNA Viruses*. Academic Press, Inc. (London), Ltd., London.

25. Palmenberg, A. C., E. M. Kirby, M. R. T. Janda, N. L. Drake, G. M. Duke, K. F. Potratz, and M. S. Collett. 1984. The nucleotide and deduced amino acid sequences of the encephalomyocarditis viral polyprotein coding region. *Nucleic Acids Res.* **12**: 2969–2985.

26. Palmenberg, A. C., M. A. Pallansch, and R. R. Rueckert. 1979. Protease required for processing picornaviral coat protein resides in the viral replicase gene. *J. Virol.* **32**:770–778.

27. Palmenberg, A. C., and R. R. Rueckert. 1982. Evidence for intramolecular self-cleavage of picornaviral replicase precursors. *J. Virol.* **41**:244–249.

28. Parks, G. D., G. M. Duke, and A. C. Palmenberg. 1986. Encephalomyocarditis virus 3C protease: efficient cell-free expression from clones which link viral 5′ noncoding sequences to the P3 region. *J. Virol.* **60**:376–384.

29. Pelham, H. R. B. 1978. Translation of encephalomyocarditis virus RNA *in vitro* yields an active proteolytic processing enzyme. *Eur. J. Biochem.* **85**:457–462.

30. Pelham, H. R. B. 1979. Synthesis and proteolytic processing of cowpea mosaic virus proteins in reticulocyte lysates. *Virology* **96**:463–477.

31. Phillips, B. A., and A. Emmert. 1986. Modulation of the expression of poliovirus proteins in reticulocyte lysates. *Virology* **148**:255–267.

32. Robertson, B. H., M. J. Grubman, G. N. Weddell, D. M. Moore, J. D. Welsh, T. Fischer, D. J. Dowbenko, D. G. Yansura, B. Small, and D. G. Kleid. 1985. Nucleotide and amino acid sequence coding for polypeptides of foot-and-mouth disease virus type A12. *J. Virol.* **54**:651–660.

33. Shih, D. S., I.-W. Park, C. L. Evans, J. M. Jaynes, and A. C. Palmenberg. 1987. Effects of cDNA hybridization on translation of encephalomyocarditis virus RNA. *J. Virol.* **61**:2033–2037.

34. Strebel, K., and E. Beck. 1986. A second protease of foot-and-mouth disease virus. *J. Virol.* **58**:893–899.

35. Svitkin, Y. V., A. E. Gorbalenya, Y. A. Kazachkov, and V. I. Agol. 1979. Encephalo-

myocarditis virus-specific polypeptide p22 possessing a proteolytic activity. *FEBS Lett.* **108:**6–9.

36. **Toyoda, H., M. J. H. Nicklin, M. G. Murray, C. W. Anderson, J. J. Dunn, F. W. Studier, and E. Wimmer.** 1986. A second virus-encoded proteinase involved in proteolytic processing of poliovirus polyprotein. *Cell* **45:**761–770.

37. **Vakharia, V. N., M. A. Devaney, D. M. Moore, J. J. Dunn, and M. J. Grubman.** 1987. Proteolytic processing of foot-and-mouth disease virus polyproteins expressed in a cell-free system from clone-derived transcripts. *J. Virol.* **61:**3199–3207.

38. **Vartapetian, A. B., A. S. Mankin, E. A. Skripin, K. M. Chumakov, V. D. Smirnov, and A. A. Bogdanov.** 1983. The primary and secondary structure of the 5′-end region of encephalomyocarditis virus RNA. *Gene* **26:**189–195.

39. **Ypma-Wong, M. F., and B. L. Semler.** 1987. *In vitro* molecular genetics as a tool for determining the differential cleavage specifities of the poliovirus 3C proteinase. *Nucleic Acids Res.* **15:**2069–2088.

40. **Ypma-Wong, M. F., and B. L. Semler.** 1987. Processing determinants required for in vitro cleavage of the poliovirus P1 precursor to capsid proteins. *J. Virol.* **61:**3181–3189.

Molecular Aspects of Picornavirus Infection and Detection
Edited by Bert L. Semler and Ellie Ehrenfeld
© 1989 American Society for Microbiology, Washington, DC 20006

Chapter 4

Molecular Biology and Genetics of Poliovirus Protein Processing

Patricia Gillis Dewalt and Bert L. Semler

The differential and temporal control of poliovirus gene expression (and that of the picornaviruses as a group) is effected by a highly regulated cascade of proteolytic processing events. Since the initial translation of the genomic viral RNA generates a polyprotein which contains the amino acid sequences of all the virus-specific proteins, the coordinate activity of the virus-encoded proteinases is necessary for the production of the mature structural and nonstructural viral proteins. The first section of this chapter will review the role of protein processing during the course of a successful poliovirus infection. (For a detailed comparison of proteolytic processing in the life cycles of picornaviruses other than poliovirus, refer to Kräusslich and Wimmer [31]).

To investigate the specific enzyme-substrate interactions of the poliovirus proteinase 3C activity, we and others have taken a molecular genetic approach to the mutagenesis of the 3C-coding region (12, 26; P. G. Dewalt and B. L. Semler, manuscript in preparation). The elucidation of the complete nucleotide sequence of the poliovirus genome (27, 46) and the isolation of infectious cDNA clones (37, 47, 55) have also led to the recovery of mutant polioviruses that bear engineered lesions elsewhere within the coding (6, 7, 32) and noncoding (24, 48, 53, 58) regions of the genome. The biochemical analysis of defined mutant polioviruses has proved to be invaluable in probing the functions of viral proteins and the nontranslated regions. The availability of engineered mutants also pro-

Patricia Gillis Dewalt and Bert L. Semler • Department of Microbiology and Molecular Genetics, California College of Medicine, University of California, Irvine, California 92717.

vides the opportunity to carry out complementation experiments with poliovirus, an approach to studying viral gene function that was previously complicated by the ill-defined nature of naturally occurring variants and chemically induced mutations. Although Bernstein et al. (6) have reported complementation of mutants containing site-directed insertions in two nonstructural regions of the genome, studies to date involving engineered poliovirus mutants have generally suffered from the lack of detailed genetic analyses. To explore this somewhat neglected aspect of systematic mutagenesis, the second section of this chapter will describe the phenotypic reversion and rescue of a poliovirus mutant bearing a single-amino-acid substitution in the 3C-coding region.

ROLE OF PROTEOLYTIC PROCESSING IN POLIOVIRUS GENE EXPRESSION

The poliovirus infection is initiated by adsorption of the virion to a specific cellular receptor that is present only on cells of primate origin. After adsorption and penetration, the capsid proteins dissociate, uncoating the viral genome. The genetic material released into the cytoplasm consists of a single strand of positive-sense RNA which is 7,441 nucleotides in length (27, 46). Because the viral RNA is of message polarity, it can immediately associate with the cellular translational machinery to direct the synthesis of virus-specific proteins. The initial translation product derives from a single large open reading frame between nucleotides 743 and 7370 of the genome that, were it not nascently and posttranslationally cleaved by two virus-encoded proteinases, would yield a giant polyprotein of 247,000 daltons (20, 23). This product, historically referred to as NCVP00, is not normally observed during the course of infection but can be detected under special conditions in which proteolytic processing is inhibited (4, 22).

The initial processing maps of poliovirus proteins were obtained by pactamycin mapping (61, 63) and tryptic peptide fingerprinting experiments (38, 51). These maps were later refined when the results of amino acid sequence data were aligned with the complete nucleotide sequence of the viral RNA (27, 46). The presently accepted genomic map is illustrated in Fig. 1. The viral proteins can be classified into three major groups according to the region of the polyprotein from which they arise. (Refer to Rueckert and Wimmer [52] for standard nomenclature.) Both limited amino acid sequence analysis and the amino acid sequence inferred from the nucleotide sequence data indicate that most of the viral proteins result from the cleavage of glutamine-glycine (Q-G) pairs (15, 33, 54, 56). In addition to the Q-G pairs, two other amino acid pairs are utilized during

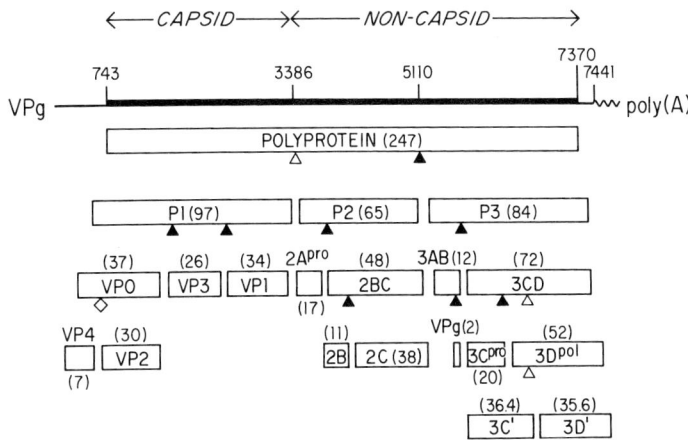

Figure 1. Genome organization and protein processing of polypeptides encoded by poliovirus RNA. The horizontal line represents the poliovirus genomic RNA. The portion of the RNA between nucleotides 743 and 7370 (bold line) denotes the polyprotein-coding region. The open rectangles below the RNA represent the virus-specific polypeptides that result from processing of the precursor polyprotein. This figure is used from reference 12 with permission. Symbols: ▲, Gln-Gly cleavage sites; △, Tyr-Gly cleavage sites; ◇, Asn-Ser cleavage site of the capsid precursor, VP0. The molecular weights (10^3) of the polypeptides are provided in parentheses.

the processing of the precursor polyprotein. Cleavage at a tyrosine-glycine (Y-G) site serves to release the capsid precursor, P1, from the nascent polypeptide chain. An additional Y-G site is cleaved to generate 3C' and 3D' from the P3 region of the genome. The cleavage of the capsid precursor VP0 into the mature capsid proteins VP4 and VP2 occurs only when the procapsid has associated with the progeny viral RNA. This event comprises the final stage of virion morphogenesis and takes place at an asparagine-serine (N-S) pair. The specific interaction of three different proteinase activities is required for the concerted processing of these sites.

The left end of the coding region yields the P1 precursor which is processed to generate the capsid proteins. The proteolytic cleavages between VP0-VP3 and VP3-VP1 occur at Q-G pairs. The previously mentioned N-S cleavage site within VP0 is buried in the interior of the virus particle and is consequently inaccessible to external proteases. Therefore, it has been proposed that the cleavage of VP0 is executed by an intramolecular protease activity (2, 19, 50). In a model describing this process, it has been proposed that the encapsidated RNA serves as a proton-abstracting base to activate a serine residue near what will become the amino terminus of VP2. The serine hydroxyl group could then autocatalytically carry out a nucleophilic attack on the peptide bond which would generate VP4 and VP2 (2).

The P2 region encodes a precursor that is cleaved into proteins 2A, 2B, 2BC, and 2C. With the exception of polypeptide 2A, the functions of these nonstructural proteins are unknown. 2A is the viral proteinase that carries out the Y-G cleavages in the poliovirus polyprotein (64). As soon as ribosomes complete translation of the 2A-coding region, the proteinase is enzymatically active and liberates the P1 precursor. All of the proteo-lytic processing of the P2 precursor takes place at Q-G sites except for the cleavage of the Y-G pair that generates the amino terminus of 2A.

The polypeptides encoded by the P3 region include VPg (the 22-amino-acid genome-linked peptide), 3AB (a membrane-bound precursor to VPg), 3C (a Q-G proteinase), and 3D (the core viral RNA polymerase). Polypeptide 3CD is a stable 72-kilodalton cleavage product that contains the amino acid sequences of both 3C and 3D. Consequently, it is conceivable that 3CD could perform some of the functions ascribed to these proteins. Support for this idea is offered by in vitro translation experiments in which it has been demonstrated that amino acid sequences downstream of the 3C carboxy terminus are required to cleave the Q-G sites in the P1 capsid precursor (68, 69). The latter section of this review will also address the possibility that 3CD may function in some aspect of viral RNA replication. All of the previously mentioned P3 proteins are generated by the cleavage of Q-G sites. Two additional proteins, 3C′ and 3D′, are the alternate cleavage products of 3CD which arise from the processing of a Y-G pair. Although they are related to 3C and 3D, the functions of 3C′ and 3D′ are unknown.

Though it had been established early on that the primary translation product of picornaviruses was extensively processed, it was unclear whether this activity was cellular or viral in origin. Translation of the viral RNA in cell-free extracts provided the first direct evidence for the existence of a virus-encoded proteinase (44, 59). Rabbit reticulocyte lysates programmed with encephalomyocarditis (EMC) virus RNA yielded processed viral polypeptides that were nearly identical to their infected-cell counterparts. Synchronized translation mixtures, blocked by the addition of emetine before translation of the middle genomic region of EMC virus RNA was complete, resulted in a precursor that was not processed further (44). These results suggested that the viral proteinase or proteinase activator responsible for the majority of the cleavages was located in the P3 region of the genome. The locus of the proteinase was later confirmed when the EMC proteinase activity present in the reticulocyte extracts was shown to copurify with the viral polypeptide p22 (3C) (41).

The proteolytic activity of the poliovirus homolog to p22 could not be demonstrated directly because of an anomaly of in vitro translation in the reticulocyte translation system. When poliovirus RNA is used to program reticulocyte lysates, translation initiates from internal AUG codons (13), a phenomenon that is not observed for EMC virus RNA. As a result, the P1 precursor which served as the substrate in the in vitro assay for EMC virus-specific proteolysis could not be synthesized in the absence of the poliovirus P3 region gene products (59). The poliovirus Q-G activity was identified by Hanecak et al. (17) in a cell-free translation system prepared from infected HeLa cells. The extracts, which contained endogenous poliovirus RNA, were capable of synthesizing viral precursors that were subsequently processed into mature viral proteins. Addition of antiserum prepared against purified 3C to the translation mixture inhibited the processing of Q-G sites. In contrast, antiserum prepared against protein 2C had no effect on proteolytic processing. Antibody inhibition of proteolytic processing in vitro was employed in a similar fashion by Toyoda et al. (64) to demonstrate that the Y-G cleavage activity was located in polypeptide 2A.

Poliovirus 3C is an unusual proteinase as manifested by its stringent specificity for Q-G amino acid pairs. The failure to identify any host cell proteins that are directly cleaved by viral proteinases in infected cells (30, 36) offers further evidence for the highly specific enzyme-substrate interactions displayed by the 3C proteinase. Of the 13 Q-G sites predicted by the nucleotide sequence of the poliovirus genome, only 9 are ever utilized as processing sites during infection. Therefore, determinants other than the mere presence of a Q-G peptide bond must be necessary for substrate recognition by 3C. A strict substrate preference is shared by the 2A proteinase, which cleaves only 2 of the possible 10 Y-G pairs.

Comparison of the amino acid sequences flanking the Q-G sites that are cleaved with those that are not processed during a poliovirus infection provides some clues concerning the specificity of the 3C activity. Although no general consensus is adhered to among the cleaved sites, there does appear to be a preference for an alanine or valine residue in the minus 4 position from the cleavage site (36). In contrast, this trend does not hold true for the uncleaved sites. Alanine and valine have small, nonpolar side chains which help to accommodate bends and close fits between regions of the polypeptide backbone. Our understanding of the determinants involved in substrate recognition by poliovirus 3C is complicated when the proteolytic processing strategies of other picornaviruses are taken into consideration. Although 3C-mediated processing of the polyprotein generates homologous mature viral proteins for all of the picornaviruses studied, in many cases the cleavage events take place at sites other than Q-G pairs (9, 34, 39, 40, 49, 60). The unconserved nature of the amino acid pairs cleaved by the various picornavirus 3C proteinases suggests that the enzyme recognizes a particular three-dimensional conformation of the substrate in addition to the Q-G bond. Additional evidence for the role of higher-order structure rather than linear sequence in proteolytic processing is provided by temperature-shift experiments. Poliovirus polyprotein synthesized at 43°C is no longer capable of being cleaved, even after the temperature is shifted down to 37°C (3). The importance of accessibility of a given Q-G site to the proteinase is supported by X-ray crystallographic analysis, which indicates that the uncleaved Q-G site within VP2 is buried deep in the interior of a β-barrel structure (19).

Since infecting virions do not contain any diffusible enzymes, the 3C activity necessary for the subsequent processing of the viral proteins must arise from the initial translation product itself. The 3C-coding region is flanked by Q-G pairs at the amino and carboxy termini. Therefore, the generation of 3C is a result of its own enzymatic activity. Early in vitro translations of EMC virus RNA demonstrated that the kinetics of cleavage of the EMC 3C proteinase (p22) from its precursor were dilution independent (42). These experiments indicated that the picornavirus 3C proteinases could autocatalytically liberate themselves from a larger polypeptide, thus freeing the infecting virus from the requirement for a virion-associated enzyme.

Bacterial expression of subgenomic poliovirus cDNA clones was successfully used to demonstrate the autocatalytic activity of poliovirus 3C. Hanecak et al. (18) showed that 3C could be produced from a bacterial lipoprotein-poliovirus fusion protein and that the proteolytic activity was present within the 3C-coding region. The experiment did not, however, distinguish whether 3C was generated by an intramolecular mechanism or by the activity of one precursor on another. This is an important consideration for the eventual understanding of the intricacies of poliovirus gene expression. The Q-G sites within the polyprotein can be cleaved in *trans* by exogenously added 3C activity (35, 69). However, the value of *cis* cleavage activity is clear in the early stages of infection when the concentration of viral proteins is not sufficient for bimolecular reactions to occur frequently. Data from the bacterial expression of subgenomic poliovirus cDNA clones and in vivo pulse-chase labeling experiments suggest that the amino-terminal cleavage of 3C proceeds very quickly and may occur in *cis* (18, 51, 57). Support for an intramolecular cleavage activity is offered by the fact that even in the presence of very high levels of anti-3C immunoglobulin G, the processing of the amino terminus of 3C is not completely inhibited (17).

All known proteolytic enzymes can be classified as members of one of four categories, depending upon their mechanism of action. These classes are the serine, aspartic acid, cysteine (thiol), and metalloproteinases. The assignment of a given enzyme to a particular class is generally made on the basis of its sensitivity to various chemical inhibitors. Initial attempts to characterize the picornavirus proteinase(s) were complicated by observations that processing was affected by agents that inhibited serine proteinases and by those that were specific for cysteine proteinases. Processing of picornavirus precursors is blocked by the substrate analogs tosyllysine chloromethyl ketone and tosylphenylalanine chloromethyl ketone (42, 62). These compounds are inhibitors of the serine proteinases (trypsin and chymotrypsin) that act by the specific alkylation of the imidazole ring of the histidine residue in the catalytic triad (10). However, diisopropylfluorophosphate, which phosphorylates the active-site serine of trypsin and other similar proteases, failed to inhibit viral protein processing. The conflicting conclusions as to the nature of the picornavirus proteolytic activity may have come about as a consequence of substrate modification by the chloromethyl ketones. These inhibitors are potentially capable of alkylating sulfhydryl groups in a nonspecific manner (62). In view of later evidence strongly supporting the 3C proteinase as a sulfhydryl enzyme, it is easy to understand how the 3C enzymatic activity could be inhibited by these agents.

Studies with other chemical inhibitors, including the thiol-reactive compounds iodoacetamide and N-ethylmaleimide (16, 28, 44) and the naturally occurring inhibitor cystatin (29), indicate that 3C is a member of the cysteine proteinase class. Most of the known cysteine proteinases belong to the papain superfamily, which includes the plant sulfhydryl enzymes (e.g., papain, ficin, bromelain, and actinidin) and the mammalian cathepsins. Comparison of the predicted amino acid sequences of representative picornavirus 3C proteins revealed no appreciable homology to any groups within the cysteine proteinase class (1). Among themselves, the picornaviruses share conserved cysteine and histidine residues in the carboxy-terminal region of the 3C polypeptide. The only type of catalytic active site consistent with the evolutionary conservation of these residues is that of a cysteine proteinase.

The catalytic mechanism of the cysteine proteinases can be considered similar, though not identical, to that proposed for the serine proteinases (14, 25, 45). The catalytic triad of the serine enzymes is made up of aspartate, histidine, and serine residues in a complex hydrogen-bonding system. The active-site cysteine of the sulfhydryl proteinase functions as an analog of the serine residue in the serine proteinase. However, the cysteine proteinases do not possess a conserved counterpart to the serine proteinase aspartate residue. In the most accepted model for the mechanism for cysteine proteinases, the histidine imidazole group serves to depolarize and deprotonate the sulfhydryl group of the catalytic-site cysteine. The resulting thiolate anion carries out a nucleophilic attack on the carbonyl carbon of the substrate, leading to the formation of a covalent thioester intermediate and the release of the first proteolytic product. The second step of the reaction consists of the transfer of the polypeptide acyl group to a water molecule. This results in the liberation of the second product and the regeneration of the sulfhydryl group.

A catalytic role for the conserved cysteine and histidine residues in poliovirus 3C has been supported by site-directed mutagenesis. Ivanoff et al. (21) showed that substitution of cysteine 147 with a serine residue or substitution of histidine 161 by a glycine residue resulted in the inability of 3C to cleave itself from a larger polypeptide in a subgenomic bacterial expression system. A four-amino-acid insertion within the putative active site of the EMC virus 3C proteinase resulted in the inhibition of proteolytic processing following translation of T7 transcripts in a reticulocyte lysate system (43). The lesion did not affect the ability of larger polypeptides containing the insertion to be cleaved by exogenously added wild-type 3C, suggesting that the inhibition of processing was the result of a defect in enzymatic activity rather than a reflection of the inability of 3C to cleave itself from a misfolded precursor.

GENETIC ANALYSIS AND RESCUE OF A SITE-DIRECTED POLIOVIRUS 3C MUTANT

In lieu of an X-ray crystallographic structure of a picornavirus 3C proteinase, we must rely entirely on the site-specific mutagenesis and in vitro expression of viral cDNAs to identify structurally and functionally important domains within the enzyme. Based upon the available amino acid sequence comparisons of the 3C-coding regions of a number of picornaviruses (1, 67), we have begun systematically scanning the poliovirus type 1 3C proteinase with mutagenic oligonucleotide cassettes. We have previously reported (12) that a conservative Val → Ala substitution at position 54 of 3C resulted in a mutant virus (Se1-3C-02) that was deficient in the production of both the mature viral polymerase 3D and the proteinase 3C. Although the virus very inefficiently cleaved the carboxy-terminal Q-G bond of 3C, the lesion had no deleterious effects on the processing of other Q-G sites within the polyprotein. The amino acid substitution is located more than 130 residues upstream of the carboxy-terminal cleavage site. Since the enzymatic activity of the mutant 3C is unaffected with respect to the processing of sites other than the 3C-3D bond, the inefficient cleavage at the carboxy-terminal Q-G pair of 3C must reflect a change in the conformation of a P3 region precursor containing the mutation.

Initial analysis of the mutant indicated that the lesion in the proteinase was also manifested by secondary effects on the replication of the mutant virus. Although Se1-3C-02 exhibits a small-plaque phenotype and produces barely detectable levels of mature 3D, it grows to nearly wild-type titers. RNA labeling experiments demonstrated that the protein processing defect of Se1-3C-02 did not severely affect the kinetics of RNA synthesis at 37°C. In contrast, when grown at 39°C the mutant attained a maximal level of RNA synthesis that was only 30% of that obtained at 37°C (Fig. 2). However, the profile of protein processing by the mutant was not noticeably different at the higher temperature (data not shown). Bellocq et al. (5), through chemical mutagenesis of the viral RNA, have isolated a mutant virus which displays a phenotype similar to that of Se1-3C-02. An Ile-74 → Thr substitution in 3C produces a small-plaque virus that also inefficiently executes the cleavage of the Q-G site between 3C and 3D. Unlike Se1-3C-02, this mutant does not appear to be temperature sensitive for RNA synthesis at 39°C and is somewhat deficient in the proteolytic processing of the P2 precursor (26).

The importance of the Val residue at position 54 for normal proteolytic processing and RNA replication was confirmed by reversion analysis of the mutation. Phenotypic revertants of Se1-3C-02 were selected by passage of the mutant virus at 39°C. Two independent plaque isolates were twice plaque purified at 39°C and expanded through passage in HeLa cell monolayers to yield high-titer revertant virus stocks (designated Se1-3C-02R1 and Se1-3C-02R2). The reversion analysis was carried out primarily to confirm that the Val-54 → Ala substitution was responsible for the observed mutant proteolytic processing and RNA synthesis phenotype. It was also of interest to see whether it was possible to select for second-site mutations that would compensate for the changes induced by the engineered mutation.

Comparison of the titers of Se1-3C-02R1 and Se1-3C-02R2 with that of the parental mutant virus, Se1-3C-02, indicated that the virus stocks selected at 39°C were no longer temperature sensitive (Table 1). Both revertant viruses displayed a large-plaque phenotype (>2.0 mm) on HeLa monolayers at 39°C. Single-step growth kinetics provided further evidence that the revertants had regained wild-type growth characteristics.

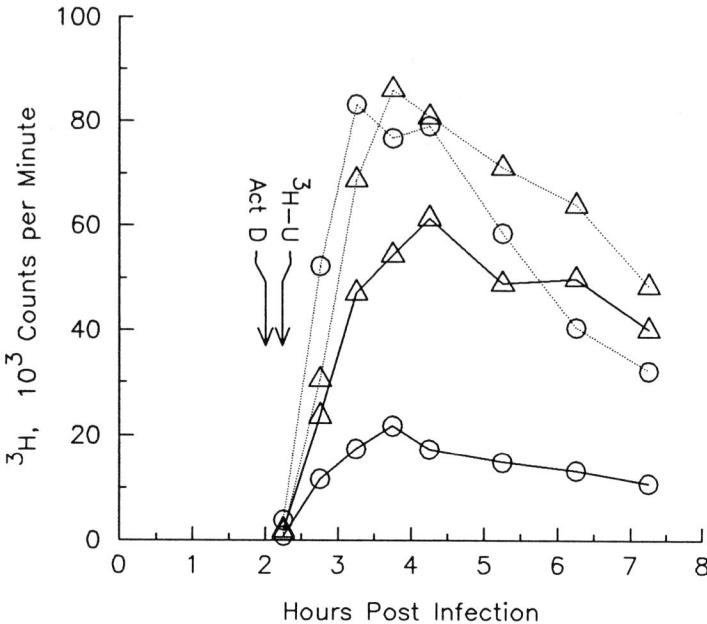

Figure 2. Kinetics of RNA synthesis by wild-type transfection-derived poliovirus (△····△, ○····○) and Se1-3C-02 (△——△, ○——○) at 37°C (△) and 39°C (○). Suspension cultures of HeLa cells were infected at a multiplicity of 25 with either wild-type virus or Se1-3C-02 and labeled with [³H]uridine (³H-U) at the indicated time postinfection. RNA synthesis was measured by liquid scintillation counting of trichloroacetic acid-precipitable material. Actinomycin D (Act D) was added to a final concentration of 5 μg/ml 15 min before labeling to inhibit the background of host cell DNA-directed RNA synthesis.

The mutant Se1-3C-02, the revertant Se1-3C-02R1, and wild-type transfection-derived virus all displayed similar growth kinetics at 37°C. At 39°C, the yield of Se1-3C-02 was 100-fold less than the revertant by 6 h postinfection (data not shown). The difference in growth kinetics between

Table 1. Titers of Wild-Type, Se1-3C-02, and Revertant Poliovirus Stocks

Virus	Titer, log₁₀ PFU/ml (avg plaque diam [mm])		Log difference (titer at 39°C/37°C)
	37°C	39°C	
Wild type	9.32 (1.5)	9.41 (2.0)	0.10
Se1-3C-02	8.89 (1.0)	7.62 (<0.5)	−1.27
Se1-3C-02R1	8.78 (1.5)	9.38 (2.5)	0.59
Se1-3C-02R2	8.17 (1.5)	9.07 (2.5)	0.90

Se1-3C-02 and the revertant was even more pronounced at early times postinfection.

Pulse-chase labeling of viral proteins with [^{35}S]methionine demonstrated that both revertant stocks had regained the wild-type protein-processing phenotype. The results of the pulse-chase labeling studies with Se1-3C-02R1 are shown in Fig. 3. The levels of 3C and 3D produced by the revertant during the course of the chase are comparable to those normally observed for wild type, indicating that Se1-3C-02R1 efficiently executes the Q-G cleavage at the carboxy terminus of 3C. The aberrant electrophoretic mobility of the Se1-3C-02 P3 region proteins also served as a marker for the mutant phenotype in this experiment. Polypeptides 3C, 3C', and 3CD, produced by the revertant, no longer exhibited retarded gel mobility, confirming that Se1-3C-02R1 did result from an actual genetic change in the parent mutant virus. The same results were obtained from pulse-chase labeling analysis of Se1-3C-02R2 (data not shown). The data resulting from RNA labeling experiments conducted at 37 and 39°C also indicated that both revertant virus stocks had regained the wild-type RNA synthesis phenotype (shown in Fig. 4 for Se1-3C-02R2).

Viral RNA was prepared for sequence analysis from both revertant virus stocks. Although revertant phenotypes can result from replacement of the mutant amino acid with the wild-type residue, it is statistically more likely that reversion to the wild phenotype is caused by a second-site mutation at a nearby locus. In fact, compensatory second-site mutations have been reported at the nucleotide level for a poliovirus containing a mutation in the 5' noncoding region of viral RNA (48). Sequencing of the region between nucleotides 5550 and 5735 revealed that, for both revertant viruses, the C at position 5598 (present in the mutant, Se1-3C-02) had been replaced by the wild-type U residue. No other changes were discovered within the 180 nucleotides that were analyzed. We concluded from the sequencing data that both of the phenotypic revertants that were selected at 39°C had undergone a primary-site reversion which restored the wild-type amino acid. It is extremely unlikely that the initial stocks of Se1-3C-02 were contaminated with wild-type virus, since these stocks were generated from single-plaque isolates, which in turn were derived from clonally amplified mutant cDNA plasmids. These results imply the existence of a strong selective pressure to maintain the Val residue at position 54.

As a whole, the data suggest that the lesion in Se1-3C-02 has two major effects on viral gene functions. The first effect is on proteolytic processing, which is evidenced by the inefficient cleavage of the Q-G pair between 3C and 3D, resulting in the production of low levels of these

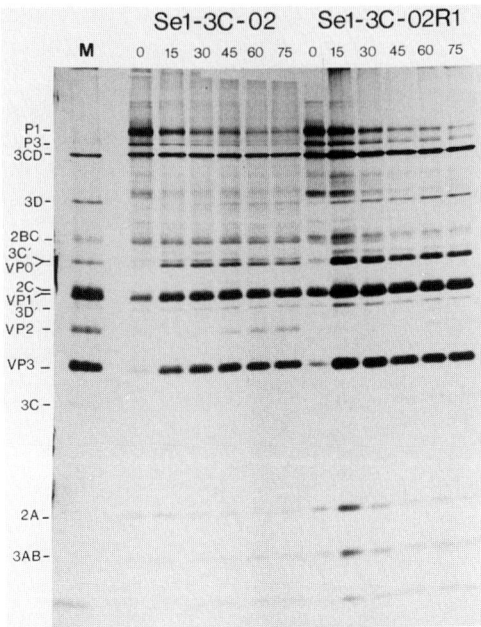

Figure 3. Pulse-chase analysis of protein processing by Se1-3C-02 and phenotypic revertant Se1-3C-02R1. Suspension cultures of HeLa cells were infected at a multiplicity of 20 with Se1-3C-02 or Se1-3C-02R1 at 37°C. At 4 h post-infection, the cells were given a 10-min pulse with [^{35}S]methionine which was followed by a chase of unlabeled methionine. Immediately after the chase, samples representing 10^6 cells were harvested at 15-min intervals, and the labeled proteins were analyzed by sodium dodecyl sulfate-polyacrylamide gel electrophoresis. Numbers above the lanes indicate the time in minutes after the chase with unlabeled L-methionine. M, Marker lane of [^{35}S]methionine-labeled poliovirus proteins (labeled from 3 to 6 h after infection of HeLa cells).

polypeptides during the course of infection. The second effect is the temperature sensitivity of RNA replication at 39°C. The problem arises of how to reconcile these two apparently different effects that result from a single-amino-acid substitution. The temperature sensitivity of RNA synthesis by Se1-3C-02 suggests that a polypeptide involved in replication is rendered nonfunctional at the nonpermissive temperature, perhaps as a result of misfolding. However, the amino acid substitution that brings about the altered phenotype is located in the 3C-coding region. This implies that a precursor containing both the 3C- and 3D-coding regions, namely, 3CD or P3, is the gene product primarily affected by the mutation. Previously described experimental evidence, including the retarded electrophoretic mobility of the mutant P3 region proteins and the elevated levels of polypeptide 4a produced by Se1-3C-02 (12), suggests that the conformation of the mutant 3C-related polypeptides is altered from the wild type. The misfolding of 3CD, which does not have a major effect on virus replication at 37°C, may be exaggerated at 39°C. The inefficient cleavage of the carboxy-terminal Q-G pair of 3C is possibly a secondary effect of the altered configuration of 3CD, rather than the primary manifestation of the mutation.

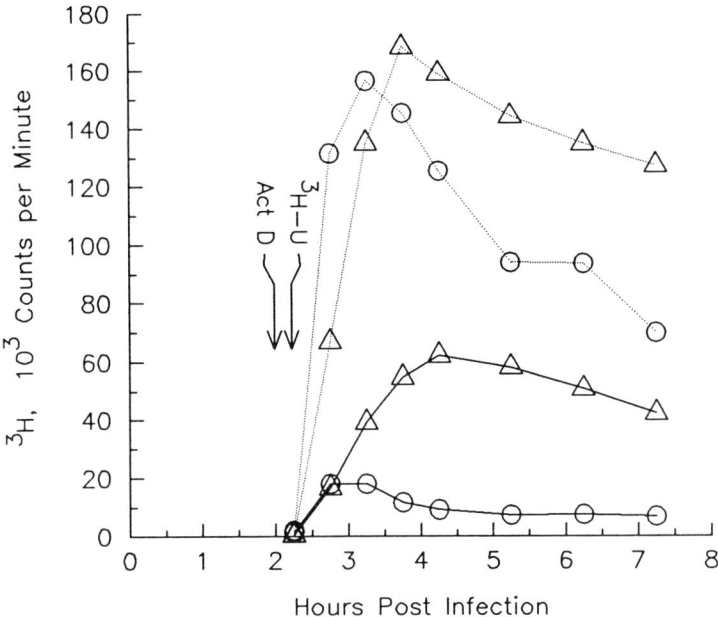

Figure 4. Kinetics of RNA synthesis by Se1-3C-02 (△——△, ○——○) and Se1-3C-02R1 (△····△, ○····○) at 37°C (△) and 39°C (○). Suspension cultures of HeLa cells were infected at a multiplicity of 20 with either Se1-3C-02 or Se1-3C-02R1. Labeling with [³H]uridine and measurement of RNA synthesis were carried out as described in the legend of Fig. 2.

It has been reported that a site-directed mutant poliovirus bearing a single-amino-acid insertion in the 3D-coding region could not be complemented significantly (as measured by increased virus yield) by coinfection with another mutant virus (6). These results were interpreted to suggest that the viral polymerase is *cis* acting. It was of interest to determine whether the RNA synthesis defect of Se1-3C-02 could be rescued by coinfection with a wild-type virus. HeLa cells were coinfected with Se1-3C-02 and a recombinant virus, PCV305, that contains a wild-type 3C-coding region. The construction of the cDNA clone that gave rise to PCV305 has been described elsewhere (24). The resulting chimeric virus contains the corresponding coxsackievirus type B3 (CB3) sequences substituted for nucleotides 1 to 627 in the 5′ noncoding region of the poliovirus genome (for a schematic diagram of the two viral genomes, refer to Fig. 5). The phenotype of this virus, with respect to the kinetics of RNA synthesis and protein processing, appears to be similar to that of wild-type poliovirus (24). Although poliovirus and CB3 share approxi-

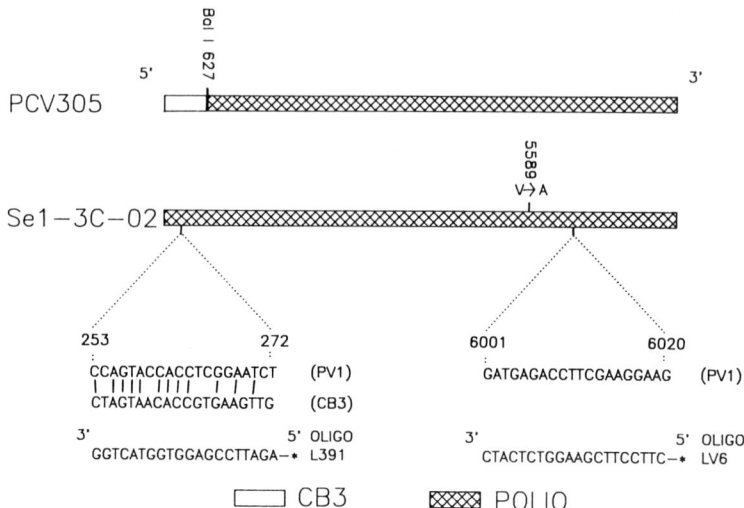

Figure 5. Strategy for detecting Se1-3C-02 RNA synthesis in cells coinfected with Se1-3C-02 and PCV305. The genomes of PCV305 and Se1-3C-02 are represented by the bars at the top of the figure. PCV305 is a recombinant picornavirus in which the 5' 627 nucleotides of the poliovirus type 1 (PV1) genome have been replaced by the corresponding sequences from CB3. Oligonucleotide L391 is complementary to poliovirus nucleotides 253 to 272, but contains eight mismatches with the corresponding CB3-derived sequence. Oligonucleotide LV6 is complementary to poliovirus nucleotides 6001 to 6020. Hybridization conditions were determined that would permit binding of L391 to Se1-3C-02 RNA but not to PCV305 RNA.

mately 70% sequence identity throughout the 5' noncoding region (65), there exist short stretches (20 nucleotides) that exhibit little sequence homology. We took advantage of these regions of dissimilarity to distinguish between the progeny RNAs generated during the coinfection.

HeLa cell monolayers were infected with PCV305 (as a negative control) or Se1-3C-02 (as a control for base-line mutant viral RNA replication) or were coinfected with both viruses at equal multiplicities. The infected cells were incubated at 39°C, and total cytoplasmic RNA was prepared after 3, 4, 5, and 6 h. The RNAs were spotted onto nitrocellulose filters which were then probed with a ^{32}P-labeled synthetic oligonucleotide (L391) complementary to nucleotides 253 to 272 of the poliovirus genome. This oligonucleotide probe contains eight mismatches with the corresponding CB3 sequence (Fig. 5) and consequently could be used to discriminate between Se1-3C-02- and PCV305-derived progeny RNAs. Hybridization conditions that would permit strong binding to Se1-3C-02 and no binding to PCV305 were determined empirically.

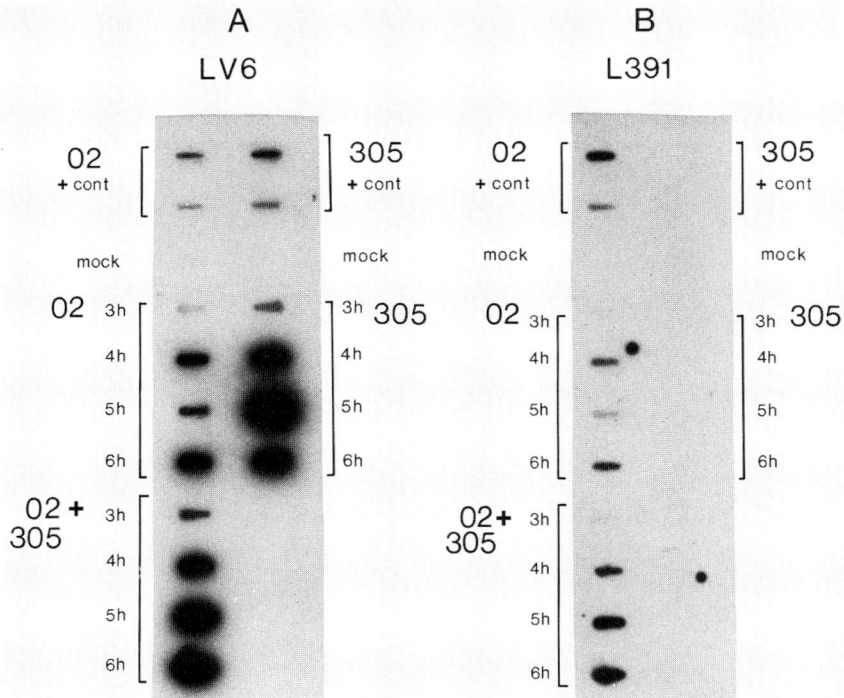

Figure 6. RNA dot-blot analysis of Se1-3C-02-specific RNA in cells coinfected with Se1-3C-02 and PCV305. Confluent HeLa cell monolayers were infected with Se1-3C-02 (02) or PCV305 (305) or coinfected with an equal number of PFU of both viruses to yield a final multiplicity of 20. At the indicated times postinfection, cytoplasmic RNA was prepared from the infected cells. A 10-μg portion of each RNA sample (or 10 μg of cytoplasmic RNA from mock-infected cells) was applied to duplicate nitrocellulose filters with a vacuum slot-blot apparatus. As a control (cont), samples containing 50 and 10 ng of purified viral RNA were included on each blot. The blots were hybridized to either the poliovirus-specific probe (LV6, panel A) or the Se1-3C-02-specific probe (L391, panel B). Note: the levels of Se1-3C-02 RNA at 5 h postinfection have been omitted from the subsequent analysis due to the loss of that sample during preparation.

Alternatively, parallel blots were hybridized with a probe (LV6) complementary to nucleotides 6001 to 6020 in the poliovirus P3 region. This probe hybridized to both viral RNAs and served as an internal control to measure levels of total RNA replication. The results of this experiment are shown in Fig. 6 and summarized in Table 2. Quantitation of the levels of Se1-3C-02 RNA by laser densitometry (Table 2) demonstrated that by 4 h after coinfection with the recombinant picornavirus, the amount of Se1-3C-02 viral RNA was approximately six times higher than in the

Table 2. Summary of Densitometric Analysis of Se1-3C-02-Specific RNA Synthesis in
Se1-3C-02 + PCV305 Coinfection

Virus	Time postinfection (h)	Peak area[a]	Peak area ratio: 6 h/4 h
Se1-3C-02	4	0.30	
Se1-3C-02	6	0.75	2.5
Se1-3C-02 + PCV305	4	1.82	
Se1-3C-02 + PCV305	6	4.40	2.4

[a] Peak area expressed in arbitrary units. The ratio of Se1-3C-02-specific RNA present in the coinfected cells versus the levels of mutant RNA synthesized in the nonrescued infection is expressed as: peak area Se1-3C-02 + PCV305 4 h/Se1-3C-02 4 h = **6.1** and peak area Se1-3C-02 + PCV305 6 h/Se1-3C-02 6 h = **5.9**.

single virus infection (see Table 2, footnote *a*). These results indicate that although coinfection with a pseudo-wild-type virus could not completely restore the RNA replication of Se1-3C-02 to wild-type levels, the temperature-sensitive lesion in RNA synthesis could be partially rescued by wild-type gene products.

It has been demonstrated that 3CD, rather than 3C, is required for efficient in vitro proteolytic processing of the P1 precursor to mature capsid proteins (68). Likewise, since it contains 3D-specific sequences, 3CD may be a replicase with properties distinct from 3D. Purified 3D is a primer-dependent enzyme that is only capable of carrying out the elongation reaction of RNA poliovirus replication in vitro (66). Poliovirus RNA polymerase preparations containing 3CD and host cell proteins, in addition to 3D, have been shown to be capable of initiating RNA synthesis de novo from exogenously added viral RNA (11). Experiments with a temperature-sensitive strain of type 3 poliovirus also indicate that 3CD is essential for the replication of viral RNA, since the proteolytic breakdown of 3CD at the nonpermissive temperature correlated with reduction in viral RNA synthesis (8). Although purified 3CD by itself has not been demonstrated to have polymerase activity in vitro, this apparent lack of activity does not rule out a role for 3CD in some other aspect of RNA replication.

The data from the coinfection with Se1-3C-02 and a recombinant virus containing a coding region from wild-type poliovirus indicate that the RNA replication defect of this mutant can be rescued by wild-type gene products provided in *trans*. It is possible that the wild-type virus is providing either 3D or proteinase activity capable of cleaving the Q-G pair between the mutant 3C and 3D, which would also serve to generate increased levels of 3D. This is an unlikely explanation for the observed rescue since RNA synthesis by the mutant is not drastically affected at 37°C even though the levels of 3D are much lower than in the wild type. A more plausible explanation is that the wild-type 3CD is in fact a necessary component of the replication complex. We are currently exploring the possibility that 3CD may be involved in the initiation step of viral RNA synthesis.

In this chapter, we have summarized some of the biochemical and genetic data that have led to our present level of understanding of picornavirus protein processing. The molecular genetic experimental approaches that we and other investigators have described should be valuable in addressing some of the key mechanistic problems regarding the regulation of picornavirus gene expression by proteolytic cleavage. These problems include (but are not limited to): (i) the mechanism of primary site recognition by the viral proteinases; (ii) the conformational contributions of precursor polypeptides in presentation of a scissile bond to the viral proteinases; (iii) the differential cleavage activities of precursor polypeptides containing proteinase amino acid sequences; (iv) the importance of *cis* versus *trans* cleavage by the viral proteinases; and (v) the resolution of the pleiotropic effects that proteinase mutations may have on both proteolytic processing and viral RNA replication.

Acknowledgments. Work described from the authors' laboratory was supported by Public Health Service grant AI22693 from the National Institutes of Health. B.L.S. is the recipient of a Research Career Development Award from the National Institutes of Health.

Literature Cited

1. **Argos, P., G. Kamer, M. J. H. Nicklin, and E. Wimmer.** 1984. Similarity in gene organization and homology between proteins of animal picornaviruses and a plant comovirus suggest common ancestry of these virus families. *Nucleic Acids Res.* **12:**7251–7267.

2. **Arnold, E., M. Luo, G. Vriend, M. G. Rossmann, A. C. Palmenberg, G. D. Parks, M. J. H. Nicklin, and E. Wimmer.** 1987. Implications of the picornavirus capsid structure for polyprotein processing. *Proc. Natl. Acad. Sci. USA* **84:**21–25.

3. **Baltimore, D.** 1971. Polio is not dead, p. 1–14. *In* M. Pollard (ed.), *Perspectives in Virology*. Academic Press, Inc., New York.

4. **Baltimore, D., M. F. Jacobson, J. Asso, and A. S. Huang.** 1969. The formation of poliovirus proteins. *Cold Spring Harbor Symp. Quant. Biol.* **34:**741–746.

5. **Bellocq, C., K. M. Kean, O. Fichot, M. Girard, and H. Agut.** 1987. Multiple mutations involved in the phenotype of a temperature sensitive small-plaque mutant of poliovirus. *Virology* **157:**75–82.

6. **Bernstein, H. D., P. Sarnow, and D. Baltimore.** 1986. Genetic complementation among poliovirus mutants derived from an infectious cDNA clone. *J. Virol.* **60:**1040–1049.

7. Bernstein, H. D., N. Sonenberg, and D. Baltimore. 1985. Poliovirus mutant that does not selectively inhibit host cell protein synthesis. *Mol. Cell. Biol.* **5**:2913–2923.

8. Bowles, S. A., and D. R. Tershak. 1978. Proteolysis of noncapsid protein 2 of type 3 poliovirus at the restrictive temperature: breakdown of noncapsid protein 2 correlates with loss of RNA synthesis. *J. Virol.* **27**:443–448.

9. Callahan, P. L., S. Mizutani, and R. Colonno. 1985. Molecular cloning and complete sequence determination of RNA genome of human rhinovirus type 14. *Proc. Natl. Acad. Sci. USA* **82**:732–736.

10. Craik, C. S., S. Roczniak, S. Sprang, R. Fletterick, and W. Rutter. 1987. Redesigning trypsin via genetic engineering. *J. Cell. Biochem.* **33**:199–211.

11. Dasgupta, A., M. H. Baron, and D. Baltimore. 1979. Poliovirus replicase: a soluble enzyme able to initiate copying of poliovirus RNA. *Proc. Natl. Acad. Sci. USA* **76**:2679–2683.

12. Dewalt, P. G., and B. L. Semler. 1987. Site-directed mutagenesis of proteinase 3C results in a poliovirus deficient in synthesis of viral RNA polymerase. *J. Virol.* **61**:2162–2170.

13. Dorner, A. J., B. L. Semler, R. J. Jackson, R. Hanecak, E. Duprey, and E. Wimmer. 1984. In vitro translation of poliovirus RNA: utilization of internal initiation sites in reticulocyte lysate. *J. Virol.* **50**:507–514.

14. Drenth, J., K. H. Kalk, and H. M. Swen. 1976. Binding of chloromethyl ketone substrate analogues to crystalline papain. *Biochemistry* **15**:3731–3738.

15. Emini, E. A., M. Elzinga, and E. Wimmer. 1982. Carboxy-terminal analysis of poliovirus proteins: termination of poliovirus RNA translation and location of unique poliovirus protein cleavage sites. *J. Virol.* **42**:194–199.

16. Gorbalenya, A. E., and Y. V. Svitkin. 1983. Encephalomyocarditis virus protease: purification and role of the SH groups in processing of the precursor of structural proteins. *Biochemistry USSR* **48**:385–395.

17. Hanecak, R., B. L. Semler, C. W. Anderson, and E. Wimmer. 1982. Proteolytic processing of poliovirus polypeptides: antibodies to polypeptide P3-7c inhibit cleavage at glutamine-glycine pairs. *Proc. Natl. Acad. Sci. USA* **79**:3973–3977.

18. Hanecak, R., B. L. Semler, H. Ariga, C. W. Anderson, and E. Wimmer. 1984. Expression of a cloned gene segment of poliovirus in E. coli: evidence for autocatalytic production of the viral proteinase. *Cell* **37**:1063–1073.

19. Hogle, J. M., M. Chow, and D. J. Filman. 1985. Three dimensional structure of poliovirus at 2.9 angstrom resolution. *Science* **229**:1358–1365.

20. Holland, J. J., and E. D. Kiehn. 1968. Specific cleavage of viral proteins as steps in the synthesis and maturation of enteroviruses. *Proc. Natl. Acad. Sci. USA* **60**:1015–1022.

21. Ivanoff, L. A., T. Towatari, J. Ray, B. D. Korant, and S. R. Petteway, Jr. 1986. Expression and site-specific mutagenesis of the poliovirus 3C proteinase in *Escherichia coli. Proc. Natl. Acad. Sci. USA* **83**:5392–5396.

22. Jacobson, M. F., J. Asso, and D. Baltimore. 1970. Further evidence on the formation of poliovirus proteins. *J. Mol. Biol.* **49**:657–669.

23. Jacobson, M. F., and D. Baltimore. 1968. Polypeptide cleavages in the formation of poliovirus proteins. *Proc. Natl. Acad. Sci. USA* **61**:77–84.

24. Johnson, V. H., and B. L. Semler. 1988. Defined recombinants of poliovirus and coxsackie virus: sequence-specific deletions and functional substitutions in the 5′-noncoding regions of viral RNAs. *Virology* **162**:47–57.

25. Kamphuis, I. G., J. Drenth, and E. N. Baker. 1985. Thiol proteases: comparative studies on the high resolution structures of papain and actinidin, and on the amino acid sequence information for cathepsins B and H, and stem bromelain. *J. Mol. Biol.* **182**:317–329.

26. **Kean, K. M., H. Agut, O. Fichot, E. Wimmer, and M. Girard.** 1988. A poliovirus mutant defective for self-cleavage at the COOH-terminus of the 3C protease exhibits secondary processing defects. *Virology* **163**:330–340.

27. **Kitamura, N., B. L. Semler, P. G. Rothberg, G. R. Larsen, C. J. Adler, A. J. Dorner, E. A. Emini, R. Hanecak, J. J. Lee, S. van der Werf, C. W. Anderson, and E. Wimmer.** 1981. Primary structure, gene organization and polypeptide expression of poliovirus RNA. *Nature* (London) **291**:547–553.

28. **Korant, B. D.** 1981. Inhibition of viral protein cleavage, p. 37–47. *In* K. K. Gauri (ed.), *Antiviral Chemotherapy: Design of Inhibitors of Viral Functions.* Academic Press, Inc., New York.

29. **Korant, B. D., J. Brzin, and V. Turk.** 1985. Cystatin, a protein inhibitor of cysteine proteinases, alters viral protein cleavages in infected human cells. *Biochem. Biophys. Res. Commun.* **127**:1072–1076.

30. **Korant, B. D., N. L. Chow, M. O. Lively, and J. C. Powers.** 1980. Proteolytic events in replication of animal viruses. *Ann. N.Y. Acad. Sci.* **343**:304–318.

31. **Kräusslich, H. G., and E. Wimmer.** 1988. Viral proteinases. *Annu. Rev. Biochem.* **15**:701–754.

32. **Kuhn, R. J., H. Tada, M. F. Ypma-Wong, J. J. Dunn, B. L. Semler, and E. Wimmer.** 1988. Construction of a mutagenesis cartridge for poliovirus VPg: isolation and characterization of viable and non-viable mutants. *Proc. Natl. Acad. Sci. USA* **85**:519–523.

33. **Larsen, G. R., C. W. Anderson, A. J. Dorner, B. L. Semler, and E. Wimmer.** 1982. Cleavage sites within the poliovirus capsid protein precursors. *J. Virol.* **41**:340–344.

34. **Najarian, R., D. Caput, W. Gee, S. J. Potter, A. Renard, J. Merryweather, G. Van Nest, and D. Dina.** 1985. Primary structure and gene organization of human hepatitis A virus. *Proc. Natl. Acad. Sci. USA* **82**:2627–2631.

35. **Nicklin, M. J. H., H. G. Krausslich, H. Toyoda, J. J. Dunn, and E. Wimmer.** 1987. Poliovirus polypeptide precursors: expression in vitro and processing by exogenous 3C and 2A proteinases. *Proc. Natl. Acad. Sci. USA* **84**:4002–4006.

36. **Nicklin, M. J. H., H. Toyoda, M. G. Murray, and E. Wimmer.** 1986. Proteolytic processing in the replication of polio and related viruses. *Biotechnology* **4**:33–42.

37. **Omata, T., M. Kohara, Y. Sakai, A. Kameda, N. Imura, and A. Nomoto.** 1984. Cloned infectious complementary DNA of the poliovirus Sabin 1 genome: biochemical and biological properties of the recovered virus. *Gene* **32**:1–10.

38. **Pallansch, M. A., O. M. Kew, B. L. Semler, D. R. Omilianowski, C. W. Anderson, E. Wimmer, and R. R. Rueckert.** 1984. The protein processing map of poliovirus. *J. Virol.* **49**:873–880.

39. **Palmenberg, A. C.** 1987. Picornaviral processing: some new ideas. *J. Cell Biochem.* **33**:191–198.

40. **Palmenberg, A. C., E. M. Kirby, M. R. Janda, N. L. Drake, G. M. Duke, K. F. Potratz, and M. S. Collett.** 1984. The nucleotide and deduced amino acid sequence of the encephalomyocarditis viral polyprotein coding region. *Nucleic Acids Res.* **12**:2969–2985.

41. **Palmenberg, A. C., M. A. Pallansch, and R. R. Rueckert.** 1979. Protease required for processing picornaviral coat protein resides in the viral replicase gene. *J. Virol.* **32**:770–778.

42. **Palmenberg, A. C., and R. R. Rueckert.** 1982. Evidence for intramolecular self-cleavage of picornaviral replicase precursors. *J. Virol.* **41**:244–249.

43. **Parks, G. D., G. M. Duke, and A. C. Palmenberg.** 1986. Encephalomyocarditis virus 3C

protease: efficient cell-free expression from clones which link viral 5' noncoding sequences to the P3 region. *J. Virol.* **60**:376–384.

44. **Pelham, H. R. B.** 1978. Translation of encephalomyocarditis virus RNA *in vitro* yields an active proteolytic processing enzyme. *Eur. J. Biochem.* **85**:457–462.

45. **Polgar, L., and P. Halasz.** 1982. Current problems in mechanistic studies of serine and cysteine proteinases. *Biochem. J.* **207**:1–10.

46. **Racaniello, V. R., and D. Baltimore.** 1981. Molecular cloning of poliovirus cDNA and determination of the complete nucleotide sequence of the viral genome. *Proc. Natl. Acad. Sci. USA* **78**:4887–4891.

47. **Racaniello, V. R., and D. Baltimore.** 1981. Cloned poliovirus complementary DNA is infectious in mammalian cells. *Science* **214**:916–919.

48. **Racaniello, V. R., and C. Meriam.** 1986. Poliovirus temperature-sensitive mutant containing a single nucleotide deletion in the 5'-non-coding region of the viral RNA. *Virology* **155**:498–507.

49. **Roberston, B. H., M. J. Grubman, G. N. Weddell, D. M. Moore, J. D. Welsh, T. Fischer, D. J. Dowbenko, D. G. Yansura, B. Small, and D. G. Kleid.** 1985. Nucleotide and amino acid sequence coding for polypeptides of foot-and-mouth disease virus type A12. *J. Virol.* **54**:651–660.

50. **Rossmann, M. G., E. Arnold, J. W. Erickson, E. A. Frankenberger, J. P. Griffith, H. J. Hecht, J. E. Johnson, G. Kamer, M. Luo, A. G. Mosser, R. R. Rueckert, B. Sherry, and G. Vriend.** 1985. The structure of a human common cold virus (rhinovirus 14) and its functional relationships to other picornaviruses. *Nature* (London) **317**:145–153.

51. **Rueckert, R. R., T. J. Matthews, O. M. Kew, M. Pallansch, C. McClean, and D. Omilianowski.** 1979. Synthesis and processing of picornaviral protein, p. 113–125. *In* R. Perez-Bercoff (ed.), *The Molecular Biology of Picornaviruses.* Plenum Publishing Corp., New York.

52. **Rueckert, R. R., and E. Wimmer.** 1984. Systematic nomenclature of picornavirus proteins. *J. Virol.* **50**:957–959.

53. **Sarnow, P., H. D. Bernstein, and D. Baltimore.** 1986. A poliovirus temperature-sensitive RNA synthesis mutant located in a noncoding region of the genome. *Proc. Natl. Acad. Sci. USA* **83**:571–575.

54. **Semler, B. L., C. W. Anderson, N. Kitamura, P. G. Rothberg, W. L. Wishart, and E. Wimmer.** 1981. Poliovirus replication proteins: RNA sequence encoding P3-1b and the sites of proteolytic processing. *Proc. Natl. Acad. Sci. USA* **78**:3464–3468.

55. **Semler, B. L., A. J. Dorner, and E. Wimmer.** 1984. Production of infectious poliovirus from cloned cDNA is dramatically increased by SV40 transcription and replication signals. *Nucleic Acids Res.* **12**:5123–5141.

56. **Semler, B. L., R. Hanecak, C. W. Anderson, and E. Wimmer.** 1981. Cleavage sites in the polypeptide precursors of poliovirus protein P2-X. *Virology* **114**:589–594.

57. **Semler, B. L., V. H. Johnson, P. G. Dewalt, and M. F. Ypma-Wong.** 1987. Site-specific mutagenesis of cDNA clones expressing a poliovirus proteinase. *J. Cell. Biochem.* **33**:39–51.

58. **Semler, B. L., V. H. Johnson, and S. Tracy.** 1986. A chimeric plasmid from cDNA clones of poliovirus and coxsackievirus produces a recombinant virus that is temperature sensitive. *Proc. Natl. Acad. Sci USA* **83**:1777–1781.

59. **Shih, D. S., C. T. Shih, O. Kew, M. Pallansch, R. Rueckert, and P. Kaesberg.** 1978. Cell-free synthesis and processing of the proteins of poliovirus. *Proc. Natl. Acad. Sci. USA* **75**:5807–5811.

60. **Stanway, G., P. J. Hughes, R. C. Mountford, P. D. Minor, and J. W. Almond.** 1984. The

complete sequence of a common cold virus: human rhinovirus 14. *Nucleic Acids Res.* **12**:7859–7874.

61. **Summers, D. F., and J. V. Maizel.** 1971. Determination of the gene sequence of poliovirus with pactamycin. *Proc. Natl. Acad. Sci. USA* **68**:2852–2856.

62. **Summers, D. F., E. N. Shaw, M. L. Stewart, and J. V. Maizel.** 1972. Inhibition of cleavage of large poliovirus-specific precursor proteins in infected HeLa cells by inhibitors of proteolytic enzymes. *J. Virol.* **10**:880–884.

63. **Taber, R., D. Rekosh, and D. Baltimore.** 1971. Effect of pactamycin on synthesis of poliovirus proteins: a method for genetic mapping. *J. Virol.* **8**:395–401.

64. **Toyoda, H., M. J. H. Nicklin, M. G. Murray, C. W. Anderson, J. J. Dunn, F. W. Studier, and E. Wimmer.** 1986. A second virus-encoded proteinase involved in proteolytic processing of poliovirus polyprotein. *Cell* **45**:761–778.

65. **Tracy, S., H. L. Liu, and N. M. Chapman.** 1985. Coxsackievirus B3: primary structure of the 5′ non-coding and capsid protein coding regions of the genome. *Virus Res.* **3**:263–270.

66. **Van Dyke, T. A., and J. B. Flanegan.** 1980. Identification of poliovirus polypeptide p63 as a soluble RNA-dependent RNA polymerase. *J. Virol.* **35**:732–740.

67. **Werner, G., B. Rosenwirth, E. Bauer, J. M. Seifert, F. J. Werner, and J. Besemer.** 1986. Molecular cloning and sequence determination of the genomic regions encoding protease and genome-linked protein of three picornaviruses. *J. Virol.* **57**:1084–1093.

68. **Ypma-Wong, M. F., P. G. Dewalt, V. H. Johnson, J. G. Lamb, and B. L. Semler.** 1988. Protein 3CD is the major poliovirus proteinase responsible for cleavage of the P1 capsid precursor. *Virology* **166**:265–270.

69. **Ypma-Wong, M. F., and B. L. Semler.** 1987. *In vitro* molecular genetics as a tool for determining the differential cleavage specificities of the poliovirus 3C proteinase. *Nucleic Acids Res.* **15**:2069–2088.

Molecular Aspects of Picornavirus Infection and Detection
Edited by Bert L. Semler and Ellie Ehrenfeld
© 1989 American Society for Microbiology, Washington, DC 20006

Chapter 5

Studies of Poliovirus RNA Polymerase Expressed in *Escherichia coli*: Attempts to Understand Viral RNA Replication

Ellie Ehrenfeld and Oliver C. Richards

Studies of poliovirus RNA replication have been conducted, somewhat intermittently, in several different laboratories during the past 25 years. Although the importance of understanding the mechanism of synthesis of RNA from an RNA template (a reaction that is unique to RNA viruses) has long been recognized, this particular aspect of picornavirus biochemistry and molecular biology is one of the least well understood of any problem in virus replication.

The first part of this chapter will attempt to provide a brief summary and overview of the kinds of approaches to the problem of poliovirus RNA replication that have been applied and are currently in progress. The aim is to examine some of the reasons for the failure to have solved the problem; thus, a comprehensive review of the many past contributions to the field has not been attempted. The subject has, however, been reviewed recently (10, 11), and the reader is directed to these discussions for a more detailed description of past work.

APPROACHES TO THE PROBLEM OF VIRAL RNA REPLICATION

Early studies of poliovirus RNA replication included attempts to characterize the structure of the viral RNA molecules isolated from

Ellie Ehrenfeld and Oliver C. Richards • Department of Biochemistry and Department of Cellular, Viral and Molecular Biology, University of Utah School of Medicine, Salt Lake City, Utah 84132.

infected cells and to characterize the structures of the replication complexes that contained the proteins involved in synthesis of viral RNA. Although much useful information was gained from these studies, little was actually learned about the mechanism of RNA replication, in part because viral RNA synthesis occurs in intracellular structures that are tightly associated with or in membranes, and these are consequently difficult to dissect or purify, and in part because procedures used to isolate RNA components from the replication complex often altered their structures. In addition, the existence of a given structure in an infected cell does not prove that it serves as an intermediate in the replication reaction. In subsequent studies the viral RNA polymerase was identified as the protein 3D (7, 14), and the important property of this enzyme, that it could catalyze formation of phosphodiester bonds in response to an RNA template but was unable to initiate synthesis of an RNA strand, was defined (6). At that time, the major interests and efforts were directed toward the question of how viral RNA synthesis was initiated.

Two basically different approaches have been pursued in studies of the poliovirus RNA chain initiation reaction. One is to isolate crude, membranous RNA replication complexes from virus-infected HeLa cells and to try to establish and analyze what reactions are occurring in the complexes. The rationale for this approach is that the isolated crude complexes synthesize in vitro all of the species of viral RNAs that are found in infected cells (5, 21), and they therefore likely contain all of the components necessary for a complete reaction. The major reaction occurring in these complexes is synthesis of new plus strands from a minus-strand template. The site of synthesis of minus strands has not been determined, and it is not known whether the reaction mechanism and the proteins required for plus- and minus-strand synthesis are the same. The crude replication complexes are difficult to work with and to analyze biochemically. The only recent consistent pursuit of this approach has been by members of Wimmer's laboratory, who have shown that there is an activity in these complexes that catalyzes the formation of a uridylylated form of VPg (VPgpU and then VPgpUpU) (23) and that at least some of these uridylylated forms of VPg can be extended into longer RNAs (21). This supported the idea, proposed earlier, that the uridylylated VPg molecule is utilized as a primer for RNA elongation. Most recently, Wimmer and co-workers have shown that the VPg uridylylation reaction is dependent upon template (22) and that viral protein 3D may be involved in the uridylylation reaction as well (24). It is not known what serves as the actual substrate for the uridylylation reaction, whether VPg itself or a precursor form of VPg, and it has not been unambiguously proven that the uridylylated protein is a true intermediate in the major reaction of RNA synthesis in vivo. The fact that the reaction occurs in a membranous complex adds to the difficulty of its study, but it also provides some confidence that the reaction might be a real participant in the overall synthesis of RNA, since the membrane complex is the site of RNA synthesis in vivo. As yet, there is no defined role for a host factor in the proposed mechanism of chain initiation, although the catalytic moiety responsible for VPg uridylylation has not been defined. In addition, the biochemical participation of other viral proteins, such as 2C, known to be involved in RNA synthesis, has not been demonstrated.

The second approach to understand the mechanism of initiation of RNA synthesis is to try to identify and purify each of the components required for a complete replication reaction and then to reconstitute each step in vitro so that each partial reaction, and ultimately the complete reaction, can be studied. These studies have also proven difficult, requiring a great deal of "brute force" biochemistry. The template invariably used in these studies is plus-strand RNA, since that is what is most easily obtained in reasonable quantities, and thus the reaction studied is the primary one, from plus to minus strand, rather than the quantitatively dominant one of synthesizing new plus strands.

Since the soluble, purified viral protein 3D was unable to initiate by itself, the first step was to search for some other protein which would allow 3D to initiate synthesis of a minus strand. Indeed, such activities can readily be found. They are found in extracts of uninfected cells and are therefore given the name "host factor" (HF). In different laboratories, purifications of 3D and HF were performed differently, in some cases yielding highly purified polymerase and relatively impure HF, whereas in other cases just the reverse was obtained: highly purified HF preparations and less pure polymerase. Not all of the data relevant to this approach will be described in this brief review. Instead, the proposed models will be presented, with the important warning that the biochemistry is not yet clean enough to know accurately what reactions are occurring in the test tube, much less to determine their physiological relevance. However, much important information has been and will no doubt continue to be accumulated, and the proposed reactions are plausible models.

Initially, both Baron and Baltimore (2) and Dasgupta (4) purified HF that may or may not have been the same and which, in conjunction with 3D, produced products that were predominantly genome-length minus strands. At least some portion of the products appeared to be attached to VPg (3, 16), which must have been present as contaminants in the polymerase preparation, and thus it is not exactly clear what the mechanisms of initiation and generation of the genome-length strands were. A protein kinase activity was also identified in the HF preparation in

Dasgupta's laboratory (15), but its participation in the RNA synthesis reaction was never proven.

In contrast, studies predominantly from the Flanegan laboratory have led to what is currently known as the "hairpin model," in which an HF is thought to act on the plus-strand template so as to enable the hydroxyl group of its 3'-terminal nucleotides to serve as primer for the synthesis of minus-strand RNA. This template self-priming produces dimer-length products, with minus strands covalently attached to the plus-strand template (26). Andrews and Baltimore (1) proposed that the HF was a terminal uridylyltransferase activity that added uridylate residues to the 3' poly(A) tract of virion RNA, which could then form a snapback structure. Subsequent elongation by 3D would generate dimer-length RNA. VPg addition to the new strand could occur subsequent to or concomitant with an endonucleolytic cleavage at the hairpin. This latter reaction has never actually been demonstrated. One problem with the interpretation of these studies is that the products made in these hairpinning reactions are always an extremely heterogeneous mixture of sizes, the vast majority being much shorter than twice template in length. Other workers (8, 12) have clearly demonstrated that heterogeneous snapback or hairpin structures can be artifactually produced by a low level of contaminating nuclease, which causes random nicking of the template with the fortuitous formation of terminal hairpins capable of self-priming. No detectable terminal uridylyltransferase activity was present in the HF preparations used in these studies. If the snapback structures produced with the Flanegan reagents (which do apparently contain terminal uridylyltransferase activity) are RNA replicative intermediates, the nature of the heterogeneous, short product RNAs must be explained as well as that of those that are longer than template.

Support for the hairpin model has been sought by looking for snapback, dimer-length poliovirus RNAs in infected cells. In fact, some fraction of the double-stranded RNAs that have been isolated from virus-infected cells always represents a population of snapback molecules (26). In our studies (18), these intracellular snapbacks also appear to result from nucleolytic nicking that generates self-priming structures, since the population of hairpin double strands is strikingly deficient in poly(A) stretches and thus does not contain the original 3' ends of plus strands.

Although the in vitro system requires further purification and definition, the hairpin model for generation of minus strands is not necessarily inconsistent with the VPg priming model proposed for the synthesis of plus strands in the membranous replication complex. The most recent report, however, on the mechanism of generation of covalently linked, dimeric RNA molecules comes from Lubinski et al. (13) in the Dasgupta laboratory, who report the synthesis of dimeric RNA containing both plus- and minus-strand sequences synthesized in vitro from a poliovirus RNA template and an oligo(U) primer. The minus strand was extended from the primer, and then it somehow snapped back to self-prime synthesis of plus strand. An endogenous nuclease was able to cleave the dimeric RNA into template-length products. Thus, although no explanation was proposed for the initiation of minus strands in the absence of oligo(U), it is suggested that new plus strands could be generated by self-priming and subsequent cleavage of the hairpin loop. Some HF activity might solve the minus-strand initiation problem, if all of the independent reactions can work together. It is not known how VPg would be added to new RNA strands produced by cleavage of dimers, or whether this reaction is thought to represent a minor reaction, producing some plus strands without VPg.

In summary, the in vitro RNA transcription systems using purified, soluble polymerase and any of several cellular proteins have not yet yielded a complete picture of how poliovirus RNA synthesis might occur in an infected cell. The biochemical role(s) of other viral proteins such as 2C is completely unknown, and the complicated architecture of the membranous replication complex may be an essential requirement for at least some of the reactions to occur efficiently. The biochemistry is difficult; there is always uncertainty as to whether purification has eliminated an essential component or included a contaminating activity, either of which can markedly affect the products formed in vitro. The most important requirement will be to collect some evidence that the reactions studied in vitro are truly a reflection of what occurs in vivo during viral RNA synthesis. Without that evidence, we can only continue to characterize properties of the enzymes, studying reactions that they are capable of catalyzing, often at low efficiencies, under the conditions of the in vitro assays.

CURRENT STUDIES IN THIS LABORATORY: PRODUCTION OF POLIOVIRUS RNA POLYMERASE IN *ESCHERICHIA COLI*

In light of the confusion currently surrounding the in vitro transcription systems described above, we have recently initiated efforts to produce poliovirus replication proteins from cloned cDNAs, either in *Escherichia coli* (19) or in recombinant baculovirus-infected insect cells. A similar approach has been instigated in Morrow's laboratory (17). It is hoped that this approach will allow us to produce large quantities of proteins such as 3D, 2C, and various forms of VPg and its precursors and to do so in the absence of each other or of host cell factors that contaminate and may confuse the activities that we study. These proteins will also be subject to genetic manipulation and thus may be appropriate for biochemical as well as enzymatic studies.

Initially, a plasmid was constructed that directed the synthesis of a fusion protein, containing the N-terminal portion of the *E. coli* TrpE protein, fused to poliovirus sequences consisting of 11 amino acids from the carboxy terminus of 3C and the entire 3D polypeptide. This plasmid, pATH-3C*D, expressed large amounts of the expected 89-kilodalton (kDa) protein, which was readily detected by staining a sodium dodecyl sulfate-polyacrylamide gel of total proteins from transformed *E. coli*. The fusion protein was electroeluted from a preparative gel and used to raise antiserum in a rabbit. After adsorption with extracts of *E. coli* expressing just the TrpE fragment, the antiserum was shown to be specific for 3D sequences: it identified 3CD, 3D, and the alternate cleavage product 3D′ in poliovirus-infected HeLa cells, but not 3C. It identified the 89-kDa fusion protein in transformed *E. coli*, as well as some characteristic smaller bands representing either proteolyzed fragments or polypeptides synthesized from fortuitous translational start sites, or both. Bacteria making the fusion protein, however, made no detectable RNA-dependent RNA polymerase activity, even though the fusion protein contained the entire 3D sequence.

We therefore attempted to generate active 3D polymerase from the fusion protein by placing in the same cells a complete 3C protease gene. Since the fusion protein contained the normal Q/G cleavage site, we reasoned that 3C activity should catalyze cleavage to active 3D, assuming that the Q/G junction was able to present itself properly and assuming that 3C can work on this junction in *trans*. This was accomplished in several ways, described below (Fig. 1).

trans Cleavages

A second plasmid, pEXC (9), which expresses 3C (fused to only an amino-terminal methionine) plus 25 amino acids of the N terminus of 3D, was introduced into bacteria harboring pATH-3C*D. The protein from pEXC had been previously shown to cleave itself at the 3C/3D junction. When expressed with the fusion proteins, in the same cells, the resulting protease activity was indeed able to cleave the fusion protein in *trans* to produce a new, 52-kDa, immunoreactive protein that comigrates with authentic 3D from poliovirus-infected cells. Only about 20% of the fusion protein underwent cleavage; nevertheless, RNA polymerase enzymatic activity was detectable.

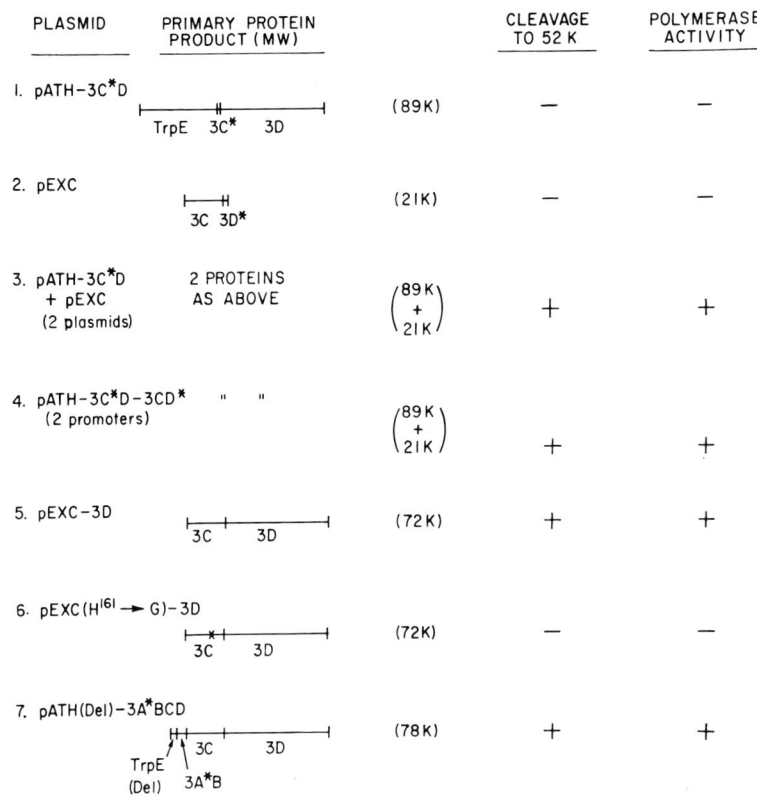

Figure 1. Properties of proteins expressed from cloned 3D cDNA sequences in *E. coli*. The indicated plasmids directed the synthesis of proteins whose structures are shown schematically and whose molecular weights (MW) are indicated in parentheses. Cleavage to the 52-kDa protein was determined by immunoblotting sodium dodecyl sulfate-polyacrylamide gels of bacterial extracts with anti-3D serum. Polymerase activity in bacterial sonic extracts was assayed with a poly(A) template and oligo(U) primer, as described previously (19).

Controlling the relative copy number of each plasmid in the above transformants was difficult, so the 3C-coding sequences from pEXC were moved into the plasmid encoding the TrpE-polymerase fusion protein such that one plasmid expressed both proteins from two different promoters. This plasmid is called pATH-3C*D-3CD* (see Fig. 1). Again, immunoblots showed the appearance of a 52-kDa polymerase, not present in cells producing fusion protein alone nor in cells producing protease alone. Approximately 50% conversion of fusion protein to polymerase occurred, and enzymatic activity was readily detectable.

Table 1. Poly(U) Polymerase Activity in Crude Extracts of *E. coli*

Plasmid	Enzyme activity[a] (U/ml)	Protein concn in extract (mg/ml)	Sp act (U/mg)
pEXC-3D	2,370	0.93	2,548
pEXC(H-G)-3D	<16	17	<0.9
pATH-3C*D	<16	17	<0.9
Infected HeLa cell cytosol[b]	2,570	2.3	1,117

[a] A unit of poly(U) polymerase activity equals 1 pmol of [³H]UMP incorporated under standard conditions. Primer-independent activity is subtracted and amounts to <0.4% of total activity for each assay.
[b] Data from Van Dyke and Flanegan (25).

cis Cleavages

A complete contiguous copy of both the 3C and 3D genes was inserted into the pEXC vector to produce pEXC-3D (Fig. 1). Self-cleavage of the expressed 3CD protein generates 3C plus 3D, similar to what happens in infected cells. Immunoblots of proteins from these bacteria reveal approximately equal amounts of the 72-kDa 3CD and the 52-kDa 3D proteins, most likely generated by *cis* cleavage. Extracts of these cells show RNA-dependent RNA polymerase activity.

Other constructs that produce proteins with additional viral sequences upstream of 3C, so that a 3B/3C cleavage site is present, always result in very efficient cleavage at the 3B/3C junction to produce 3CD, just as is seen in vivo. Thus, cleavage at the 3B/3C junction is preferred over 3CD, which occurs incompletely and more slowly.

The conclusion from all of the proteins we have produced in *E. coli* is that cleavage of fusion proteins containing the 3D sequences to form the 52-kDa polymerase correlates with the appearance of enzymatic activity. That cleavage of 3CD is required to generate enzymatically active protein is suggested by the results obtained with a mutant 3CD protein. Wild-type 3C sequences in the pEXC-3D plasmid were replaced with a point mutant that converts His-161 of 3C to Gly, which has been shown to abolish self-cleavage (9). Immunoblot analysis with antipolymerase serum confirmed that the 72-kDa 3CD was not cleaved to produce a 52-kDa protein (Fig. 1). In addition, no polymerase activity was detected (Table 1). We interpret this to mean that uncleaved 3CD does not have polymerase activity, although it does have protease activity (see other chapters in this volume). It is, of course, possible that the point mutation in 3C affects the overall folding of the protein such that the polymerase active site is inactivated. Additionally, uncleaved 3CD may function in the RNA replication reaction in some way other than chain elongation, which is all we have measured here. Failure to detect polymerase activity in 3CD supports the biochemical evidence presented by Van Dyke and Flanegan (25) that polymerase activity copurified with 3D from infected cells and not with 3CD.

The poliovirus polymerase synthesized in *E. coli* manifests a specific activity in crude extracts that is significantly higher than that found at peak times in infected HeLa cells (Table 1), although it varies markedly, depending upon the physiological state of the bacterial culture (20). Assay conditions are selected so that incorporation is linear for at least 60 min and the rate is proportional to enzyme concentration. Incorporation is absolutely dependent upon both template and primer. The enzyme can be purified from *E. coli*, essentially by the same procedures used to purify it from infected HeLa cells, and it shows all of the same biochemical properties and assay characteristics. It will use poliovirus RNA as a template, if provided with a primer, and generates genome-length, minus-strand product. In the absence of primer, no product is made at all, indicating that there is no apparent HF activity in *E. coli*.

At present, Cara Burns, a graduate student in our laboratory, is preparing mutant 3D genes to try to identify a catalytic site or substrate recognition site domain in the resulting proteins. In addition, mutant 3D proteins that are active in vitro, but which lead to an RNA⁻ phenotype when inserted back into an infectious viral RNA, might indicate other functional domains in the protein.

Literature Cited

1. **Andrews, N. C., and D. Baltimore.** 1986. Purification of a terminal uridylyltransferase that acts as host factor in the *in vitro* poliovirus replicase reaction. *Proc. Natl. Acad. Sci. USA* **83**:221–225.
2. **Baron, M. H., and D. Baltimore.** 1982. Purification and properties of a host cell protein required for poliovirus RNA replication *in vitro*. *J. Biol. Chem.* **257**:12351–12358.
3. **Baron, M. H., and D. Baltimore.** 1982. Anti-VPg antibody inhibition of the poliovirus replicase reaction and production of covalent complexes of VPg-related proteins and RNA. *Cell* **30**:745–752.
4. **Dasgupta, A.** 1983. Purification of host factor required for *in vitro* transcription of poliovirus RNA. *Virology* **128**:245–251.
5. **Etchison, D., and E. Ehrenfeld.** 1981. Comparison of replication complexes synthesizing poliovirus RNA. *Virology* **111**:33–46.
6. **Flanegan, J. B., and D. Baltimore.** 1977. Poliovirus-specific primer-dependent RNA polymerase able to copy poly(A). *Proc. Natl. Acad. Sci. USA* **74**:3677–3680.

7. **Flanegan, J. B., and D. Baltimore.** 1979. Poliovirus polyuridylic acid polymerase and RNA replicase have the same viral polypeptide. *J. Virol.* **29:**352–360.
8. **Hey, T. D., O. C. Richards, and E. Ehrenfeld.** 1987. Host factor-induced template modification during synthesis of poliovirus RNA in vitro. *J. Virol.* **61:**802–811.
9. **Ivanoff, L. A., T. Towatari, J. Ray, B. D. Korant, and S. R. Pettaway.** 1986. Expression and site-specific mutagenesis of the poliovirus 3C protease in *E. coli. Proc. Natl. Acad. Sci. USA* **83:**5392–5396.
10. **Koch, G., and F. Koch.** 1985. *The Molecular Biology of Poliovirus.* Springer-Verlag, Vienna.
11. **Kuhn, R. J., and E. Wimmer.** 1987. The replication of picornavirus, p. 17–51. *In* D. J. Rowlands, B. W. J. Mahy, and M. Mayo (ed.), *The Molecular Biology of Positive Strand RNA Viruses.* Academic Press, Inc. (London), Ltd., London.
12. **Lubinski, J. M., G. Kaplan, V. R. Racaniello, and A. Dasgupta.** 1986. Mechanism of in vitro synthesis of covalently linked dimeric RNA molecules by the poliovirus replicase. *J. Virol.* **58:**459–467.
13. **Lubinski, J. M., L. R. Ransone, and A. Dasgupta.** 1987. Primer-dependent synthesis of covalently linked dimeric RNA molecules by poliovirus replicase. *J. Virol.* **61:**2997–3003.
14. **Lundquist, R. E., E. Ehrenfeld, and J. V. Maizel.** 1974. Isolation of a viral polypeptide associated with poliovirus RNA polymerase. *Proc. Natl. Acad. Sci. USA* **71:**4773–4777.
15. **Morrow, C. D., G. F. Gibbons, and A. Dasgupta.** 1985. The host protein required for *in vitro* replication of poliovirus is a protein kinase that phosphorylates eukaryotic initiation factor-2. *Cell* **40:**913–921.
16. **Morrow, C. D., M. Navab, C. Peterson, J. Hocko, and A. Dasgupta.** 1984. Antibody to poliovirus VPg precipitates *in vitro* synthesized RNA attached to VPg-precursor polypeptide(s). *Virus Res.* **1:**89–100.
17. **Morrow, C. D., B. Warren, and M. R. Lentz.** 1987. Expression of enzymatically active poliovirus RNA-dependent RNA polymerase in *E. coli. Proc. Natl. Acad. Sci. USA* **84:**6050–6054.
18. **Richards, O. C., T. D. Hey, and E. Ehrenfeld.** 1987. Poliovirus snapback double-stranded RNA isolated from infected HeLa cells is deficient in poly(A). *J. Virol.* **61:**2307–2310.
19. **Richards, O. C., L. A. Ivanoff, K. Bienkowska-Szewczyk, B. Butt, S. Pettaway, M. A. Rothstein, and E. Ehrenfeld.** 1987. Formation of poliovirus RNA polymerase 3D in *E. coli* by cleavage of fusion proteins expressed from cloned viral cDNA. *Virology* **161:**348–356.
20. **Rothstein, M. A., O. C. Richards, C. Amin, and E. Ehrenfeld.** 1988. Enzymatic activity of poliovirus RNA polymerase synthesized in *Escherichia coli* from viral cDNA. *Virology* **164:**301–308.
21. **Takeda, N., R. J. Kuhn, C.-F. Yang, T. Takegami, and E. Wimmer.** 1986. Initiation of poliovirus plus-strand RNA synthesis in a membrane complex of infected HeLa cells. *J. Virol.* **60:**43–53.
22. **Takeda, N., C.-F. Yang, R. J. Kuhn, and E. Wimmer.** 1987. Uridylylation of the genome-linked protein of poliovirus *in vitro* is dependent upon an endogenous RNA template. *Virus Res.* **8:**193–204.
23. **Takegami, T., R. J. Kuhn, C. W. Anderson, and E. Wimmer.** 1987. Membrane-dependent uridylylation of the genome-linked protein VPg of poliovirus. *Proc. Natl. Acad. Sci. USA* **80:**7447–7451.
24. **Toyoda, H., C.-F. Yang, A. Nomoto, and E. Wimmer.** 1987. Analysis of RNA synthesis

of type 1 poliovirus by using an in vitro molecular genetic approach. *J. Virol.* **61:**2816–2822.

25. **Van Dyke, T. A., and J. B. Flanegan.** 1980. Identification of poliovirus polypeptide P63 as a soluble RNA-dependent RNA polymerase. *J. Virol.* **35:**732–740.

26. **Young, D. C., D. M. Tuschall, and J. B. Flanegan.** 1985. Poliovirus RNA-dependent RNA polymerase and host cell protein synthesize product RNA twice the size of poliovirus RNA in vitro. *J. Virol.* **54:**256–264.

Molecular Aspects of Picornavirus Infection and Detection
Edited by Bert L. Semler and Ellie Ehrenfeld
© 1989 American Society for Microbiology, Washington, DC 20006

Chapter 6

A Large Segment of Poliovirus 5' Noncoding Region Allows Cap-Independent Translation of Downstream Sequences in Mammalian Cells

Didier Trono, Raul Andino, and David Baltimore

INTRODUCTION

The genome of poliovirus is a single-stranded molecule of plus-strand RNA, approximately 7,500 bases long. Its 5' end is linked to a 22-amino-acid peptide, VPg, and its 3' end is connected to a stretch of poly(A), 40 to 100 nucleotides long (8, 11). A total of more than 800 untranslated nucleotides flank a 6,528-base-long reading frame: 742 upstream of the AUG used to initiate translation (3) and 65 preceding the poly(A) tract. Poliovirus mRNA not only has an unusually long 5' noncoding region, but it is also not capped at its 5' end: it terminates in pUp instead of the "capping group" $m^7G(5')pppN...$, found on almost all other mRNAs (6, 10). Poliovirus RNA must therefore be translated in a cap-independent manner; this characteristic is of primary importance, because poliovirus specifically inhibits all cap-dependent translation in the infected cell (4, 14).

The 5' noncoding region must serve some crucial functions in the

Didier Trono and Raul Andino • Whitehead Institute for Biomedical Research, Nine Cambridge Center, Cambridge, Massachusetts 02142. **David Baltimore** • Whitehead Institute for Biomedical Research, Nine Cambridge Center, Cambridge, Massachusetts 02142, and Department of Biology, Massachusetts Institute of Technology, Cambridge, Massachusetts 02139.

virus life cycle, a hypothesis reinforced by the high sequence similarity extending through the first 650 nucleotides of the three poliovirus serotypes (16). These functions could be initiation of protein synthesis (binding of ribosomes and initiation factors), initiation of synthesis of plus-strand RNA, stabilization of the RNA through secondary structure formation, interaction with the viral capsid proteins during packaging of the RNA, or binding to presumptive regulatory molecules controlling RNA replication or translation.

The 5' noncoding region appears to have a critical role in neurovirulence. Neurovirulent revertants of the Sabin 3 serotype have been isolated from patients with vaccine-associated poliomyelitis and found to have a single-base substitution at position 472 as compared with the parental strain (5). In addition, in vitro experiments using recombinants between wild-type and attenuated strains strongly suggest that a major determinant of neurovirulence is located in the 5' untranslated region of poliovirus (9, 18).

In an attempt to elucidate its functions, we have created a number of mutations in the 5' noncoding region of poliovirus type 1 (Mahoney) RNA, using an infectious cDNA copy of the genome (12). As a first step, we have studied the biological behavior of several phenotypically recognizable viral strains generated by some of these mutations. On that basis, we have been able to delineate at least two functional regions, one primarily involved in the synthesis of the viral RNA and the other in its translation. This last function seemed to be accomplished by an extensive sequence of nucleotides, suggesting the role of an RNA "superstructure" in the initiation of translation.

In a second set of experiments, we have demonstrated that a foreign gene cloned downstream of the poliovirus 5' noncoding region acquires the ability to be translated in a cap-independent manner in mammalian cells. We have also extended the information obtained from the analysis of the 5' noncoding mutants by identifying major and minor determinants of cap-independent translation.

MUTAGENESIS OF POLIOVIRUS 5' NONCODING REGION

Our initial experimental approach consisted in generating viruses with mutations in the 5' noncoding region (designated 5NC mutants) so as to study the biological characteristics of those showing a recognizable phenotype. For this, multiple nucleotides were inserted or deleted at various convenient sites, using a cDNA copy of poliovirus RNA (a complete description of materials and methods used is presented elsewhere [17]). RNA was made from these mutated clones and used to

Table 1. Mutagenesis

Clone	Wild-type sequence	Mutant sequence	Phenotype[a]
pPN-1	$AGT_{52}ACT$	AGT_{52}CCCGGGACT	D
pPN-2	$G_{66}GTACC$	G_{66}....C	D
pPN-3	$GGTAC_{70}C$	$GGTAC_{70}$GTACC	M (5NC-11)
pPN-4	$GGTAC_{70}C$	$GGTAC_{70}$GGAATTCCGTACC	M (5NC-111)
pPN-5	$CTTA_{108}G$	$CTTA_{108}$TTΔG	WT
pPN-6	$CTTA_{108}G$	$CTTA_{108}$GGAATTCCTTAG	WT
pPN-7	$CTTA_{108}G$	$CTTA_{108}$GGAATTAATTCCTTAG	WT
pPN-8	$GGATC_{224}C$	$GGATC_{224}$GATCC	M (5NC-13)
pPN-9	$GGATC_{224}C$	$GGATC_{224}$GGAATTCCGATCC	D
pPN-10	$GAAT_{270}C$	$GAAT_{270}$AATC	WT
pPN-11	$GAAT_{270}C$	$GAAT_{270}$GGAATTCCAATC	M (5NC-114)
pPN-12	$GAGT_{325}C$	$GAGT_{325}$AGTC	D
pPN-13	$CCATG_{392}G$	$CCATG_{392}$CATGG	WT
pPN-14	$CCATG_{392}G$	$CCATG_{392}$GGAATTCCCATGG	M (5NC-116)
pPN-15	$CCATG_{392}G$	$CCATG_{392}$GGAATTAATTCCCATGG	M (5NC-1116)
pPN-16	$GAAT_{443}C$	$GAAT_{443}$GGAATTCCAATC	D
pPN-17	$GAATG_{460}CGGC$	$GAATG_{460}$..GC	D
pPN-18	$GAATG_{460}CGGC$	$GAAT_{460}$..GGAATTCCGC	D
pPN-19	$CAG_{499}TGATTG$	CAG_{499}...TTG	D
pPN-20	$CAG_{499}TGATTG$	CAG_{499}...GGAATTCCTTG	D

[a] D, Dead (lethal mutation); WT, wild type (silent mutation); M, mutant. Parentheses give mutant designation.

transfect HeLa cells, and the recovery of infectious virus particles was tested by plaque assay. Among 21 different mutations, three kinds of results were observed (Table 1 and Fig. 1).

First, nine mutations involving seven sites, distributed evenly over much of the 5' noncoding region, were "lethal," meaning that no infectious virus was recovered from transfected cells. Most of the 5' noncoding region, therefore, appears necessary to viral growth.

Second, six mutations at four different sites, including a deletion extending from nucleotides 630 to 723, were silent, at least as assessed by plaque assay. Thus, some regions can be altered without significantly inhibiting viral growth, and one region is dispensable for in vitro growth.

Third, six mutations, affecting four different sites, generated infectious viruses with a phenotype easily distinguishable from that of the wild type, as follows.

(i) Mutants 5NC-11 and 5NC-111 have a 4-base and a 12-base insertion, respectively, at nucleotide 70. Both generate minute plaques and are temperature sensitive.

(ii) Mutant 5NC-13 has a four-nucleotide insertion at position 224, is

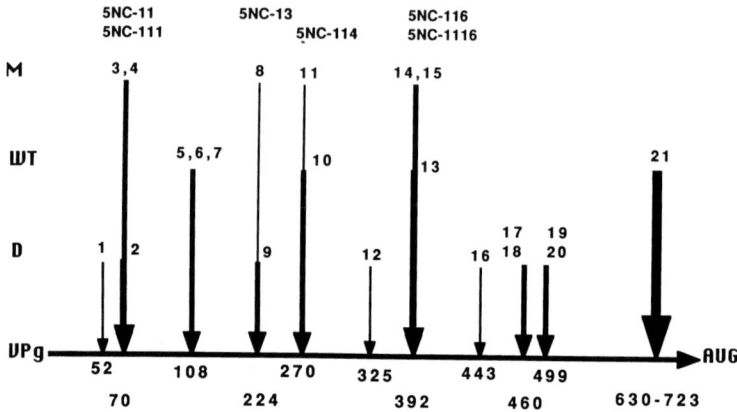

Figure 1. Mutagenesis. The result obtained from transfection of HeLa cells with RNA made from the 21 mutated clones is shown. Numbers at the top of each arrow correspond to the constructs described in Table 1 (i.e., 1 is pPN-1). Construct pPN-21 has a deletion of the sequence indicated on the bottom line. Short arrows indicate lethal mutations (D), medium-length arrows denote silent mutations (WT), and long arrows point to mutations responsible for a recognizable viable phenotype (M). Names of the mutant viruses are shown at the top. Numbers on bottom lines indicate the nucleotide position of the mutation.

also temperature sensitive, and generates plaques that are smaller than those of the wild type.

(iii) Mutant 5NC-114: an 11-base insertion at nucleotide 270 creates a mutant with a small-plaque phenotype and a slight degree of temperature sensitivity.

(iv) Mutants 5NC-116 and -1116 result from the insertion of 12 and 16 bases, respectively, at nucleotide 392; they have a phenotype grossly similar to that of 5NC-114.

REPLICATION OF VIRUSES WITH MUTATIONS IN THE 5' NON-CODING REGION RESULTING IN A RECOGNIZABLE PHENOTYPE

Assuming that slightly different modifications of the same site would affect the same function, mutants 5NC-111 and 5NC-1116 were not further characterized. Several aspects of the life cycle of the four other 5' noncoding mutants were studied.

One-step growth curves showed that all of these mutants replicated more slowly than the wild-type parental strain and produced a final progeny yield that was lower (data not shown). This was, however, the only aspect studied that was common to all mutants, since the analysis of viral RNA synthesis clearly distinguished two groups, as follows.

(i) Of the first group, mutant 5NC-11 was primarily deficient in RNA synthesis, making less than 1% of the wild-type amount of RNA, at all temperatures (Fig. 2A). A mutant with a deletion of nucleotide 10, having a closely similar phenotype, has been described (13). This suggests that the far 5' end of the genome is likely to be primarily involved in one or more steps of RNA replication.

(ii) Forming a second group, mutants 5NC-13, -114, and -116 all synthesized a significant amount of RNA, 30 to 60% of the wild-type level, both plus and minus strand, however with some delay (Fig. 2B and C). Although these three mutants were somewhat defective in RNA synthesis, this clearly was not their primary defect, as outlined by a comparison of progeny virion yield and RNA synthesis (Fig. 2D): for instance, at 6 h postinfection, 5NC-114 (and the others similarly) made 50% of the normal yield of RNA, but produced only 0.3% of the normal yield of progeny.

The analysis of protein synthesis in cells infected with viral mutants 5NC-13, -114, and -116 revealed two fundamental characteristics of these mutants: they made very little protein, and they did not inhibit host cell translation (Fig. 3 and 4). Translational shutoff in poliovirus-infected cells correlates with the cleavage of one of the components of the cap-binding complex, eucaryotic initiation factor 4F, also called p220 (2, 6, 14). The three mutants failed to induce this cleavage (data not shown).

GENETIC ANALYSIS OF THE 5NC MUTANTS

To define the primary defects of mutants 5NC-13, -114, and -116, we studied their ability to complement and be complemented by other well-defined poliovirus mutants. In such experiments, cells are infected with one virus alone or with two viruses together, and the yields of single and mixed infections are compared. If the growth of one of the partners is enhanced in mixed infection, it is said to be defective in a function that can be provided in *trans*, and it can also be concluded that the other partner accomplishes this function normally. The existence of poliovirus defective interfering particles (7) demonstrates that the capsid proteins can be provided in *trans*. In addition, it was shown recently that two nonstructural proteins, 2A and 3A, involved in translational shutoff and RNA synthesis, respectively, carry out a *trans* rather than a *cis* activity; other nonstructural functions (2B, 3D, 3' noncoding region) seemingly act only in *cis* (1).

In view of this, the behavior of the 5NC mutants in complementation experiments was of the highest interest, as they presented a phenotype that was closely similar to that of mutants with lesions in the 2A protein

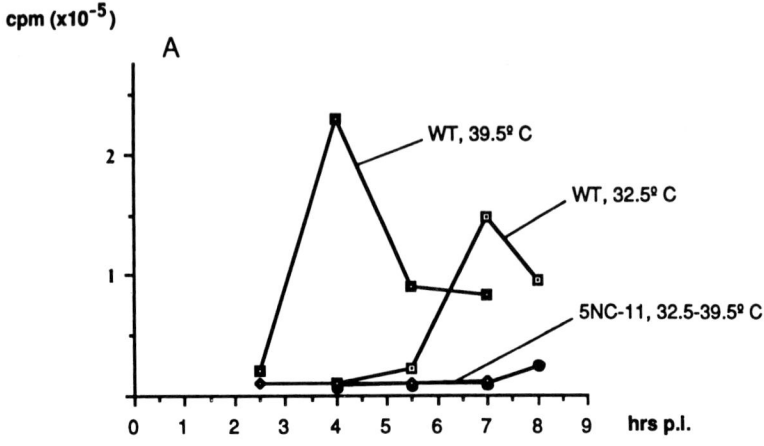

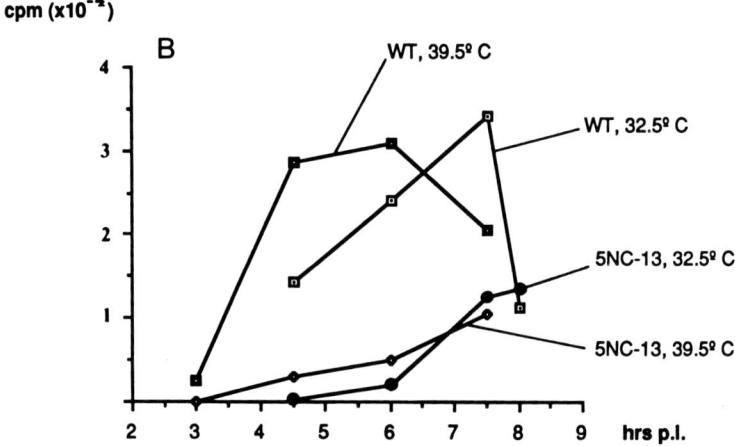

Figure 2. RNA synthesis in virus-infected cells. Analysis was done by dot-blot as described (17). Only the result of the positive strand is shown because the ratio of the positive to negative strand was indistinguishable from that of the wild type with all mutants. (A) 5NC-11 at 32.5 and 39.5°C; (B) NC-13 at 32.5 and 39.5°C; (C) 5NC-114 and -116 at 37°C; (D) 5NC-114, progeny yield (as determined by plaque assay) and RNA synthesis, comparison with wild type.

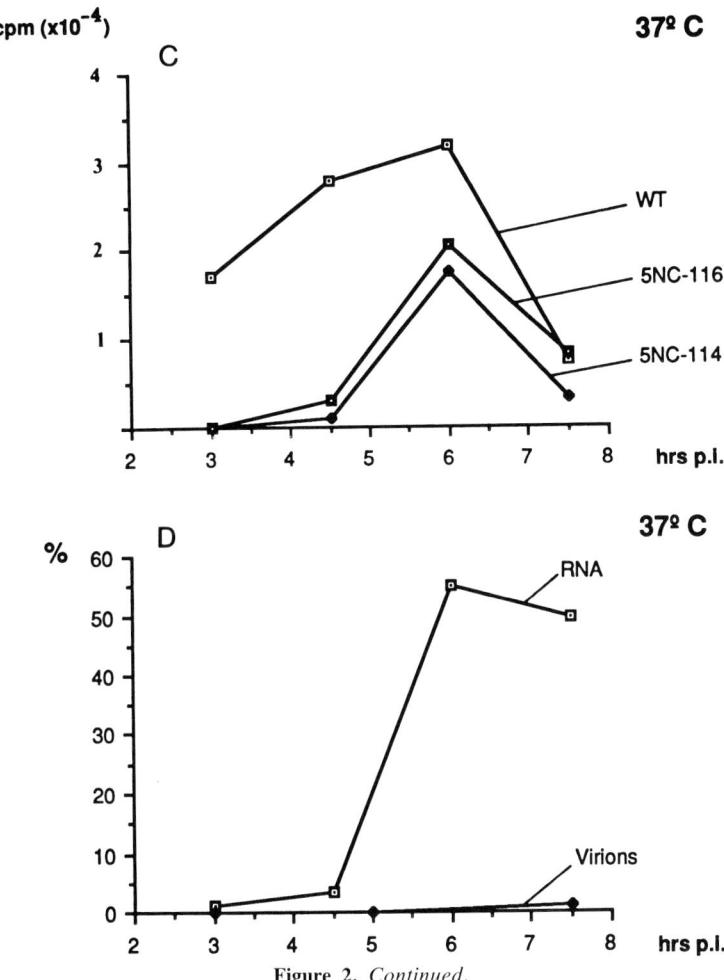

Figure 2. *Continued.*

(2): reduced level of viral protein synthesis and failure to inhibit host cell translation. Was the absence of translational shutoff a consequence of the decreased amount of viral proteins, namely, of an insufficient level of 2A, or was it a direct effect of lesions in the 5' noncoding region, independently of 2A? One could have imagined a *trans*-acting function of the 5' noncoding region, for instance ribosomal modification.

The mutants used in these experiments and the results of the complementation experiments are shown in Fig. 5 and Table 2, respectively. First, mutants 5NC-13, -114, and -116 were found to belong to the

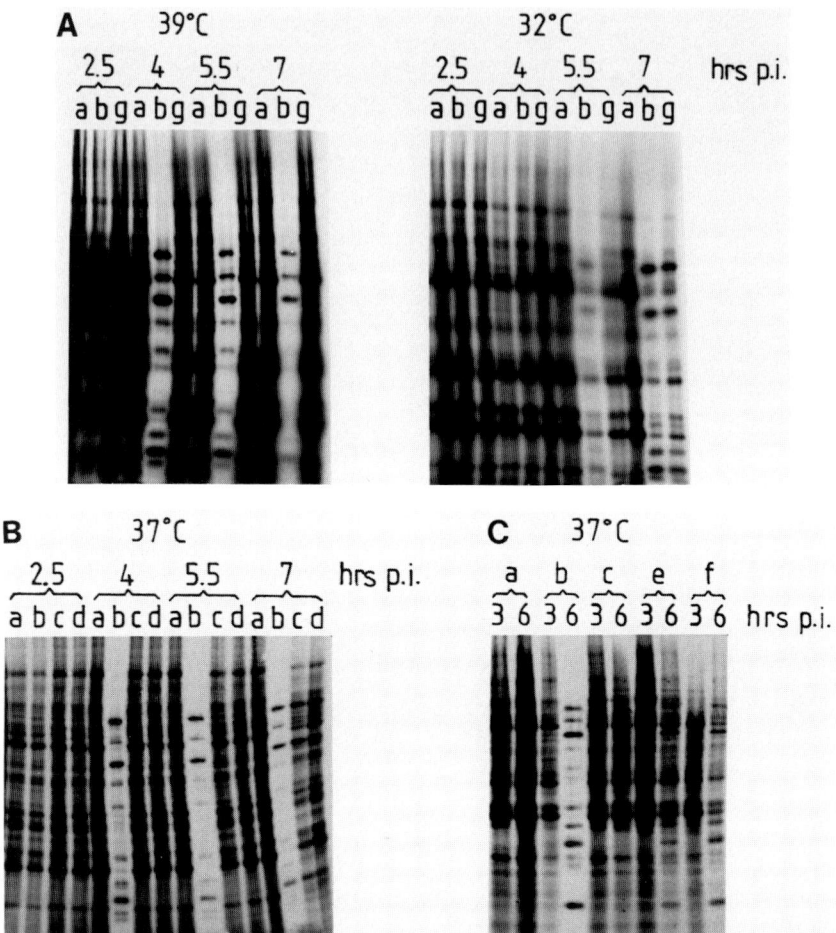

Figure 3. Protein synthesis in virus-infected cells. Cells were pulsed at different times after infection with [³⁵S]methionine, and extracts were fractionated by electrophoresis through a 12.5% polyacrylamide gel which was then autoradiographed. Initial multiplicity of infection was 10 PFU/cell unless specified. (A) 5NC-13 at 39 and 32°C; (B) 5NC-114 and -116 at 37°C; (C) effect of increasing multiplicities of 5NC-114. Lanes in all panels: a, mock-infected cells; b, wild-type virus; c, 5NC-114; d, 5NC-116; e, 5NC-114 (50 PFU/cell); f, 5NC-114 (100 PFU/cell); g, 5NC-13.

same complementation group, as they did not complement each other (Table 2, lines 1, 2, and 3). Second, they were efficiently complemented by mutants having defects mapping in the capsid region, the replicase gene, and the 3' noncoding region (Table 2, lines 5 to 10). Third and most

Figure 4. Analysis of virus-specific protein synthesis. Radiolabeled cytoplasmic extracts were immunoprecipitated with a polyclonal antivirion antibody and analyzed by electrophoresis through a 12.5% polyacrylamide gel. Lanes: v, [^{35}S]methionine-radiolabeled virion; a, mock-infected cells; b, wild-type virus; c, 5NC-114; d, 5NC-116; e, 5NC-13.

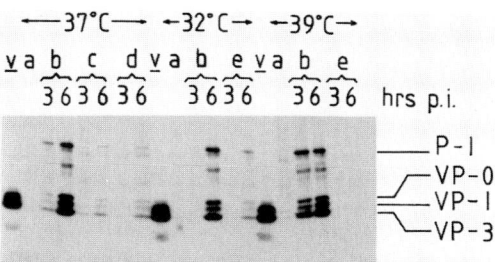

importantly, they were not complemented efficiently by mutants with lesions in 2A (Table 2, lines 12, 13, 15, and 16).

It can therefore be concluded that 5NC-13, -114, and -116 map in a region of the 5' noncoding segment of the poliovirus genome primarily involved in viral protein synthesis. We call this "region P." The absence of complementation by 2A mutants is key to understanding the role of region P, because 2A is a function easily complemented by all mutants other than the 5' noncoding mutants (1) (Table 2, lines 11 and 14). It appears that region P mutants, due to a global decrease in viral protein synthesis, make so little of protein 2A that they are unable to shut off host cell translation, a defect which becomes the limiting factor of viral infection. When 2A is provided to region P mutants, the cellular ribosomes, no longer attracted to cellular mRNA, can better translate poliovirus RNA with region P lesion, and thus the mutants grow much more efficiently. Indeed, as region P mutants make significant yields of RNA without complementation, their particle yield must be a direct function of the amount of capsid protein they make: any increase in protein synthesis will be a yield enhancement.

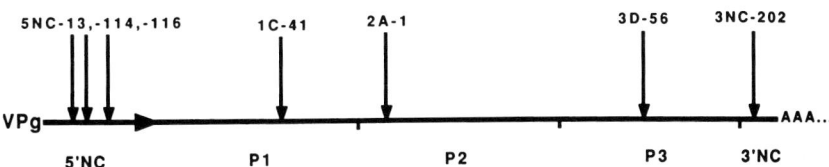

Figure 5. Mutants used in complementation experiments. The approximate genomic location of the mutants described in the text is shown.

Table 2. Complementation Experiments

Virus pair[a]	Complementation index[b] at time postinfection:			
	3 h		6 h	
1. 5NC-13 + 5NC-114	1	(total)	1	(total)
	1	(5NC-114)	1	(5NC-114)
	15	(5NC-13)	6	(5NC-13)
2. 5NC-114 + 5NC-116	1.2	(total)	1	(total)
3. 5NC-13 + 5NC-116	1	(total)	1.5	(total)
4. 3NC-202 + 3D-56	1.3		1	
5. 5NC-13 + 3NC-202	120		140	
6. 5NC-114 + 3NC-202	150		43	
7. 5NC-13 + 3D-56	180		50	
8. 5NC-114 + 3D-56	40		20	
9. 5NC-13 + 1C-41	90		ND	
10. 5NC-114 + 1C-41	40		ND	
11. 3NC-202 + 2A-1[c]	50		ND	
12. 5NC-13 + 2A-1[c]	1.2	(total)	ND	
	20	(5NC-13)		
13. 5NC-114 + 2A-1[c]	2	(total)	ND	
14. 3D-56 + R2-2A-2	40		ND	
15. 5NC-13+ R2-2A-2	2	(total)	ND	
16. 5NC-114 + R2-2A-2	4	(total)	ND	

[a] The mutant complemented is underlined.
[b] Complementation index calculated as described in reference 17. ND, Not determined. Parentheses indicate the mutant complemented.
[c] Done on CV1 cells.

THE POLIOVIRUS 5′ NONCODING REGION CAN DIRECT IN MAMMALIAN CELLS THE CAP-INDEPENDENT TRANSLATION OF FOREIGN SEQUENCES PLACED DOWNSTREAM

To better understand the translational properties of the poliovirus 5′ noncoding region, we used transfected mammalian cells to analyze its potential to direct the cap-independent translation of foreign sequences, and the influence of various mutations on that potential.

For that purpose, the gene encoding chloramphenicol acetyltransferase (CAT) was cloned at position 630 of the poliovirus 5′ noncoding region, either from the wild-type sequence or from one of the mutated clones described earlier. In all cases, the constructs were placed under the control of the simian virus 40 early promoter. These constructs, as well as a pSV2-CAT control, were introduced in parallel into COS cells by electroporation. Forty hours later, a fraction of the cells that had been electroporated with each one of the constructs was harvested for CAT assay. At the same time, another fraction was infected with poliovirus at a multiplicity of infection of 100 in the presence of guanidine (to inhibit poliovirus replication and therefore prevent early death of the cell) and actinomycin D (to stop any further accumulation of mRNA in all cases). A third fraction was mock treated, with addition of guanidine and actinomycin D to the medium. At various times later, an equal number of cells from each one of these two fractions were assayed for CAT activity.

If the hypothesis based on our previous results were true, cells electroporated with constructs able to carry on translation in a cap-independent manner should show an enhancement of CAT activity after poliovirus infection, thanks to the disappearance of most competitor mRNA. In contrast, cells containing constructs able to translate only in a cap-dependent manner should stop any further synthesis of CAT after infection.

The results of this experiment (Fig. 6 and Table 3) are best considered in four ways, as follows.

(i) First, poliovirus infection of cells electroporated with a pSV2-CAT plasmid resulted in the abolition of further CAT synthesis (Fig. 6; Table 3, line 1).

(ii) Second, CAT activity was stimulated by poliovirus infection in cells transfected with a construct in which the wild-type poliovirus 5′ noncoding region preceded the CAT encoding sequence (Fig. 6; Table 3, line 2). The same result was obtained when the leader sequence was the one of mutant 5NC-11, shown to contain a mutation without effect on the translation of the viral RNA but impairing its replication (Table 3, line 3).

(iii) Third, constructs corresponding to viable poliovirus mutants with a decreased translational potential (i.e., mutants 5NC-13, -114, and -116) showed a reduced base-line level of CAT activity, but kept the ability to translate in a cap-independent manner, that is, were stimulated by poliovirus infection (Fig. 6C; Table 3, lines 4, 6, and 7).

(iv) Finally, constructs containing mutations that had been previously identified as lethal (see Table 1 and Fig. 1) were found to have lost the potential to translate in a cap-independent manner (Table 3, lines 5 and 8 to 11).

We therefore have confirmed the results described earlier, but also disclosed what appear to be regions crucial to a cap-independent translation. The sequence extending from nucleotides 460 to 500 seems critical in that regard.

These experiments also demonstrate that the poliovirus 5′ noncoding region can direct in mammalian cells the cap-independent translation of foreign sequences. As a consequence, it provides a unique method to program mammalian cells to express only a single gene of interest.

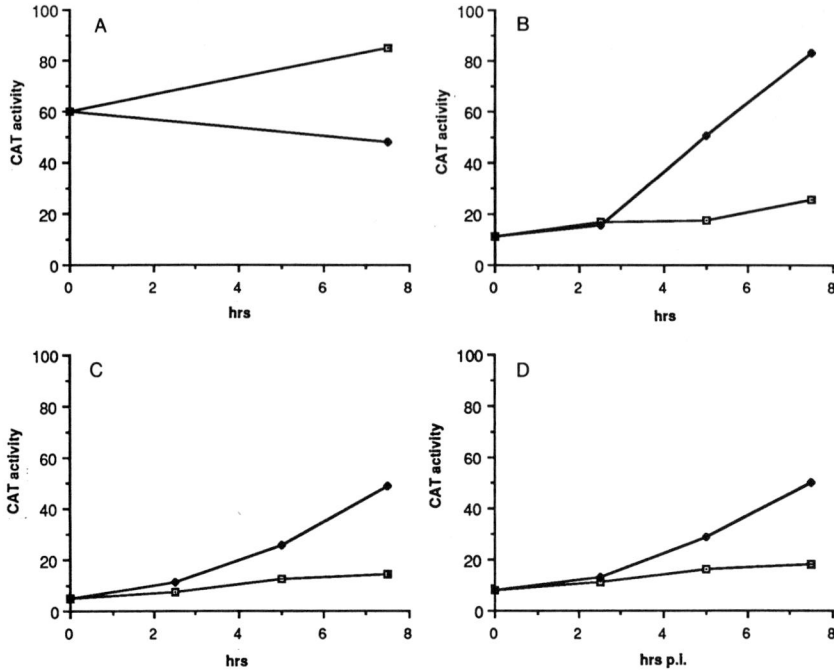

Figure 6. Effect of poliovirus infection on the expression of a gene placed downstream of different 5′ noncoding regions. COS cells were electroporated with constructs containing the CAT-coding sequence downstream of 5′ noncoding regions (see Table 3) and either mock-treated (□) or infected with poliovirus 40 h later (time 0) (◆). CAT activity was assayed at various times after infection. (A) pSV2-CAT; (B) pWT-CAT; (C) pPN8-CAT; (D) pPN11-CAT.

THE UNITARY FUNCTION OF REGION P: ROLE IN NEUROVIRULENCE

The cumulative data from both sets of experiments show that an extensive RNA sequence is responsible for allowing translation of the viral RNA. It extends at least from nucleotides 220 to 500, and we are currently defining its exact limits. We have found major and minor determinants of poliovirus RNA translation and demonstrated that minimal changes in some crucial regions (for instance, a 2- or a 3-base deletion at position 460 or 499, respectively) completely abolish its ability to be translated in a cap-independent manner, explaining why no viable virus containing such mutations could be obtained. The precise function of region P is still to be defined, but providing an internal ribosome-binding site is a likely possibility.

Table 3. Translation of 5NC-CAT Constructs into COS Cells

Construct[a]	CAT activity after electroporation[b]	Fold increase in CAT activity 7 h after:	
		Mock treatment	Poliovirus infection
1. pSV2-CAT	6.5	1.3	0.8
2. pWT-CAT	1	1.4	2.4
3. pPN3-CAT	0.9	1.4	3.1
4. pPN8-CAT	0.33	2	2.7
5. pPN9-CAT	0.38	1.9	1.3
6. pPN11-CAT	0.73	1.2	2
7. pPN14-CAT	0.75	1.4	2.2
8. pPN17-CAT	0.56	1.3	0.7
9. pPN18-CAT	0.21	1.2	1.1
10. pPN19-CAT	0.29	1.6	0.9
11. pPN20-CAT	0.53	1.3	1.2

[a] The CAT gene was cloned at position 630 of the wild-type or mutated poliovirus noncoding region; the nature of the mutation is given by pPN designation, corresponding to the mutated clones described in Table 1.
[b] CAT activity 40 h after electroporation; the activity obtained with pWT-CAT, which contains a wild-type poliovirus 5' noncoding region, is taken as reference (line 2).

If the whole sequence contained within the limits of region P belongs to the same functional unit, it will include nucleotides that have been found to be major determinants of the attenuation of type 3 (5, 18) and, to a lesser extent, type 1 neurovirulence (5). This would suggest that attenuation in vaccine strains comes from a specific inability of motor neurons, as opposed to enterocytes, to translate the viral genome. It has actually been shown that vaccine strains translate more poorly than their wild-type parents, at least in some in vitro systems (15). If such a model is true, defining how region P accomplishes its function may highlight some specific characteristics of the translational machinery in neurons. Also, and most importantly, the 5' noncoding region of poliovirus should be used as a target to engineer new vaccines with full immunogenic properties but better stability than the ones currently in use.

Acknowledgments. We are grateful to V. R. Racaniello for mutant R2-2A-2, to J. P. Li for mutant 1C-41, and to I. Edery for anti-p220 antiserum.

This work was supported by Public Health Service grant AI 22346 to D.B. from the National Institutes of Health. D.T. is the recipient of a Fellowship from the Fondation Suisse de Bourses en Medecine et Biologie. R.A. is the recipient of a Fellowship from the Consejo Nacional de Investigaciones en Ciencia y Tecnica-Argentina.

Literature Cited

1. **Bernstein, H. D., P. Sarnow, and D. Baltimore.** 1986. Genetic complementation among poliovirus mutants derived from an infectious cDNA clone. *J. Virol.* **60:**1040–1049.

2. **Bernstein, H. D., N. Sonenberg, and D. Baltimore.** 1985. Poliovirus mutant that does not selectively inhibit host cell protein synthesis. *Mol. Cell. Biol.* **5:**2913–2923.

3. **Dorner, A. J., L. F. Dorner, G. R. Larsen, E. Wimmer, and C. W. Anderson.** 1982. Identification of the initiation site of poliovirus polyprotein synthesis. *J. Virol.* **42:**1017–1028.

4. **Etchison, D., S. C. Milburn, I. Edery, N. Sonenberg, and J. W. B. Hershey.** 1982. Inhibition of HeLa cell protein synthesis following poliovirus infection correlates with the proteolysis of a 220,000 dalton polypeptide associated with eukaryotic initiation factor 3 and a cap-binding protein complex. *J. Biol. Chem.* **257:**14806–14810.

5. **Evans, D. M. A., G. Dunn, P. D. Minor, G. C. Schild, A. J. Cann, G. Stanway, J. W. Almond, K. Currey, and J. V. Maizel, Jr.** 1985. Increased neurovirulence associated with a single nucleotide change in a noncoding region of the Sabin type 3 poliovaccine genome. *Nature* (London) **314:**548–550.

6. **Hewlett, M. J., J. K. Rose, and D. Baltimore.** 1976. 5′-Terminal structure of poliovirus polyribosomal RNA is pUp. *Proc. Natl. Acad. Sci. USA* **73:**327–330.

7. **Huang, A. S., and D. Baltimore.** 1970. Defective viral particles and viral disease processes. *Nature* (London) **226:**325–327.

8. **Kitamura, N., B. Semler, P. G. Rothberg, G. R. Larsen, C. J. Adler, A. J. Dorner, E. A. Emini, R. Hanecak, J. J. Lee, S. van der Werf, C. W. Anderson, and E. Wimmer.** 1981. Primary structure, gene organization and polypeptide expression of poliovirus RNA. *Nature* (London) **291:**547–553.

9. **Nomoto, A., M. Kohara, S. Kuge, N. Kawamura, M. Arita, T. Komatsu, S. Abe, B. L. Semler, E. Wimmer, and H. Itoh.** 1987. Study on virulence of poliovirus type 1 using in vitro modified viruses. *UCLA Symp. Mol. Cell. Biol.* **24:**437–452.

10. **Nomoto, A., Y. F. Lee, and E. Wimmer.** 1976. The 5′ end of poliovirus mRNA is not capped with m^{7}G(5′)ppp(5′)NP. *Proc. Natl. Acad. Sci. USA* **73:**375–380.

11. **Racaniello, V. R., and D. Baltimore.** 1981. Molecular cloning of poliovirus cDNA and determination of the complete nucleotide sequence of the viral genome. *Proc. Natl. Acad. Sci. USA* **78:**4887–4891.

12. **Racaniello, V. R., and D. Baltimore.** 1981. Cloned poliovirus complementary DNA is infectious in mammalian cells. *Science* **214:**916–919.

13. **Racaniello, V. R., and C. Meriam.** 1986. Poliovirus temperature-sensitive mutant containing a single nucleotide deletion in the 5′-noncoding region of the viral RNA. *Virology* **155:**498–507.

14. **Sonenberg, N.** 1987. Regulation of translation by poliovirus. *Adv. Virus Res.* **33:**175–204.

15. **Svitkin, Y. V., S. V. Maslova, and V. I. Agol.** 1985. The genomes of attenuated and virulent poliovirus strains differ in their *in vitro* translation efficiency. *Virology* **147:**243–252.

16. **Toyoda, H., M. Kohara, Y. Kataoka, T. Suganuma, T. Omata, N. Imura, and A. Nomoto.** 1984. Complete nucleotide sequences of all three poliovirus serotype genomes: implication for genetic relationship, gene function and antigenic determinants. *J. Mol. Biol.* **174:**562–585.

17. **Trono, D., R. Andino, and D. Baltimore.** 1988. An RNA sequence of hundreds of nucleotides at the 5′ end of poliovirus RNA is involved in allowing viral protein synthesis. *J. Virol.* **62:**2291–2299.

18. **Westrop, G. D., D. Evans, M. Skinner, M. Ferguson, D. Magrath, G. Schild, P. Minor, and J. W. Almond.** 1987. Investigation of the molecular basis of attenuation in the Sabin type 3 vaccine using novel recombinant polioviruses constructed from infectious cDNA, p. 56–60. *In* D. J. Rowlands, B. W. J. Mahy, and M. Mayo (ed.), *The Molecular Biology of Positive Strand RNA Viruses.* Alan R. Liss, Inc., New York.

Part II

VIRION STRUCTURE AND CELL SURFACE INTERACTIONS

Molecular Aspects of Picornavirus Infection and Detection
Edited by Bert L. Semler and Ellie Ehrenfeld
© 1989 American Society for Microbiology, Washington, DC 20006

Chapter 7

Structural Basis for Serotypic Differences and Thermostability in Poliovirus[*]

James M. Hogle, David J. Filman, Rashid Syed, Marie Chow, and Philip D. Minor

Crystallographic studies have provided considerable insight into the architecture, evolution, assembly, and immune recognition of picornaviruses (5, 6, 10; D. J. Filman, R. Syed, M. Chow, P. D. Minor, and J. M. Hogle, manuscript in preparation). We have now solved the structures of two strains of poliovirus: the Mahoney strain of type 1 poliovirus (Mahoney 1) (5) and the Sabin (attenuated vaccine) strain of type 3 poliovirus (Sabin 3) (Filman et al., in preparation). Comparison of the structures of the Sabin 3 and the Mahoney 1 strains has provided the first detailed view of the structural differences between serotypes of related spherical viruses. In addition, the availability of the two independently derived atomic models provides constraints for attempts to model the effect of strain- and serotype-dependent sequence differences on the

* Publication no. 5444-MB from the Molecular Biology Department of the Research Institute of Scripps Clinic.

James M. Hogle, David J. Filman, and Rashid Syed • Department of Molecular Biology, Research Institute of Scripps Clinic, 10666 North Torrey Pines Road, La Jolla, California 92037.   **Marie Chow** • Department of Biology, Massachusetts Institute of Technology, Cambridge, Massachusetts 02139.   **Philip D. Minor** • National Institute for Biological Standards and Control, Blanche Lane, South Mimms, Potters Bar, Hertfordshire EN6 3QG, United Kingdom.

structures of other poliovirus strains. Such modeling has taken on increased importance in light of the recent success of several investigators in constructing intertypic hybrids of poliovirus, as well as the potential use of these antigenic chimeras of poliovirus as vaccines (2, 6a, 7a, 8; see also J. Bradley, M. G. Murray, R. J. Kuhn, H. Tada, X.-F. Yang, C. Mirzayan, and E. Wimmer [this volume] and M. Girard, A. Martin, T. Couderc, R. Crainic, and C. Wychowski [this volume]). Finally, the comparison of the Mahoney strain with Sabin 3 has suggested a plausible structural explanation for the temperature sensitivity of the Sabin 3 strain.

GENERAL INTRODUCTION TO POLIOVIRUS STRUCTURE

We have previously described the three major capsid proteins (VP1, VP2, and VP3) of Mahoney 1 as having similar conserved cores (eight-stranded antiparallel beta barrels shaped roughly like triangular wedges), dissimilar extensions at their amino and carboxyl termini, and dissimilar loops connecting the regular secondary structural elements of the cores (Fig. 1). In the virions the cores form the closed shell of the particle, packing together with the narrow ends of the cores of VP1 pointing toward the particle fivefold axes and the narrow ends of the cores of VP2 and VP3 alternating around the particle threefold axes. The tilts of the cores produce prominent radial protrusions at the fivefold and threefold axes, giving the particle the shape of the geometric solid shown in Fig. 2. The amino-terminal extensions form a network on the internal surface of the protein shell which may function to direct the assembly of the virus. The formation of this network must be preceded by proteolytic processing of the capsid protein precursors to free the chain termini. The connecting loops and carboxyl termini contribute to the major features exposed on the external surface of the virion (Fig. 3).

COMPARISON OF THE STRUCTURES OF THE MAHONEY 1 AND SABIN 3 STRAINS

The Sabin 3 and Mahoney 1 strains of poliovirus share 83% sequence identity at the amino acid level in the capsid protein region. VP4 is the most conserved with 92% sequence identity; VP1 is the least conserved with only 76% identity. Consistent with the high degree of sequence homology between the two strains, the structures of Sabin 3 and Mahoney 1 are strikingly similar. The root mean square difference in alpha carbon position is 0.78 Å (0.078 nm) for 846 sequentially equivalent residues (though the root mean square difference is only 0.034 nm when the 19 pairwise discrepancies greater than 0.2 nm are omitted). This structural

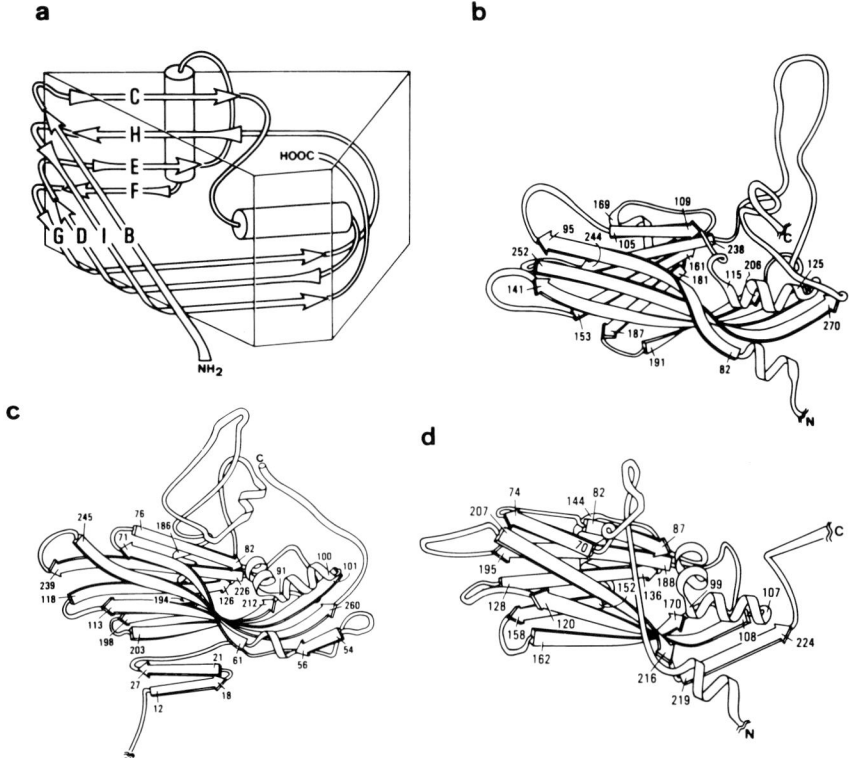

Figure 1. (a) Schematic representation of the conserved wedge-shaped eight-stranded antiparallel beta-barrel cores of VP1, VP2, and VP3. Individual beta strands are indicated by arrows and labeled alphabetically as they proceed from the amino to the carboxyl terminus. Helices are shown as cylinders. (b to d) Ribbon diagrams of (b) VP1, (c) VP2, and (d) VP3. The amino and carboxyl termini of VP1 and VP3 have been truncated for clarity. The ribbon diagrams were drawn by Elizabeth Getzoff, Research Institute of Scripps Clinic.

similarity is especially pronounced in the cores of the capsid protein subunits, but is also seen in the amino-terminal extensions and in many of the connecting loops. In several instances similarity of the main chain structure is maintained despite significant local sequence differences. In such instances, side chain differences are accommodated either through localized adjustments of side chain conformations or by compensating changes in neighboring side chains.

Significant conformational differences in the main chain are confined to the exposed loops and to the chain termini of the virus (see Fig. 4). These structural differences fall into three general categories: (i) differ-

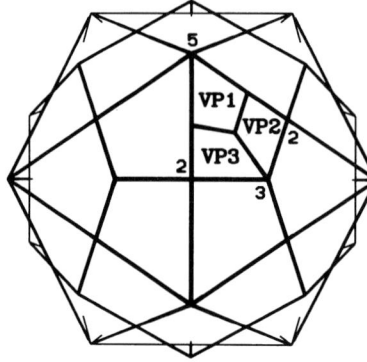

Figure 2. Geometric representation of the outer surface of poliovirus. The geometric figure is formed by the intersection of an icosahedron and a dodecahedron. The particle fivefold axes are located at the apices of the icosahedron (one such axis is indicated by a 5). The particle threefold axes (labeled 3) are located at the apices of the dodecahedron. The twofold axes (labeled 2) are located at the intersection of the icosahedron and the dodecahedron. The positions and general shapes of VP1, VP2, and VP3 from one promoter are indicated. The figure emphasizes the prominent radial protrusions at the particle fivefold axes and the somewhat smaller radial protrusions at the particle threefold axes.

ences in loop conformation due to insertions in one strain relative to the other; (ii) loops with several sequence differences **including the replacement of a proline residue**; and (iii) differences observed at points of transition between ordered and disordered structure. All insertions occur in exposed loops (VP1 221, 289; VP2 138) or in the disordered amino terminus of VP1. The observable insertions/deletions cause limited, highly localized structural perturbations which have little effect further than one or two residues on either side of the insertion (as shown in Fig. 4).

The most significant conformational difference between the Sabin 3 and Mahoney 1 structures occurs in the loop connecting the B and C strands of the core of VP1 (residues 95 to 104), where the difference between equivalent alpha carbon positions is as large as 0.8 nm (Fig. 4). Of the six sequence differences in this loop, the most obvious causes of the conformational difference are the substitution of a Glu (Sabin 3) for a Pro (Mahoney 1) at position 95 and the substitution of a Pro (Sabin 3) for a Ser (Mahoney 1) at position 97. A large structural difference associated with the substitution of a proline in an exposed loop with considerable local sequence divergence is also seen in the B-C loop of VP3 and in the H-I loop of VP2. In contrast, the E-F loop of VP2 (residues 160 to 170, part of antigenic site 2), which is an area of high sequence variability but which does not involve the substitution of a proline, is nearly identical in structure in the two poliovirus strains.

It may be relevant that significant structural differences occur in all three of the major antigenic sites of the virion (5, 9). In particular, structural changes are seen in the B-C loop of VP1 (which constitutes a major portion of antigenic site 1), the insertion in the G-H loop of VP1

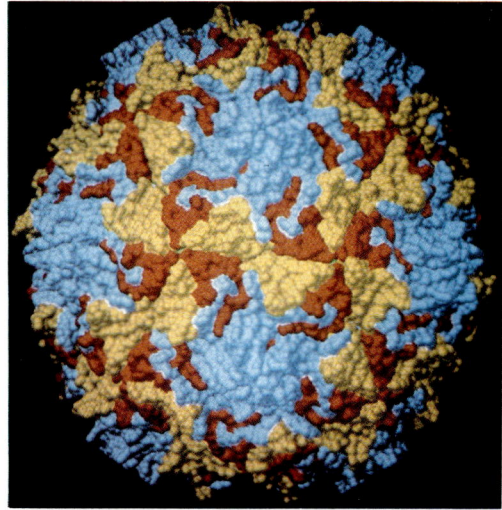

Figure 3. Space-filling model of the surface of the poliovirion. Blue, VP1; yellow, VP2; red, VP3. The orientation of the particle is the same as the orientation of the geometric representation shown in Fig. 2.

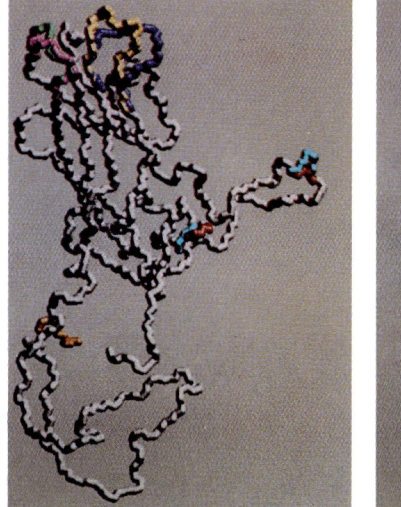 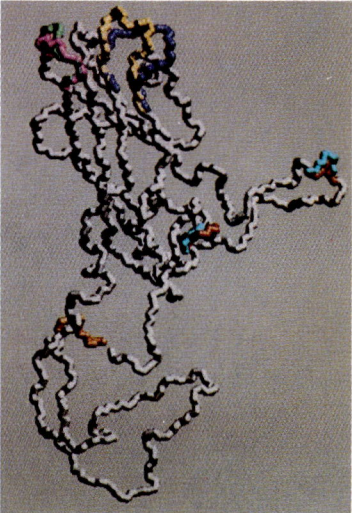

Figure 4. Significant differences in the main chain conformation of VP1 between Mahoney 1 and Sabin 3. In this stereo pair, structurally conserved main chain atoms are white. Differences, which are observed only in the loops and terminal extensions of the capsid proteins, are shown in color. The B-C loop (yellow for Mahoney 1 and dark blue for Sabin 3) involves the replacement of proline residues in a loop having substantial sequence differences. The D-E loops (magenta for Mahoney 1 and green for Sabin 3) are significantly different in the two structures despite their general similarity in sequence. The highly localized structural changes due to one-residue insertions or deletions are indicated in red for Mahoney 1 and cyan for Sabin 3. The residues indicated in orange represent a portion of the amino-terminal extension (residues 20 to 23) which is ordered in Mahoney 1, but disordered in Sabin 3.

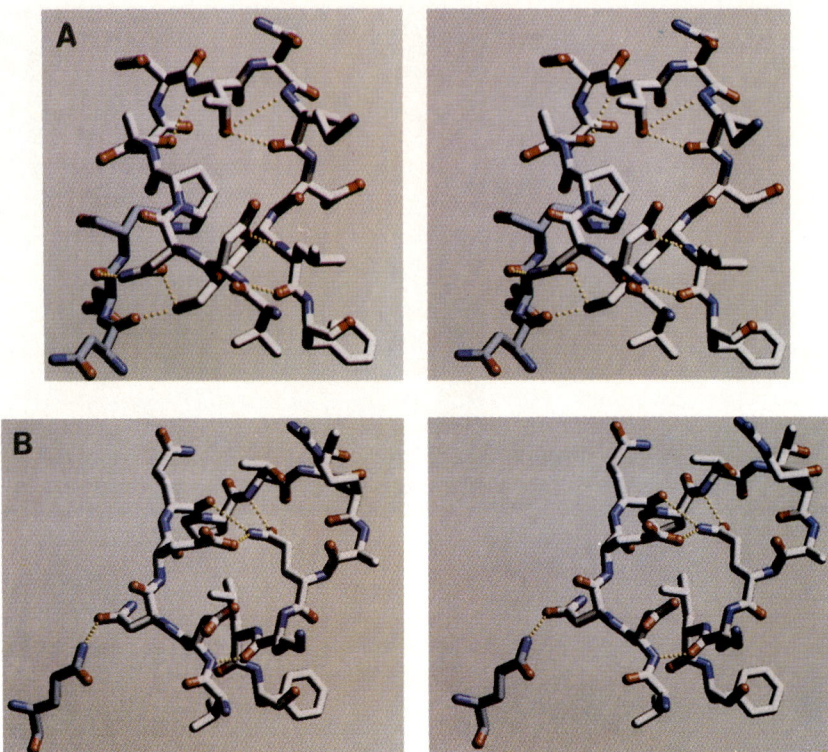

Figure 5. Atomic structure of the B-C loop of VP1 in (A) Mahoney 1 and (B) Sabin 3. In the stereo representations, oxygen atoms are red and nitrogens are dark blue. The carbon atoms shown in white belong to the B-C loop, while those shown in light blue belong to adjacent portions of the protein which make structurally important interactions with the loop. The yellow dotted lines represent hydrogen bonds which stabilize the B-C loop and which appear to play a significant role in determining its conformation.

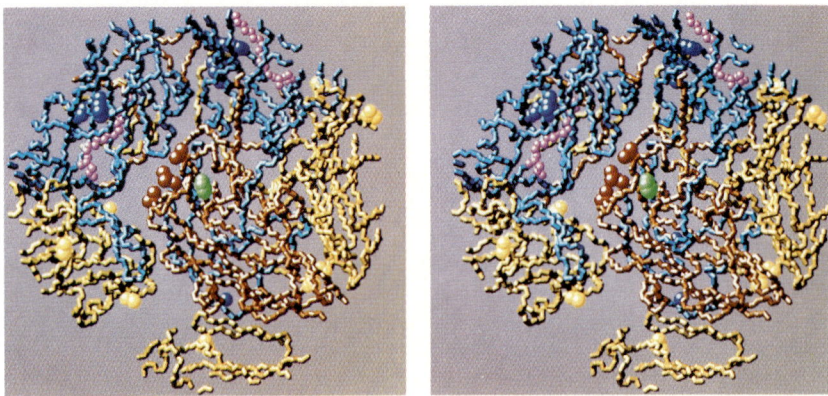

Table 1. Consensus Sequences for the B-C Loop of VP1 in Poliovirus[a]

Poliovirus type	Amino acid at position:														
	91	92	93	94	95	96	97	98	99	100	101	102	103	104	105
Type 1	Thr	Val	Asp	Asn	<u>Ser</u>	Ala	Ser	<u>Thr</u>	<u>Thr</u>	<u>Ser</u>	Lys	Asp	Lys	Leu	Phe
Type 2	Glu	Val	Asp	Asn	Asp	Ala	Pro	Thr	Lys	Arg	Ala	Ser	<u>Lys</u>	Leu	Phe
Type 3	Glu	Val	Asp	Asn	Glu	Gln	Pro	Thr	<u>Thr</u>	<u>Arg</u>	Ala	Gln	Lys	Leu	Phe
Consensus		Val	Asp	Asn				Thr					Lys	Leu	Phe

[a] The consensus sequences for the B-C loop of VP1 of the three serotypes of poliovirus are shown. The consensus sequences are based on sequence information from 11 strains of type 1 poliovirus, 2 strains of type 2 poliovirus, and 13 strains of type 3 poliovirus. For type 1 and type 3 poliovirus, the underlined residues indicate positions where there are two or more substitutions in the known strains. For type 2 the two sequences differ only at amino acid 103 (underlined), which is a Lys in the Lansing strain and an Arg in the Sabin strain. The sequence in Mahoney 1 is identical to the type 1 consensus sequence except at position 95 which is a Pro and at position 100 which is an Asn. The sequence of Sabin 3 is identical to the type 3 consensus sequence (adapted from reference 7).

(site 2), the insertion at position 289 of VP1 (site 3A), and the large structural difference in the B-C loop of VP3 (site 3B). This suggests that three-dimensional structural differences, as well as simple sequence changes, might play an important role in determining serotypes.

Because of its importance as an antigenic site, the B-C loop of VP1 has been the focus of considerable attention. The sequence of this loop has been determined for a number of strains of all three serotypes of poliovirus (7; Table 1). Recently, three research groups have reported the construction of viable intertypic chimeras in which the B-C loop from one strain of poliovirus has been replaced with the corresponding sequence from a different serotype. These chimeras include two examples in which a type 3 loop (2, 8) and one example in which a type 2 loop (6a) has been constructed into a type 1 background. In every case, the hybrid displays the appropriate mosaic antigenicity and is able to induce neutralizing antibodies against both parental serotypes. In the type 2/type 1 chimera, replacement of the B-C loop in the primate-specific Mahoney 1 strain with

Figure 6. Mutations observed in non-temperature-sensitive revertants of type 3 poliovirus. The stereo representations show the portion of the Sabin 3 structure which is within 4.0 nm of Phe-91 of VP3 and includes atoms from three different protomers. Main chain atoms are depicted as thin tubes. Large spheres indicate the side chains which have been found to mutate in one or more of the revertants. Green, Phe-91; blue, VP1; yellow, VP2 and VP4; red and brown, VP3. The extended hydrocarbon which occupies the "drug-binding pocket" in the center of the VP1 beta barrel is magenta. The model has been expanded so as to separate individual protomers. Note that most of the mutations are located in the interfaces between protomers or in the hydrocarbon-binding site.

the corresponding loop from the mouse-adapted poliovirus type 2, Lansing strain (Lansing 2), produces a hybrid which is able to cause fatal paralysis in mice (6a). This demonstrates that the replacement of this 10-amino-acid loop (which includes only six sequence changes) is sufficient to confer mouse adaptation on the Mahoney 1 strain. The type 2/type 1 chimera grows to very high titer in cultured cells. We have received seed stocks of this virus from Marc Girard (Institut Pasteur, Paris) and have produced crystals that are highly suitable for high-resolution crystallographic study. This structure has recently been solved by molecular replacement methods, and a preliminary atomic model has been built to fit the electron density map. The 95-104 loop has a significantly different conformation from that observed in either the Mahoney 1 or the Sabin 3 structure (T. Yeates, D. J. Filman, T. Critchlow, R. Syed, D. Jacobson, A. Martin, C. Wychowski, M. Girard, and J. M. Hogle, manuscript in preparation).

MOLECULAR MODELING OF LOOP STRUCTURES
IN POLIOVIRUS

The biological significance, the availability of sequence information from a large number of strains, and the large structural differences between strains make the B-C loop of VP1 a particularly attractive candidate for computational modeling studies designed to predict the effects of sequence variation on the conformation of the loop. These studies are analogous to previous calculations, in which the structure of the complementarity-determining loops of an antibody has been predicted from sequence information by assuming structural conservation in the beta-barrel core of the Fv region of an Fab (1, 3, 4, 12). In some ways, the loop-modeling calculations for poliovirus are more ambitious, because many of the loops, and in particular the B-C loop of VP1, are longer than the loops which have been modeled in Fabs. On the other hand, the structures of the beta-barrel cores of the polioviruses are better conserved than the beta barrels of the Fv regions of Fabs.

In anticipation of more rigorous modeling calculations, we have begun to examine the structures and sequences of the B-C loops of type 1 and type 3 polioviruses in an effort to define specific features ("signatures") in the sequences which are responsible for strain- and serotype-specific differences in the structure of the loop. The loop begins with the sequence Val-Asp-Asn and ends with the sequence Lys-Leu-Phe. Both sequences are highly conserved in all known naturally occurring strains of poliovirus (Table 1). Between these highly conserved sequences there are eight residues which are highly variable. In both the Mahoney 1 and the

Sabin 3 structures the two ends of the loop are connected by a main chain-main chain hydrogen bond. In the Mahoney 1 loop (shown in Fig. 5A) this hydrogen bond is donated by the amide nitrogen of Asp-93 and accepted by the carbonyl oxygen of Leu-104. Interestingly, in Sabin 3 the analogous hydrogen bond is accepted by the carbonyl oxygen of Lys-103 (Fig. 5B), reflecting the conformational differences in the loop.

The conserved sequence Val-Asp-Asn constitutes the last three residues of the B strand of the beta barrel of VP1. It is, therefore, not surprising that their main chain structure is highly conserved. The side chains of Val-92 and Asn-94 also display very similar conformations. In the Sabin 3 structure the side chain of Asn-94 is hydrogen bonded to the side chain of Gln-153 (from the amino end of the E strand of the barrel of VP1 [Fig. 5B]), while in the Mahoney 1 structure, the side chain of Asn-94 is hydrogen bonded to the side chain of Lys-103 and to the main chain carbonyl oxygen of amino acid 247 in the H-I loop of VP1 (Fig. 5A). Asn-94 thus appears to serve as an anchor for the carboxyl end of the B strand of the beta barrel of VP1 in both type 1 and type 3 polioviruses. In contrast, the side chain of Asp-93 is oriented very differently in the two structures. In Mahoney 1 one of the carboxylate oxygens of Asp-93 is hydrogen bonded to the amide nitrogen of Leu-104 (Fig. 5A), whereas in Sabin 3 the Asp side chain is rotated away from the top surface of the loop, apparently owing to the proximity of the acidic side chain of Glu-95 (Fig. 5B).

Although the regular pattern of main chain-main chain hydrogen bonding that is typical of antiparallel beta sheet does not extend beyond amino acid 94, residue 95 and the amide nitrogen of residue 96 are in a nearly extended conformation in the Sabin 3 structure (Fig. 5B). In the Mahoney 1 structure, however, the proline at position 95 is unable to adopt the main chain torsions required for an extended structure. Interestingly, the proline at amino acid 95 is not conserved among type 1 polioviruses (Table 1), occurring in only two of the known type 1 sequences (7; M. Chow, personal communication). This raises the possibility that the structure observed in the B-C loop of the Mahoney strain is not typical of most type 1 polioviruses. Fortunately, we have been able to grow crystals of the Sabin strain of poliovirus type 1, which has the consensus serine at position 95. Crystallographic studies of the Sabin 1 strain may eventually provide an alternative structure for this loop which could prove to be more representative of type 1 polioviruses.

The conserved sequence Lys-Leu-Phe is located at the carboxyl end of the B-C loop. Phenylalanine 105 is the first residue in the C strand of the beta-barrel core of VP1. In both the Sabin 3 and the Mahoney structures, the aromatic side chain is buried in a hydrophobic pocket between the C strand and the loop that connects the E and F strands of VP1 (Fig. 5), thus providing a conserved structural anchor. Unlike Phe-105, however, the sequentially conserved Lys-103 and Leu-104 are found in very different environments in the two structures. In the Mahoney 1 structure the side chain of Lys-103 extends from the lower surface of the B-C loop to make a hydrogen bond with the side chain of Asn-94 and with the (main chain) carbonyl oxygen of Asn-246, while the side chain of Leu-104 is on the upper surface of the loop (Fig. 5A). In Sabin 3, however, the side chain of Lys-103 is highly exposed on the upper surface of the loop, participating in an extensive network of charged side chains, while the side chain of Leu-104 is on the lower surface of the loop (Fig. 5B). The magnitude of the structural differences in the B-C loop and the dissimilarity in the conformations of highly conserved residues indicate that structural predictions based on simple interactive model building (which does not make explicit use of energy minimization and molecular dynamics) are likely to be seriously in error.

In each of the structures a number of specific interactions are observed that appear to be important for the stability of the loop conformation. In Mahoney 1 these interactions include a beta turn between residues 96 and 99, a hydrogen bond between the hydroxyl oxygen of Thr-99 and the amide nitrogen and carbonyl oxygen of Lys-101, and the previously mentioned hydrogen bond between the carboxylate oxygen of Asp-93 and the amide nitrogen of Leu-104 (Fig. 5A). It should be noted that Thr-99 is not conserved in the known type 1 poliovirus sequences (Table 1), suggesting once again that the B-C loop in the Mahoney strain might not be representative of type 1 polioviruses.

In Sabin 3 the side chain of Gln-102 is involved in several very interesting stabilizing interactions. The amide nitrogen of the Gln side chain donates a hydrogen bond to the carbonyl oxygen of Gln-96 and to one of the side chain carboxylate oxygens of Glu-95, while its carbonyl oxygen accepts a hydrogen bond from the amide nitrogen of Thr-98 (Fig. 5B). This network of hydrogen bonds is expected to contribute significantly to the stability of the loop structure. Furthermore, Gln-102 is highly conserved among the known sequences of naturally occurring poliovirus type 3 isolates (Table 1). The only known exception is a strain which has a His at 102 (7). Interactive modeling suggests that His could make only one of the two side chain-main chain hydrogen bonds. The conservation in type 3 sequences of Gln-102, of the acidic side chain at position 95, and of Pro-97 suggests that the structure of this loop in the Sabin 3 model is probably typical for type 3 polioviruses.

TEMPERATURE SENSITIVITY IN THE SABIN 3 STRAIN

The Sabin 3 strain is temperature sensitive, growing well at 37°C but not at 40°C. While our crystallographic studies were in progress, the temperature sensitivity phenotype of the Sabin 3 strain was shown to be due to the substitution of phenylalanine (Sabin 3) for serine (Leon) at amino acid 91 of VP3 during the derivation of the Sabin 3 strain from its neurovirulent and non-temperature-sensitive parent poliovirus type 3, Leon strain (Leon 3) (13; P. D. Minor, G. Dunn, D. M. A. Evans, D. I. Magrath, A. John, J. Howlett, A. Phillips, G. Westrop, K. Wareham, J. W. Almond, and J. M. Hogle, manuscript in preparation). In addition, we have found that exposure of Sabin 3 virions to brief incubations at elevated temperatures results in irreversible conformational changes, suggesting that the temperature sensitivity of Sabin 3 is due in part to thermolability of the virion itself (C. E. Fricks, T. Critchlow, and J. M. Hogle, unpublished observations).

In both the Sabin 3 and the Mahoney 1 structures, residue 91 of VP3 is located in a turn of helix at the carboxyl end of the C strand of the beta barrel. The loop connecting the G and H strands of the beta barrel folds up against this turn of helix, trapping a chain of three buried solvent molecules in a pocket. In Mahoney 1 (which, like Leon 3, has a serine at position 91 of VP3) the serine side chain points downward into this pocket, making a hydrogen bond with one of the trapped water molecules.

In Sabin 3, however, the phenylalanine side chain points upward and is fully exposed to solvent at the base of the deep depression or "canyon" which surrounds the fivefold axes of the particle (Fig. 6). Indeed, the Sabin 3 electron density maps contain clear indications for several ordered solvent molecules adjacent to edge and to one of the faces of the aromatic group. Considering that the solvation of the side chain is energetically unfavorable, we have postulated that the substitution of Phe for Ser at VP3 91 causes an increased thermolability of the Sabin 3 particle by promoting thermally induced conformational changes which permit the phenylalanine to be buried (Filman et al., in preparation).

Phil Minor and his colleagues at the National Institute for Biological Standards and Control, London, in collaboration with Jeffrey Almond and his colleagues at Reading University, Reading, England, have characterized a number of second-site non-temperature-sensitive revertants of Sabin 3 (Minor et al., in preparation). We have mapped those sequence differences which occurred in the capsid region, and found that·the mutations tend to occur in the interfaces between protomers. In particular, a striking cluster of mutations is seen in the interface between fivefold-related protomers close to the position of Phe-91 (Fig. 6). These second-site mutations might restore thermal stability simply by stabilizing these interfaces against conformational rearrangements. A notable exception to the general observation that the second-site suppressors occur in the interfaces between protomers is a mutation involving a substitution of a leucine for a phenylalanine at residue 134 of VP1. This mutation has been observed in three of four non-temperature-sensitive revertants of virus produced from a construct in which the coding sequence for Phe at VP3 91 has been substituted into the Leon 3 genome. The side chain of residue 134 of VP1 interacts with an as-yet-unidentified hydrophobic molecule which is bound between the two sheets of the beta barrel of VP1 in both the Sabin 3 and Mahoney 1 structures (Filman et al., in preparation; see Fig. 6). Interestingly, the hydrocarbon-binding site is nearly identical to the site which binds a class of antiviral drugs, exemplified by WIN 51711 (Sterling-Winthrop), in rhinovirus 14 (11). The binding of these drugs is known to stabilize the virus against thermally induced conformational changes. We are currently engaged in a series of X-ray crystallographic and theoretical modeling studies of both types of revertants to understand the structural factors that govern thermal stability in polioviruses (Filman et al., in preparation).

Acknowledgments. This work was supported by Public Health Service grants AI20566 and GM38794 from the National Institutes of Health (to J.M.H.) and by the World Health Organization.

Literature Cited

1. **Bruccoleri, R. E., and M. Karplus.** 1987. Prediction of the folding of short polypeptide segments by uniform conformational sampling. *Biopolymers* **26:**137–168.
2. **Burke, K. L., G. Dunn, M. Ferguson, P. D. Minor, and J. W. Almond.** 1988. Antigenic chimaeras of poliovirus as potential new vaccines. *Nature* (London) **332:**81–82.
3. **Chothia, C., A. M. Lesk, M. Levitt, A. G. Amit, R. A. Mariuzza, S. E. V. Phillips, and R. J. Poljak.** 1986. The predicted structure of immunoglobulin D1.3 and its comparison with the crystal structure. *Science* **233:**755–758.
4. **Fine, R. M., H. Wang, P. S. Shenkin, D. L. Yarmush, and C. Levinthal.** 1986. Predicting antibody hypervariable loop conformations. II. Minimization and molecular dynamics studies of MCPC603 from many randomly generated loop conformations. *Proteins Struct. Funct. Genet.* **1:**342–362.
5. **Hogle, J. M., M. Chow, and D. J. Filman.** 1985. Three-dimensional structure of poliovirus at 2.9 Å resolution. *Science* **229:**1358–1365.
6. **Luo, M., G. Vriend, G. Kamer, I. Minor, E. Arnold, M. G. Rossmann, U. Boege, D. C. Scraba, G. M. Duke, and A. C. Palmenberg.** 1987. The atomic structure of mengo virus at 3.0 Å resolution. *Science* **235:**182–191.
6a.**Martin, A., C. Wychowski, T. Couderc, R. Crainic, J. M. Hogle, and M. Girard.** 1988. Engineering a poliovirus type 2 antigenic site on a type 1 capsid results in a chimaeric virus which is neurovirulent for mice. *EMBO J.* **7:**2839–2847.
7. **Minor, P. D., M. Ferguson, A. Phillips, D. I. Magrath, A. Huovialainen, and T. Hovi.**

1987. Conservation in vivo of protease cleavage sites in antigenic sites of poliovirus. *J. Gen. Virol.* **68**:1857–1865.

7a. **Murray, M. G., J. Bradley, X.-F. Yang, E. Wimmer, E. G. Moss, and V. R. Racaniello.** 1988. Poliovirus host range is determined by a short amino acid sequence in neutralization antigenic site 1. *Science* **241**:213–215.

8. **Murray, M. G., R. J. Kuhn, M. Arita, N. Kawamura, A. Nomoto, and E. Wimmer.** 1988. Poliovirus type 1/type 3 antigenic hybrid virus constructed in vitro elicits type 1 and type 3 neutralizing antibodies in rabbits and monkeys. *Proc. Natl. Acad. Sci. USA* **85**:3203–3207.

9. **Page, G. S., A. G. Mosser, J. M. Hogle, D. J. Filman, R. R. Rueckert, and M. Chow.** 1988. Three-dimensional structure of the poliovirus serotype 1 neutralizing determinants. *J. Virol.* **62**:1781–1794.

10. **Rossmann, M. G., E. Arnold, J. W. Erickson, E. A. Frankenberger, J. P. Griffith, H.-J. Hecht, J. E. Johnson, G. Kamer, M. Luo, A. G. Mosser, R. R. Rueckert, B. Sherry, and G. Vriend.** 1985. Structure of a human common cold virus and functional relationship to other picornaviruses. *Nature* (London) **317**:145–153.

11. **Smith, T. J., M. S. Kremer, M. Luo, G. Vriend, E. Arnold, G. Kamer, M. G. Rossmann, M. A. McKinlay, G. D. Diano, and M. J. Otto.** 1986. The site of attachment in human rhinovirus 14 for antiviral agents that inhibit uncoating. *Science* **233**:1286–1293.

12. **Snow, M. E., and L. M. Amzel.** 1986. Calculating three-dimensional changes in protein structure due to amino-acid substitutions: the variable region of immunoglobulins. *Proteins Struct. Funct. Genet.* **1**:267–279.

13. **Westrop, G. D., D. M. A. Evans, P. D. Minor, D. Magrath, G. C. Schild, and J. W. Almond.** 1987. Investigation of the molecular basis of attenuation in the Sabin type 3 vaccine using novel recombinant polioviruses constructed from infectious cDNA, p. 53–60. *In* D. J. Rowlands, B. W. J. Mahy, and M. A. Mayo (ed.), *The Molecular Biology of the Positive Strand RNA Viruses.* Academic Press, Inc. (London), Ltd., London.

Molecular Aspects of Picornavirus Infection and Detection
Edited by Bert L. Semler and Ellie Ehrenfeld
© 1989 American Society for Microbiology, Washington, DC 20006

Chapter 8

Conformational Adaptations by Picornaviruses to Antiviral Agents and pH Changes

Michael G. Rossmann

GENERAL

Atomic resolution structures of a few simple spherical RNA plant and animal viral capsids have been determined in the past decade (Table 1). In each case the essential shell domain consists of an eight-stranded anti-parallel β barrel (Fig. 1). In general, the quaternary organization of the virus icosahedral shells is remarkably alike. The character of the three-dimensional viral structures is sufficiently similar to make it most probable that many simple spherical RNA viruses have evolved from a common evolutionary precursor (22). The similarity extends to the more complex icosahedral DNA adenoviruses (27) and to the bacillus-shaped RNA alfalfa mosaic virus (10; I. Fita and M. G. Rossmann, unpublished data). It has also been shown that the core protein of Sindbis virus is homologous to VP3 of foot-and-mouth disease virus (FMDV) and may, therefore, also have the same tertiary and quaternary structure (11). Thus observations made on structural properties of picornaviruses might often be applicable to many other kinds of viruses. However, the relative simplicity of picornaviruses makes them particularly suitable for careful structural analyses.

Michael G. Rossmann • Department of Biological Sciences, Purdue University, West Lafayette, Indiana 47907.

Table 1. Viruses Whose Capsid Proteins Are Similarly Folded and Assembled

Virus	Capsid		Reference
	Symmetry[a]	Diam (nm)	
Plant RNA viruses			
Tomato bushy stunt virus	$T = 3$	35.0	13
Southern bean mosaic virus	$T = 3$	30.0	1
Satellite tobacco necrosis virus	$T = 1$	19.0	20
Turnip crinkle virus	$T = 3$	35.0	15
Cowpea mosaic virus	$P = 3$	30.0	36
Alfalfa mosaic virus	Bacillus	19.0	10; Fita and Rossmann, unpublished results
Animal RNA viruses			
HRV14	$P = 3$	30.0	30
Poliovirus	$P = 3$	30.0	14
Mengovirus	$P = 3$	30.0	22
Insect RNA virus			
Black beetle virus	$T = 3$	30.0	16
Animal DNA virus			
Adenovirus hexon			27

[a] T (triangulation) = 3 implies that there are 3×60 capsid proteins in the icosahedral viral shell. P (pseudo) = 3 implies that there are 60 copies of each of three different major proteins in the icosahedral viral shell. However, the fold of the three different polypeptide chains is very similar.

CONFORMATIONAL CHANGES OF MENGOVIRUS DUE TO pH CHANGES

Alteration in structure of picornaviruses due to attachment to cellular receptors can include conformational changes of the capsid proteins, loss of VP4, and loss of RNA (33). When virions that have complexed with receptors are eluted, then the virions are irreversibly changed and will not reattach to cells (6). Virion modification, analogous to that due to receptor attachment, is observed upon treatment with heat, irradiation with UV light, proteolytic digestion, or organic solvents (19, 24). These changes can also be accompanied by alterations in antigenicity, isoelectric point, and sedimentation.

Mengovirus can also exist in two pI forms (pH 4.4 and about 7.0 for mengovirus type M [4, 5]), analogous to poliovirus and rhinovirus. Mengovirus (23) and mouse Elberfeld virus (8) dissociate into 13.4S pentamers in the presence of Cl^- or Br^- ions (Fig. 2). The dissociation is

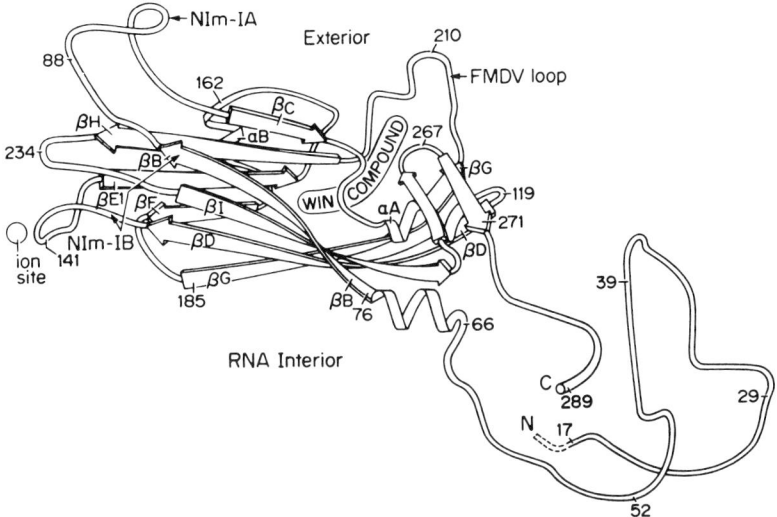

Figure 1. The eight-stranded antiparallel β barrel as found in VP1 of HRV14, showing site of attachment of antiviral agent WIN 51711 in the hydrophobic internal pocket (35).

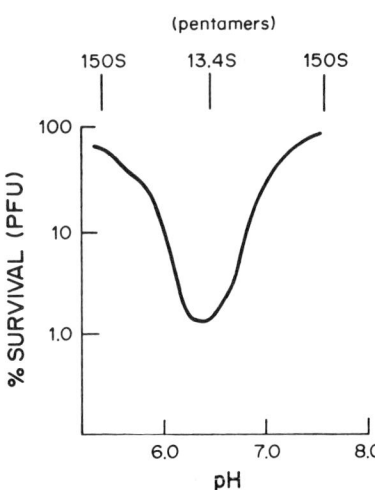

Figure 2. Stability of mengovirus in the presence of 150 mM Cl⁻ at various pH values (adapted from reference 23).

Table 2. Conformational Changes of Mengovirus

pH	Buffer	Changes at Cl^- concn:			
		0 mM	25 mM	150 mM	500 mM
7.4	Phosphate	VP3 loop disordered; FMDV loop standard; PO_4^{2-} present			
6.2	Phosphate or acetate	VP3 loop disordered; FMDV loop moved; PO_4^{2-} present	VP3 loop disordered; FMDV loop moved; PO_4^{2-} present	VP3 loop ordered; FMDV loop moved; PO_4^{2-} small occupancy	VP3 loop ordered; FMDV loop moved; PO_4^{2-} absent
4.6	Acetate	VP3 loop ordered; FMDV loop moved; PO_4^{2-} absent			

dependent on time and halide ion concentration. The structure of mengovirus was determined to atomic resolution at pH 7.4 (22). The structure of mengovirus at pH 6.2 and 4.6 has been examined in various buffers and Cl^- concentrations.

Orthorhombic mengovirus crystals were prepared as described by Luo et al. (22). The native crystals were then soaked in various buffers at a variety of pH values and halide ion concentrations (Table 2). The lengths of soaks depended on crystal stability. Thus, soaking at 150 mM Cl^- caused rapid crystal disintegration, making it essential to use only a very short soaking time, followed by immediate exposure to X rays. The X-ray diffraction data were collected to 3-Å (0.3-nm) resolution at the Cornell High Energy Synchrotron Source, using previously described techniques (22, 30). Structure amplitudes were locally scaled to the native structure amplitudes in about 20 resolution shells. Differences between low-pH and pH 7.4 diffraction data were used to compute difference electron density maps (Fig. 3). In addition, the amplitudes were used to compute phases for the structurally altered virions, using the 60-fold redundancy of the viral particle in the $P2_12_12_1$ space group. These phases were used to compute electron density maps, with respect to which the new structure was interpreted.

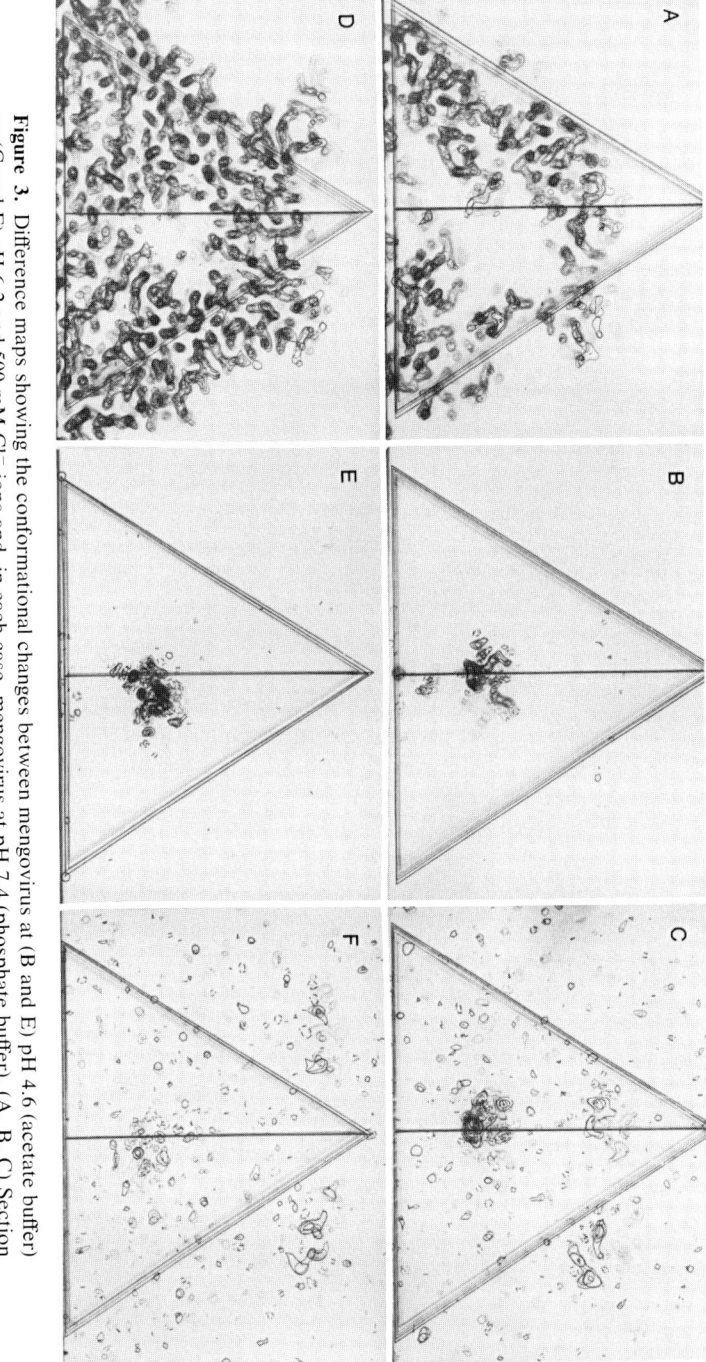

Figure 3. Difference maps showing the conformational changes between mengovirus at (B and E) pH 4.6 (acetate buffer) or (C and F) pH 6.2 and 500 mM Cl⁻ ions and, in each case, mengovirus at pH 7.4 (phosphate buffer). (A, B, C) Section 14.3 to 13.7 nm. (D, E, F) Section 13.6 to 13.1 nm. These same sections for the native virus are shown in panels A and D, respectively. Note how the conformational changes line the pit, or putative receptor attachment site.

The conformational changes induced on the mengo virion by pH changes or the presence of Cl⁻ were large, but confined entirely to the "pit" area (the putative receptor attachment site). The major changes were as follows.

(i) The VP3 loop between residues 3176 and 3182 became ordered. (The convention used here for numbering amino acids is to add 1000, 2000, 3000, or 4000 to the sequential numbers in VP1, VP2, VP3, and VP4, respectively.) The previously visible ends of the loop (residues Gly-3174 to Gly-3175 and Gly-3183 to Gly-3185) also underwent slight conformational changes.

(ii) The FMDV loop in VP1 (residues 1204 to 1214) underwent a conformational change of up to 0.1 nm in the main-chain position. The largest side chain movement occurred between Gly-1204 and Gly-1212, including His-1205, which alters its hydrogen bonding from Asp-1208 to Asp-2138.

(iii) A well-occupied phosphate ion site, immediately below the FMDV loop in VP1, but mostly associated with VP2, became unoccupied. The combination of the FMDV loop movement and the ordering of the VP3 loop makes this site inaccessible from the outside at low pH. Ligands to the phosphate site included His-2209 on VP2.

Conformational changes are frequently involved in permitting interactions between proteins and ligands such as DNA, metals, coenzymes, and substrates. Order-disorder conformational changes have been shown to be essential in the function of many proteins (17). Thus, the exceedingly localized area on the surface of mengovirus where the conformational changes occur as a result of a pH drop suggests that this region on the viral surface plays an important functional role. This same area had previously been associated with the possible site for receptor attachment (22).

The narrow pH effect of halide ions on the stability of mengovirus shows that there are at least two titratable groups at around pH 6.2. That one of these is likely to be a histidine was noted by Mak et al. (23). Indeed, the FMDV loop contains His-1205, which alters its hydrogen-bonding arrangement during the conformational changes. Furthermore, His-2209 is a ligand to the bound phosphate which is displaced at lower pH or high Cl⁻ ion concentration. Thus, it is probable that these two residues control the stability of the virion. Unfortunately, it is not yet clear how the various conformational changes that occur control the disassembly into pentamers, nor how cell attachment might bring about these same conformational changes. Nevertheless it is not surprising that receptor recognition is associated with flexible components of the structure to permit an "induced fit."

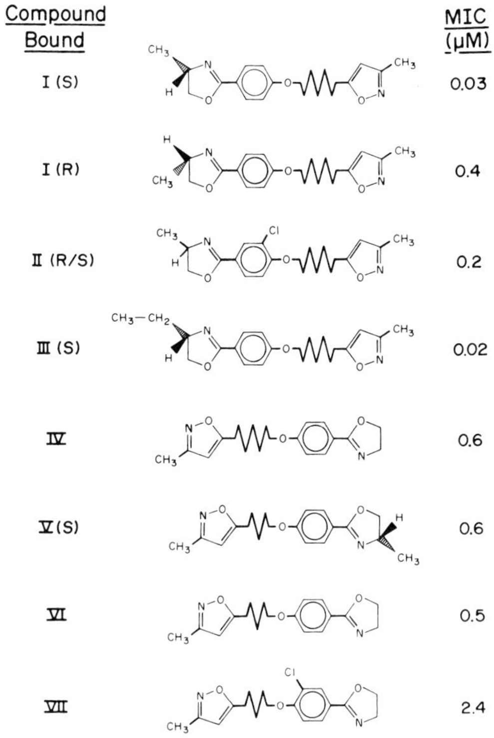

Figure 4. Formulae of antiviral compounds used in the studies reported here. Shown also are their in vitro activity against HRV14 measured in terms of the concentration required to reduce the plaque counts by a factor of 2 (MIC).

INHIBITION OF RHINOVIRUS UNCOATING

A series of structurally related antiviral compounds (Fig. 4) synthesized at the Sterling-Winthrop Research Institute (7) has been found to inhibit picornaviral replication by preventing uncoating (9), as have some other structurally unrelated compounds (21, 25). My colleagues and I have examined the binding of some of these compounds to human rhinovirus 14 (HRV14) in an extension of the work reported by Smith et al. (35). All of these compounds bind into a hydrophobic pocket beneath the "canyon floor" (Fig. 1), accompanied by large (up to 0.55-nm) conformational changes in the HRV14 structure from the native conformation in the vicinity of the binding site. Here I report and contrast the structures of eight virus-drug complexes (Fig. 4). I also report the structure of a drug-resistant mutant.

Cubic HRV14 crystals of 0.25 to 0.45 mm in diameter (2) were soaked in solutions of the WIN compounds dissolved in dimethyl sulfoxide. The

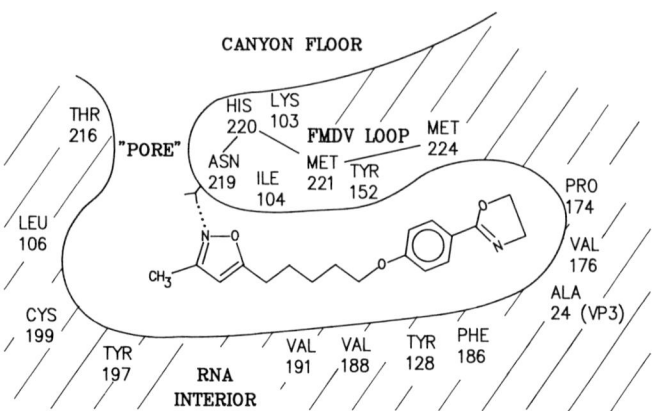

Figure 5. Diagrammatic representation of compound VI bound in the WIN pocket. Note that the orientation is opposite to that of compound I. Residues lining the pocket were selected as having atoms within 0.36 nm of any atom in compound I(*S*).

X-ray diffraction data were collected primarily at the Cornell High Energy Synchrotron Source and also (compound III) at the Brookhaven Laboratory's National Light Source. The resulting films were processed and scaled as previously described (28, 31). Phase angles were initially obtained from the native HRV14 structure determination (30) and then refined using 20-fold noncrystallographic redundancy. Further computational details are given by Smith et al. (35) and Badger et al. (2a).

The binding site for all the WIN antiviral agents is located at the same general area in the hydrophobic pocket beneath the canyon floor. This pocket can be entered by way of a pore on the floor of the canyon (Fig. 5). The compounds with a seven-membered aliphatic chain and an aliphatic group (methyl or ethyl) on the oxazoline group, examined thus far, bind in one orientation, whereas the other compounds examined here bind in the opposite direction (Fig. 4). Indicators of orientation are the relative size of the denser aromatic groups at either end of the compound, a constriction in the bigger density showing separation of the oxazoline and phenyl group, and the site of the denser chlorine atoms in compound II and compound VII (Fig. 6). Previously it had been assumed that compound I(*S*) and compound IV were oriented in an identical way (35). Improved analytical procedures (2a) show that the predominant orientation of compound IV is in the opposite direction from that of compound I(*S*), with the positions of the isoxazole group and oxazoline-phenyl group exchanged. The environments of compound II and compound VII (Fig. 7) represent the two major alternative and opposite binding modes, depending on the particular compound.

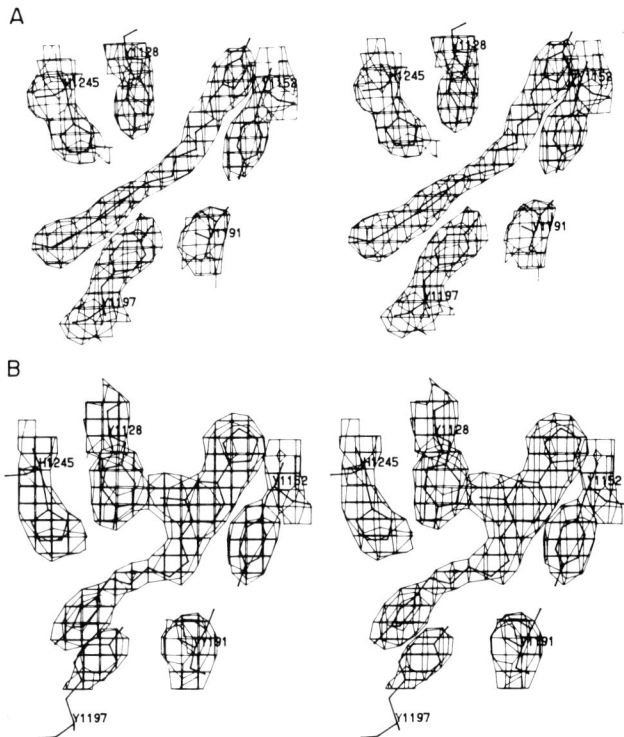

Figure 6. Electron density showing (A) compound I(*S*) and (B) compound VII. Note that the chlorine atom of compound VII readily defines the compound's orientation, although the larger volume of density for the phenyloxazoline group compared to the isoxazole group is also abundantly clear in both cases. The two compounds differ in the length of their aliphatic chains.

When the WIN compounds are bound, three stretches of the VP1 polypeptide chain occupy conformations distinctly different from the native rhinovirus structure. The residues between 1213 and 1224 (part of the chain leading from the FMDV loop to βH, as in Fig. 1) are pushed out of the drug-binding site by distances of up to 0.45 nm in C_α positions. A smaller but still considerable displacement of residues 1151 to 1159 is also evident. Even smaller changes are found for the region between residues 1100 and 1110. The conformational changes were found to be largely independent (less than 0.02 nm for any main-chain atom) of the particular compound bound to the virus. The three displaced chain segments are associated with the β sheet forming one side of the VP1 β barrel. Hence, the conformational changes induced by drug binding are a concerted effect, causing the systematic movement of a β sheet.

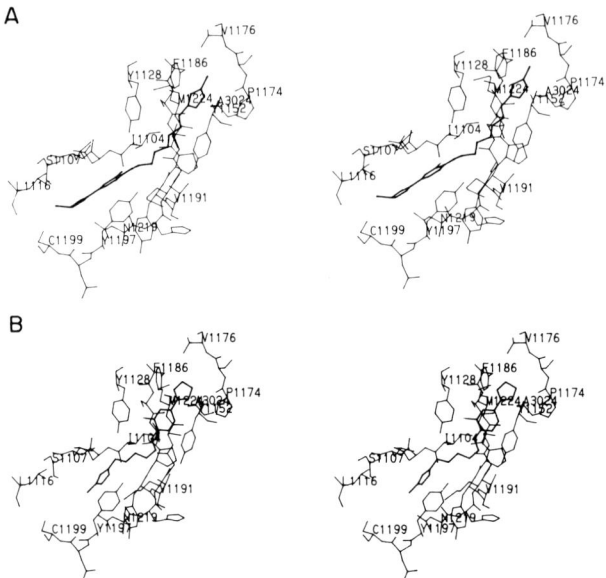

Figure 7. The protein environment of the bound compounds (A) compound II and (B) compound VII. Both these compounds contain chlorine atoms, but they differ in the lengths of their aliphatic chain, in the presence of a methyl substituent on the oxazoline group of compound II, and in their orientation within the binding pocket.

Two classes of drug-resistant mutants have been identified: a high-resistance class, which replicated efficiently in high concentrations of drug (18 μg/ml), and a low-resistance class, which could replicate only in low concentrations of drug (0.3 μg/ml). The frequencies of the high-resistance mutants (1 per 10^4 PFU) were consistent with expectations for single-step mutations. These rates correspond to the frequency of mutations seen in the search for mutants that escape neutralization by antibodies (34). Indeed, they may be considered as escape mutants to the drug. Thus, it was anticipated that these mutants would have single amino-acid substitutions.

High-resistance mutants were selected in the presence of compound IV and compound I. The sites of mutations were determined by sequencing the RNA in the region encoding the amino acids that line the WIN-binding site. All mutants so far characterized have base substitutions in codons specifying amino acids 1188 (Val to Leu or Met) or 1199 (Cys to Trp, Arg, or Tyr). Clone SW12, selected using compound IV (Fig. 4), has leucine substituted for Val-1188. This mutant propagated well and crystallized easily. A difference electron density map (Fig. 8) confirmed

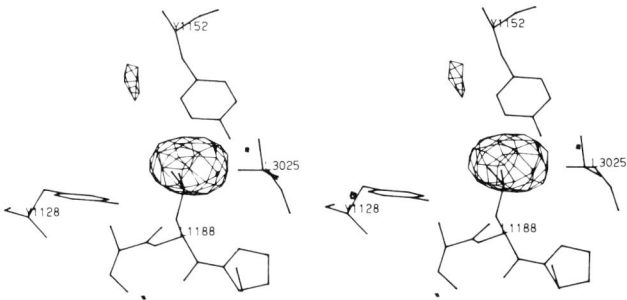

Figure 8. Electron density difference map between mutant SW12 (Val-1188 → Leu) and wild-type virus. The positive peak at site 1188 had a height of 470 relative electron density units. The highest noise peak had a height of 186 units, and the negative peak where the valine used to be had a value of −126 units.

the identity of the mutated amino acid and also showed that there were no other visible mutations or conformational differences.

High MICs (low activity; defined in legend to Fig. 4) of compound IV were observed for mutant SW-12 of HRV14 and wild-type strains HRV2, HRV39, and HRV49, all of which have a leucine at the position homologous to 1188 in HRV14 (Table 3). However, compound I(S) is more active with respect to the mutated HRV14 virus. It is, therefore, of interest that the aliphatic chain of compound I would have relatively little steric hindrance (nearest approach is 0.29 nm), with a leucine at position 1188 due to the opposite orientation of compound I versus compound IV. Similarly, compound I has much greater activity than compound IV against HRV2, HRV39, and HRV49, demonstrating that the opposite orientations of compounds IV and I observed crystallographically are roughly consistent with their observed antiviral activities.

Table 3. Effect of Mutations on MICs

Virus	Residue		MIC (μM) of compound:					
	1199	1188	I(S) (n = 7)	I(R) (n = 7)	IV (n = 7)	VI (n = 5)	V (n = 5)	VII (n = 5)
HRV14 wild type	Cys	Val	0.03	0.4	0.6	0.5	0.6	2.4
HRV14 mutant 2-9[a]	Trp	Val	3.6	6.5	9.1	2.5	1.5	2.9
HRV14 mutant SW-12	Cys	Leu	1.6	>10	>20	3.1	>10	>20
HRV2	Met	Leu	0.1	>10	4.1	1.5	0.1	0.1
HRV39	Met	Leu	1.0	>10	>20	2.9	0.6	0.4
HRV49	Met	Leu	0.3	3.1	4.4	2.0	0.3	0.2

[a] Results are consistent with 19 other independently isolated Cys → Trp mutations.

Smith et al. (35) proposed two possible mechanisms for the activity of the WIN compounds: (i) binding of the compounds might inhibit the flow of ions through the WIN pocket into the virus interior, which would cause swelling and disassembly; or (ii) binding of the compounds would stiffen VP1 and, therefore, inhibit the necessary flexibility for uncoating. Although the diffusion of heavy metal ions into the virus interior does occur (30), systematic analysis of the refined HRV14 structure does not reveal any special channels (32). Furthermore, the interior of the pocket, where the $n = 5$ compounds bind, would not be an optimal site for stopping ion flow. The second hypothesis has some confirmation in light of the observation that the SW-12 mutant (Val-1188 to Leu) is significantly more resistant to thermal inactivation in the presence of drugs than wild-type virus (F. J. Dutko, unpublished results). Filling the pocket with WIN compounds or WIN compounds plus larger amino acid side chains can improve stability and, hence, decrease uncoating potential.

APPLICATION TO OTHER TYPES OF VIRUSES

The presence of the hydrophobic pocket (into which the WIN compounds bind in enteroviruses) in the typical viral capsid is probably not fortuitous. Some degree of flexibility is required in the assembly and disassembly process. This can be provided by the loosely packed internal hydrophobic pocket of the standard viral capsid protein. Indeed, the requirement for this function in a protein that can also assemble into an icosahedral particle may, in part, be the cause for the retention of the same protein fold in the evolution of so many viral capsid structures. The design of a suitable antiviral agent that inhibits uncoating will then depend on the knowledge of the precise structure of the targeted virus capsid protein. The agent must be sufficiently flexible to enter the pocket through an available pore on the capsid's exterior, sufficiently hydrophobic to be retained by the pocket, and of suitable size to fit into the pocket.

The eight-stranded antiparallel β-barrel motif of viral capsid proteins has not been found in other classes of proteins (29). Thus, compounds like WIN 51711 (compound IV in Fig. 4), which have been particularly adapted to bind to a specific viral capsid protein, are unlikely to bind well to other types of proteins with different folds. They are, therefore, less likely to be toxic. In contrast, antiviral agents targeted at, for instance, viral proteases or polymerases have to have greater specificity in order not to interfere with essential metabolic processes that are dependent on proteins which have similar function and therefore are probably also of a similar polypeptide fold.

The greatest conservation between different picornaviruses occurs in the internal residues, whereas the greatest variability occurs on the antigenic surface (32). The high level of surface variability accounts for the large number of serologically distinct viruses which, nevertheless, bind to the same receptor (30). Thus, antiviral agents such as WIN 51711 have a relatively large range covering not only most rhinoviruses but also many enteroviruses. Hence, antiviral agents targeted at the hydrophobic pocket of a capsid protein are less likely to become ineffective as the virus mutates.

Compounds like WIN 51711 are rather hydrophobic and can usually be solubilized in dimethyl sulfoxide. Their hydrophobic character is essential for their binding to the hydrophobic pocket in VP1 of picornaviruses. However, their hydrophobic character is likely to allow them to be absorbed and transported across viral membranes. They could also be adsorbed onto the capsid during assembly. Thus, variations of these compounds might also be useful to inhibit uncoating of enveloped viruses such as Sindbis virus or human immunodeficiency virus.

The major core protein for human immunodeficiency virus is the *gag* protein p24. Although its amino acid sequence is known (26), there is little detectable homology to the sequences of simple icosahedral RNA viral capsid proteins. Of interest is the highly conserved sequence PPGAP in picornaviruses (30), which can also be spotted in p24 at an appropriate distance along the polypeptide chain. However, viral capsid proteins generally do not show amino acid sequence homology, although they all have exceedingly similar three-dimensional structures. The size of p24 roughly corresponds to the size of simple capsid proteins. For instance, the molecular weight of VP3 in picornaviruses is about 26,000. In human immunodeficiency virus the p24 protein forms a bacillus- or cone-shaped capsid containing the RNA (about $2 \times 9,600$ bases) (12). A proportional comparison with similarly bacillus-shaped particles of alfalfa mosaic virus (10, 18) suggests that the p24 core capsid could, for instance, have icosahedral ends with $T = 3$ symmetry (3), a 30-nm diameter, and a 95-nm-long cylindrical central component. Not only would such an envelope be of sufficient size to contain the RNA, but it also corresponds very roughly with the observed dimensions. In light of these general properties and the observations that many virus capsid proteins have similar tertiary structure and have a similar capsid organization, it does not seem impossible that p24 might also have the eight-stranded antiparallel β-barrel structure typical of other viruses. It is possible, therefore, that p24 might be a suitable antiviral target for compounds that would inhibit viral uncoating. Furthermore, the efficacy of these compounds is less likely to be affected by rapid change in amino acid sequences of the p24 protein.

Acknowledgments. The results on the conformational changes of mengovirus were obtained in a collaborative endeavor between Ulrike Boege and Doug Scraba (University of Alberta) and Ming Luo, Sangsoo Kim, S. Krishnaswamy, Iwona Minor, and Thomas J. Smith (Purdue University). The results on the binding of antiviral agents to HRV14 were obtained in a collaborative endeavor between John Badger, Iwona Minor, Marcia J. Kremer, Marcos A. Oliveira, Thomas J. Smith, S. Krishnaswamy, and Ming Luo (Purdue University), Mark A. McKinlay, Guy D. Diana, Frank J. Dutko, and Marilyn Fancher (Sterling-Winthrop Research Institute), and Roland R. Rueckert and Beverly A. Heinz (University of Wisconsin). I am greatly indebted to Mark McKinlay for discussions on considering other viruses as possible targets for WIN-type compounds. It was Roland Rueckert who suggested the collaborative work on HRV14 which in turn has led to all the topics (as well as others) discussed in this chapter. I would also like to extend my deep appreciation to all members of the groups at Purdue University, the University of Wisconsin, and Sterling-Winthrop Research Institute for numerous stimulating discussions.

The work was supported by grants from the National Institutes of Health, the National Science Foundation, and the Sterling-Winthrop Research Institute.

Literature Cited

1. **Abad-Zapatero, C., S. S. Abdel-Meguid, J. E. Johnson, A. G. W. Leslie, I. Rayment, M. G. Rossmann, D. Suck, and T. Tsukihara.** 1980. Structure of southern bean mosaic virus at 2.8 Å resolution. *Nature* (London) **286:**33–39.
2. **Arnold, E., J. W. Erickson, G. S. Fout, E. A. Frankenberger, H. J. Hecht, M. Luo, M. G. Rossman, and R. R. Rueckert.** 1984. Virion orientation in cubic crystals of the human common cold virus HRV14. *J. Mol. Biol.* **177:**417–430.
2a.**Badger, J., I. Minor, M. J. Kremer, M. A. Oliveira, T. J. Smith, J. P. Griffith, D. M. A. Guerin, S. Krishnaswamy, M. Luo, M. G. Rossmann, M. A. McKinlay, G. D. Diana, F. J. Dutko, M. Fancher, R. R. Rueckert, and B. A. Heinz.** 1988. Structural analysis of a series of antiviral agents complexed with human rhinovirus 14. *Proc. Natl. Acad. Sci. USA* **85:**3304–3308.
3. **Caspar, D. L. D., and A. Klug.** 1962. Physical principles in the construction of regular viruses. *Cold Spring Harbor Symp. Quant. Biol.* **27:**1–24.
4. **Chlumecka, V., P. D'Obrenan, and J. S. Colter.** 1973. Electrophoretic studies of three variants of Mengo encephalomyelitis virus. *Can. J. Biochem.* **51:**1521–1526.
5. **Chlumecka, V., P. D'Obrenan, and J. S. Colter.** 1977. Isoelectric focusing studies of Mengo virus variants, their protein structure units and constituent polypeptides. *J. Gen. Virol.* **35:**425–437.
6. **Crowell, R. L., and L. Philipson.** 1971. Specific alterations of coxsackievirus B3 eluted from HeLa cells. *J. Virol.* **8:**509–515.
7. **Diana, G. D., M. A. McKinlay, M. J. Otto, V. Akullian, and C. Oglesby.** 1985. [[(4,5-Dihydro-2-oxazolyl)phenoxy]alkyl]isoxazoles. Inhibitors of picornavirus uncoating. *J. Med. Chem.* **28:**1906–1910.

8. **Dunker, A. K., and R. R. Rueckert.** 1971. Fragments generated by pH dissociation of ME-virus and their relation to the structure of the virion. *J. Mol. Biol.* **58**:217–235.

9. **Fox, M. P., M. J. Otto, and M. A. McKinlay.** 1986. Prevention of rhinovirus and poliovirus uncoating by WIN 51711, a new antiviral drug. *Antimicrob. Agents Chemother.* **30**:110–116.

10. **Fukuyama, K., S. S. Abdel-Meguid, J. E. Johnson, and M. G. Rossmann.** 1983. Structure of a *T*=1 aggregate of alfalfa mosaic virus coat protein seen at 4.5 Å resolution. *J. Mol. Biol.* **167**:873–894.

11. **Fuller, S. D., and P. Argos.** 1987. Is Sindbis a simple picornavirus with an envelope? *EMBO J.* **6**:1099–1105.

12. **Gelderblom, H. R., E. H. S. Hausmann, M. Özel, G. Pauli, and M. A. Koch.** 1987. Fine structure of human immunodeficiency virus (HIV) and immunolocalization of structural proteins. *Virology* **156**:171–176.

13. **Harrison, S. C., A. J. Olson, C. E. Schutt, F. K. Winkler, and G. Bricogne.** 1978. Tomato bushy stunt virus at 2.9 Å resolution. *Nature* (London) **276**:368–373.

14. **Hogle, J. M., M. Chow, and D. J. Filman.** 1985. Three-dimensional structure of poliovirus at 2.9 Å resolution. *Science* **229**:1358–1365.

15. **Hogle, J. M., A. Maeda, and S. C. Harrison.** 1986. Structure and assembly of turnip crinkle virus. I. X-ray crystallographic structure analysis at 3.2 Å resolution. *J. Mol. Biol.* **191**:625–638.

16. **Hosur, M. V., T. Schmidt, R. C. Tucker, J. E. Johnson, T. M. Gallagher, B. H. Selling, and R. R. Rueckert.** 1987. Structure of an insect virus at 3.0 Å resolution. *Proteins* **2**:167–176.

17. **Huber, R., and W. S. Bennett, Jr.** 1983. Functional significance of flexibility in proteins. *Biopolymers* **22**:261–279.

18. **Hull, R.** 1969. Alfalfa mosaic virus. *Adv. Virus Res.* **15**:365–433.

19. **Korant, B. D., K. Lonberg-Holm, F. H. Yin, and J. Noble-Harvey.** 1975. Fractionation of biologically active and inactive populations of human rhinovirus type 2. *Virology* **63**:384–394.

20. **Liljas, L., T. Unge, T. A. Jones, K. Fridborg, S. Lövgren, U. Skoglund, and B. Strandberg.** 1982. Structure of satellite tobacco necrosis virus at 3.0 Å resolution. *J. Mol. Biol.* **159**:93–108.

21. **Lonberg-Holm, K., L. B. Gosser, and J. C. Kauer.** 1975. Early alteration of poliovirus in infected cells and its specific inhibition. *J. Gen. Virol.* **27**:329–342.

22. **Luo, M., G. Vriend, G. Kamer, I. Minor, E. Arnold, M. G. Rossmann, U. Boege, D. G. Scraba, G. M. Duke, and A. C. Palmenberg.** 1987. The atomic structure of Mengo virus at 3.0 Å resolution. *Science* **235**:182–191.

23. **Mak, T. W., J. S. Colter, and D. G. Scraba.** 1974. Structure of the Mengo virion. II. Physicochemical and electron microscopic analysis of degraded virus. *Virology* **57**:543–553.

24. **Mandel, B.** 1971. Characterization of type 1 poliovirus by electrophoretic analysis. *Virology* **44**:554–568.

25. **Ninomiya, Y., C. Ohsawa, M. Aoyama, I. Umeda, Y. Suhara, and H. Ishitsuka.** 1984. Antivirus agent, Ro 09-0410, binds to rhinovirus specifically and stabilizes the virus conformation. *Virology* **134**:269–276.

26. **Ratner, L., W. Haseltine, R. Patarca, K. J. Livak, B. Starcich, S. F. Josephs, E. R. Doran, J. A. Rafalski, E. A. Whitehorn, K. Baumeister, L. Ivanoff, S. R. Petteway, Jr., M. L. Pearson, J. A. Lautenberger, T. S. Papas, J. Ghrayeb, N. T. Chang, R. C. Gallo, and F. Wong-Staal.** 1985. Complete nucleotide sequence of the AIDS virus, HTLV-III. *Nature* (London) **313**:277–283.

27. **Roberts, M. M., J. L. White, M. G. Grütter, and R. M. Burnett.** 1986. Three-dimensional structure of the adenovirus major coat protein hexon. *Science* **232:**1148–1151.

28. **Rossmann, M. G.** 1979. Processing oscillation diffraction data for very large unit cells with an automatic convolution technique and profile fitting. *J. Appl. Crystallogr.* **12:**225–238.

29. **Rossmann, M. G., and P. Argos.** 1981. Protein folding. *Annu. Rev. Biochem.* **50:**497–532.

30. **Rossmann, M. G., E. Arnold, J. W. Erickson, E. A. Frankenberger, J. P. Griffith, H. J. Hecht, J. E. Johnson, G. Kamer, M. Luo, A. G. Mosser, R. R. Rueckert, B. Sherry, and G. Vriend.** 1985. Structure of a human common cold virus and functional relationship to other picornaviruses. *Nature* (London) **317:**145–153.

31. **Rossmann, M. G., A. G. W. Leslie, S. S. Abdel-Meguid, and T. Tsukihara.** 1979. Processing and post-refinement of oscillation camera data. *J. Appl. Crystallogr.* **12:**570–581.

32. **Rossmann, M. G., and A. C. Palmenberg.** 1988. Conservation of the putative receptor attachment site in picornaviruses. *Virology* **164:**373–382.

33. **Rueckert, R. R.** 1976. On the structure and morphogenesis of picornaviruses, p. 131–213. *In* H. Fraenkel-Conrat and R. R. Wagner (ed.), *Comprehensive Virology*, vol. 6. Plenum Publishing Corp., New York.

34. **Sherry, B., and R. R. Rueckert.** 1985. Evidence for at least two dominant neutralization antigens on human rhinovirus 14. *J. Virol.* **53:**137–143.

35. **Smith, T. J., M. J. Kremer, M. Luo, G. Vriend, E. Arnold, G. Kamer, M. G. Rossmann, M. A. McKinlay, G. D. Diana, and M. J. Otto.** 1986. The site of attachment in human rhinovirus 14 for antiviral agents that inhibit uncoating. *Science* **233:**1286–1293.

36. **Stauffacher, C. V., R. Usha, M. Harrington, T. Schmidt, M. V. Hosur, and J. E. Johnson.** 1987. The structure of cowpea mosaic virus at 3.5 Å resolution, p. 293–308. *In* D. Moras, J. Drenth, B. Strandberg, D. Suck, and K. Wilson (ed.), *Crystallography in Molecular Biology*. Plenum Publishing Corp., New York.

Molecular Aspects of Picornavirus Infection and Detection
Edited by Bert L. Semler and Ellie Ehrenfeld
© 1989 American Society for Microbiology, Washington, DC 20006

Chapter 9

Neutralization of Picornaviruses: Support for the Pentamer Bridging Hypothesis

*Anne G. Mosser, Donna M. Leippe, and
Roland R. Rueckert*

The mechanisms by which antibodies neutralize viruses have been studied extensively (5, 7, 8, 16) but inconclusively, perhaps because antiserum contains a variety of antibodies which act in different ways. The availability of monoclonal antibodies offers a solution to this problem and has stimulated new activity in this field.

Some antibodies neutralize virus by reducing the infective efficiency of a single virus particle, a process we call disablement. It has been suggested that antibodies neutralize virus by preventing a conformational change necessary for penetration and uncoating (15) or by inducing some lethal change such as RNA unpackaging (3). One measure of conformational change is a change in the isoelectric point of virus particles, which has frequently been associated with neutralization (9, 10, 15), but more quantitative studies revealed no regular relationship between the quantity of antibody needed to cause an isoelectric shift and that needed to neutralize (4). This absence of a relationship may be due in part to the method generally used for examining the isoelectric point, focusing with ampholines, which represents a highly unphysiological condition (low ionic strength, unusual dipolar ions, and none of the ions characteristic of

Anne G. Mosser, Donna M. Leippe, and Roland R. Rueckert • Institute for Molecular Virology and Department of Biochemistry, University of Wisconsin, Madison, Wisconsin 53706.

physiological fluids). No serious effort to measure isoelectric conversion of neutralized virus under more nearly physiological conditions has yet been reported. Thus, the relationship between isoelectric conversion and neutralization, if any, is still controversial.

Another mechanism of neutralization is aggregation (2, 25). Purely aggregating antibodies act by cross-linking virus particles to each other and thereby reduce the concentration of infectivity units in the virus suspension without damaging the intrinsic infectivity of the virus particle. The neutralizing effect of aggregation has been ignored or discounted in most studies. Icenogle et al. (12) encountered extensive aggregation in their studies but removed aggregates on sucrose gradients and studied antibody-complexed single virions. They showed that gradient-purified, antibody-complexed monomeric virions did not reaggregate, implying that both arms of the immunoglobulin G (IgG) molecule were stably complexed to the same particle. Moreover, the specific infectivity of these virions declined by a factor of roughly one-fourth for each antibody molecule added to the virus particle. This study provided the first direct evidence that antibodies are indeed able to reduce the intrinsic infectivity of a virus particle. Recently Wetz and co-workers, who measured the stoichiometry of antibodies to virus particles in immune precipitates made with rabbit antiserum, concluded that a single antibody molecule can neutralize a polio virion (28).

Thomas et al. (25) studied 13 murine monoclonal antibodies against poliovirus type 1 (PV-1) produced in three different laboratories and found that all generated aggregates when mixed with virus. From virus-antibody mixtures they separated virus monomers, dimers, and higher aggregates on sucrose gradients and concluded that aggregation was the primary basis of neutralization for these antibodies and probably all antibodies. A more recent study reports that different antibodies show very different aggregating abilities, and the ability to aggregate often does not parallel neutralization efficiency (24).

We have studied 15 neutralizing antibodies against PV-1 and 32 against human rhinovirus 14 (HRV-14) to discover what factors determine the different neutralization mechanisms. Here we report on the utility of titration curves in gathering information on the neutralizing properties of monoclonal antibodies.

LOCATION OF ANTIBODY-BINDING SITES

The molecular structures of both PV-1 (11) and HRV-14 (22) are known. Thus, it has been possible to map the target sites of all of our antibodies to three binding sites on the surface of PV-1 and four on

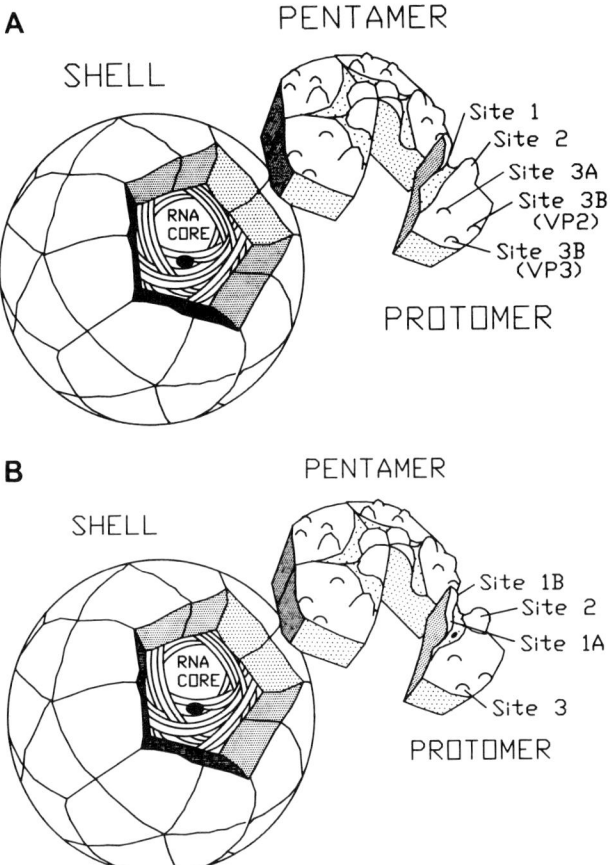

Figure 1. Schematic drawing of the antibody-binding sites on PV-1 (A) and on HRV-14 (B). The approximate shape of the viral surface is diagrammed for one pentamer, and binding sites are labeled on a single protomer.

HRV-14, using a combination of cross-resistance and escape mutation analysis (21, 23). Figure 1 shows diagrammatically the positions of these sites on the surfaces of PV-1 (panel A) and HRV-14 (panel B).

HRV-14 site 1A is located on the plateau at the fivefold axis of symmetry, including capsid protein VP1 amino acids 91 and 95. Site 1B is also located on this plateau and includes VP1 amino acids 83, 85, 138, and 139. We have not found evidence for PV-1 sites which correspond to either site 1A or 1B of HRV-14. Other laboratories have reported the isolation of neutralizing anti-PV-1 monoclonal antibodies which interact with a site corresponding to HRV-14 site 1A (19, 27, 29), but this site is immunorecessive for PV-1 (13). For poliovirus type 3 this is the dominant site (17, 18, 20).

Site 2 is located on a large protrusion from the virus surface. In PV-1, changes are located in VP2, amino acids 164 through 170 and 270, and in VP1, amino acids 221 through 226. Site 2 is essentially the same for HRV-14, except that there is apparently a smaller contribution from VP1 residues than there is in PV-1.

PV-1 site 3A consists of a small loop of VP3 which protrudes from the surface at amino acids 58 through 60. We have found no evidence of a corresponding site in HRV-14. PV-1 site 3B consists of a segment including the surface residue of VP2, amino acid 72, and another segment containing VP3, amino acid 76. These two segments are separated on the surface of the protomer by a distance of 30 Å (3 nm); interestingly, however, the VP2 segment of one protomer is separated from the VP3 segment on the adjacent protomer by only 1.2 nm (J. Hogle, personal communication). (See two circled pairs of mutation sites lying on the pentamer-pentamer boundary in Fig. 5.) This situation suggests that antigenic site 3B is actually made up from segments of two different pentamers; if so, then it might be expected that site 3B is not fully formed until the two pentamers join to form shells, a testable prediction.

Minor et al. (18) and Blondel et al. (1) have found escape mutants with changes in VP3 amino acid 71, and Diamond et al. (6) have found changes in VP3 amino acid 73. These two residues lie roughly midway between our sites 3A and 3B. These mutants are not neutralized by antibodies which we would classify as site 3A, those that elicit escape mutations at VP3 amino acids 58 through 60.

HRV-14 site 3 includes VP3 amino acids 72, 75, 78, and 203 and the carboxy terminus of VP1. It corresponds to PV-1 site 3B, but all of these residues lie in close apposition on one protomer. This site apparently lies entirely within a single protomer for HRV-14; we have found no evidence for mutations in an adjacent protomer.

CHARACTERIZATION OF ANTIBODIES BY NEUTRALIZATION CURVES

To study the interaction of our antibodies with their binding sites, we have constructed titration curves for 32 antibodies against HRV-14 and for all 15 antibodies against PV-1. These curves represent a simple titration of the ability of antibody to neutralize a constant amount of virus. Figure 2 shows the profiles of several antibodies interacting with site 3A

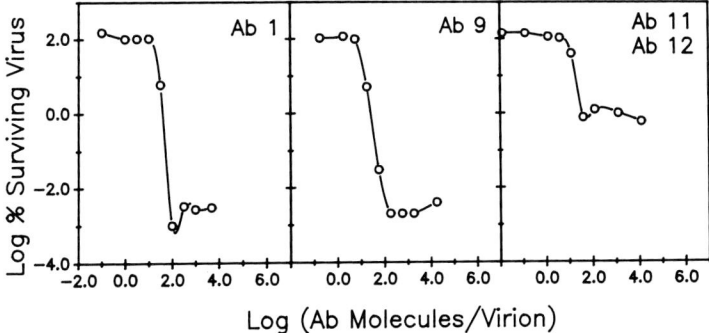

Figure 2. Neutralization profiles of the site 3A anti-PV-1 antibodies (Ab). Sabin type 1 virus (2×10^8 PFU/ml) and dilutions of antibodies purified from mouse ascites fluids by affinity chromatography were mixed and incubated for 1 h at room temperature and then overnight at 4°C. Surviving infectious virus was determined by plaque assay.

of PV-1. For antibodies 1 and 9, the infectivity falls very steeply for 4 or 5 logs before leveling off. Plaques picked from tissue culture dishes infected with PV-1 treated with an excess of these antibodies show that this residual virus consists primarily of antibody-resistant mutants (data not shown).

This is the kind of profile one might expect to find for a strongly neutralizing antibody. Thus, we were surprised to find that only these 2 of 15 antipoliovirus antibodies and only 10 of the 32 antirhinovirus antibodies gave profiles like these. The other two antibodies that interact with site 3A displayed a 1,000-fold higher fraction of persistent infectivity at high antibody concentration, about 1% of input virus.

The neutralization profiles of antibodies against polioviral site 3B are shown in Fig. 3. All of our site 3B antibodies (Fig. 3) give neutralization curves that are similar to site 3A antibodies 11 and 12; the percentage of surviving virus falls rapidly but levels off at 0.1 to 1% of surviving infectivity.

The profiles of the seven anti-PV-1 antibodies against site 2 are shown in Fig. 4. Three of these neutralization curves are characterized by a U-shaped profile (antibodies 3, 5, and 8). At high antibody-to-virus ratios, an increase in antibody actually increases the amount of surviving virus. We think these antibodies neutralize the virus primarily, if not solely, by aggregation. Similar kinds of neutralization profiles are observed with antibodies against HRV-14; for some of these anti-HRV-14 antibodies, we have shown that the dip in this curve represents precipitation of virus-antibody complexes, while the rise corresponds to resolubilization of the precipitate to form single virions saturated with excess

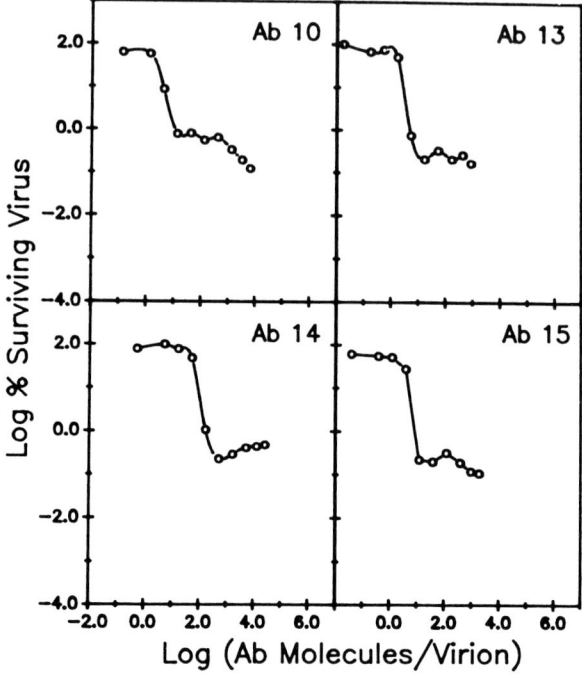

Figure 3. Neutralization profiles of the site 3B anti-PV-1 antibodies (Ab). Sabin type 1 virus $(2 \times 10^8$ to 2×10^9 PFU/ml) and dilutions of antibodies purified from mouse ascites fluids by affinity chromatography were mixed and incubated as described in the legend to Fig. 2. Surviving infectious virus was determined by plaque assay.

antibody. The rise in infectivity demonstrates that virions saturated with these antibodies are still infectious, albeit only 3 to 30% as infectious as free virus. Antibodies 2, 4, 6, and 7 against PV-1 do not show such an increase in virus infectivity with increasing antibody concentration at antibody excess. These may correspond to antibodies for which immune aggregates are not solubilized even at very high antibody-to-virus ratios or for which antibody-coated virus is not particularly infectious.

Thus, for the anti-PV-1 antibodies (Table 1), all strongly neutralizing antibodies interact with site 3A. Antibodies that bind to site 2 frequently give U-shaped curves, which we believe is a characteristic of those antibodies whose primary or perhaps sole mechanism of neutralization is aggregation.

Similar experiments have been done with 32 neutralizing antibodies against HRV-14 (Table 1). All of the strongly neutralizing antibodies are targeted to site 1A, although 5 of the 15 antibodies reacting with this site

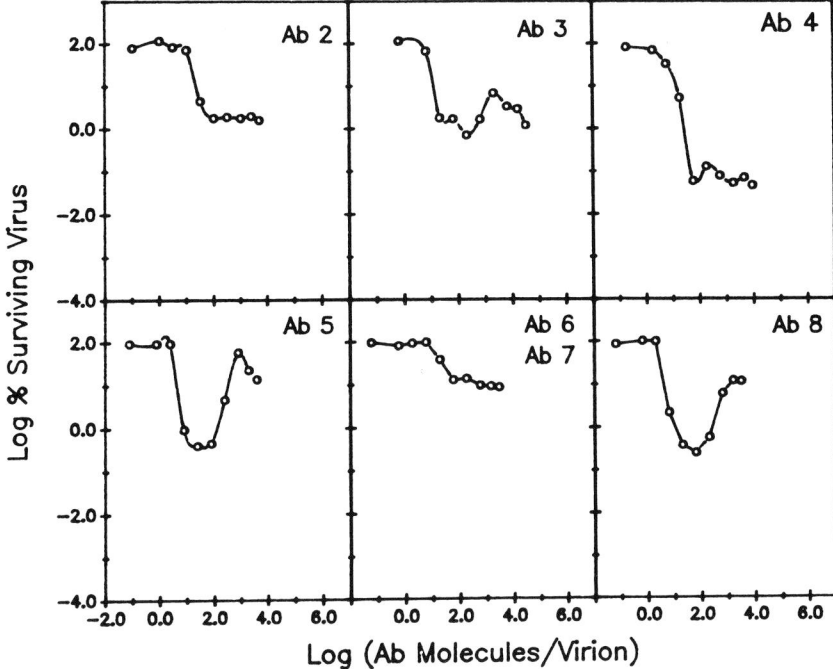

Figure 4. Neutralization profiles of the site 2 anti-PV-1 antibodies (Ab). Sabin type 1 virus (2×10^8 to 2×10^9 PFU/ml) and dilutions of antibodies purified from mouse ascites fluids by affinity chromatography were mixed and incubated as described in the legend to Fig. 2. Surviving infectious virus was determined by plaque assay.

are less efficient. Two of these site 1A antibodies give neutralization profiles with U-shaped curves. Site 1B and site 2 antibodies tend to give U-shaped profiles (two of three and three of four tested antibodies, respectively). Four of the 10 site 3 antibodies give U-shaped profiles, and the remaining 6 give profiles that resemble those of PV-1 site 3B antibodies.

POSSIBLE FACTORS DETERMINING MECHANISM OF NEUTRALIZATION

The different shapes of the neutralization profiles evidently reflect different mechanisms of neutralization and perhaps also different affinities between antibodies and their binding sites. Antibodies associated with only one of the PV-1 sites and one of the HRV-14 sites give evidence of

Table 1. Classification of Neutralization Curves for Neutralizing Antibodies against PV-1 and HRV-14

Virus	Site	Approx spanning distance (nm)[a]	No. of neutralization curves of type[b]		
			Strong	U-shaped	Intermediate
PV-1	2	6 or 15	0	3	4
	3A	8 or 11	2	0	2
	3B	7, 12, or 14	0	0	4
HRV-14	1A	11 or 14	10	2	3
	1B	13.5 or 16	0	2	1
	2	6 or 14	0	3	1
	3	6.5 or 12	0	4	6

[a] The spanning distances are estimated between pairs of identical sites related by a twofold axis of symmetry.
[b] For explanation of neutralization curve type, see text.

efficient neutralization, with little or no residual viral infectivity at high antibody concentrations. The antibody studied by Icenogle et al. (12) is one of the two efficient PV-1 antibodies, and for this antibody, bivalent binding to a single virion was determined to be important to its ability to neutralize that virion. Therefore, we speculate that bivalent binding of an antibody to a single virion may be important for other antibodies generating this sort of neutralization profile. The probability of such bivalent binding depends on the relative ease with which its second arm can bind to a partner site on the same virion (thus linking pentamers) compared with the ease with which it binds another virion (causing aggregation). This in turn is probably determined by the spanning distance between two identical sites related by twofold symmetry and the orientation with which the antibody binds to the first of these sites.

SPANNING DISTANCES BETWEEN SITES RELATED BY TWOFOLD SYMMETRY

Figure 5 depicts the symmetry relationship of antigenic sites on PV-1. It highlights the neutralization sites in four viral protomers, represented by the kite shapes, and shows the approximate spanning distances across a twofold axis for all three sites. These distances range from approximately 6 to 16 nm. Several authors have suggested that the two arms on an IgG molecule can span distances ranging from 9 to 15 nm (12), but the true range limits are not known with certainty. We propose that the site 2 pair (Fig. 5, hatched with closely spaced lines), at 6 nm apart, are too close together for antibody spanning, while the pair hatched with widely

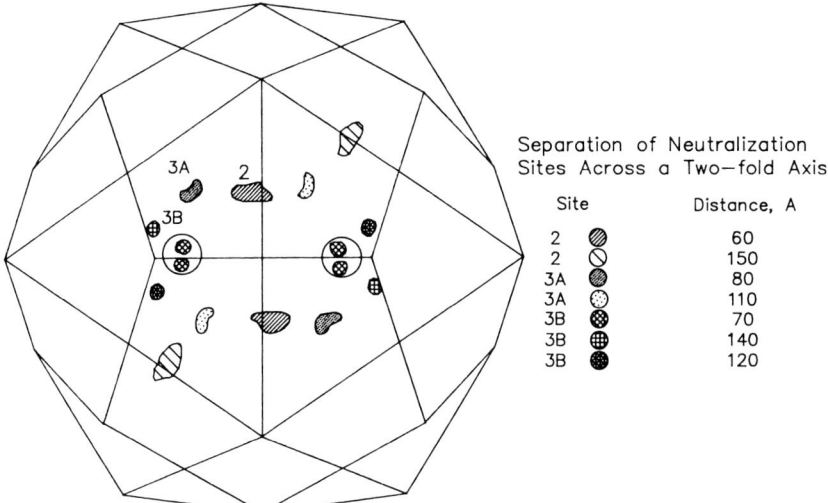

Separation of Neutralization
Sites Across a Two—fold Axis

Site		Distance, A
2		60
2		150
3A		80
3A		110
3B		70
3B		140
3B		120

Figure 5. Geometrical diagram of PV-1 showing positions of antibody-binding sites and distances between pairs of sites related by a twofold axis of symmetry. The kite-shaped segments represent the morphogenetic units or protomers and consist of the four capsid proteins VP1, 2, 3, and 4. Circles enclose the two segments of site 3B which flank a protomer-protomer boundary.

spaced lines, at 15 nm apart, are a bit too far apart. For the other sites with intermediate spanning distances, for example, the site 3A pairs with distances of 8 and 11 nm, bivalent binding might be possible, and this indeed is the only target site at which we have found strongly neutralizing antibodies.

Figure 6 is a similar schematic of HRV-14 binding sites showing spanning distances across a twofold axis. These distances are averages of the distances between symmetrically related pairs of those amino acids which were substituted in antibody escape mutants. The distance between alpha carbons was determined by using the FRODO software (14) and atomic coordinates for HRV-14 kindly provided by Michael Rossmann. Again, site 1A, the only site with strongly neutralizing antibodies, is the site for which the spanning distance (11 nm) is most likely to be ideal for a bivalent antibody.

IMPORTANCE OF ORIENTATION OF ANTIBODY BINDING

Based on previous estimates (12), the spanning distance of 12 nm associated with site 3 in HRV-14 should also be easily spannable by an

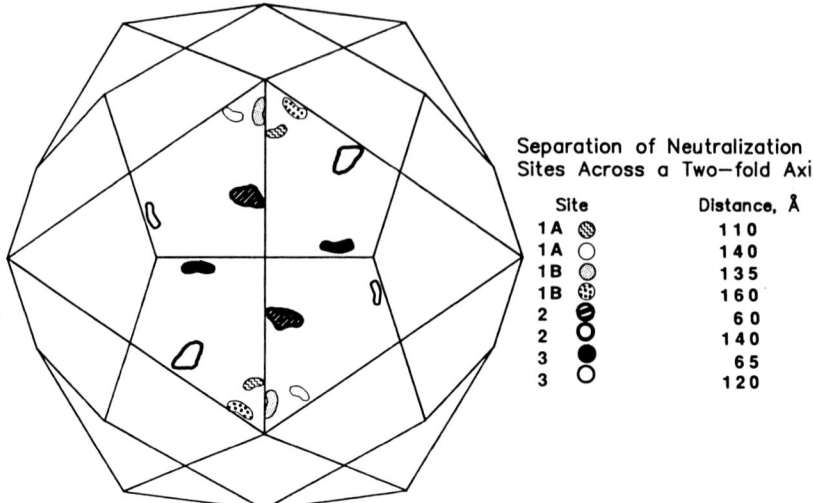

Separation of Neutralization
Sites Across a Two—fold Axis

Site		Distance, Å
1 A	◉	1 1 0
1 A	○	1 4 0
1 B	◎	1 3 5
1 B	⊕	1 6 0
2	◉	6 0
2	●	1 4 0
3	●	6 5
3	○	1 2 0

Figure 6. Geometrical diagram of HRV-14 showing positions of antibody-binding sites and distances between pairs of sites related by a twofold axis of symmetry.

antibody, yet none of the antibodies associated with this site show efficient neutralization curves. It may be that antibody binding bivalently to a curved surface such as a virus particle may not be able to span even 12 nm, but another interpretation is also possible. Not all antibodies that

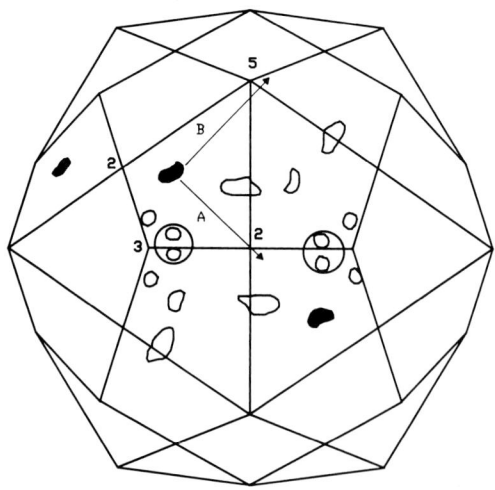

Figure 7. Geometrical diagram of PV-1 showing positions of antibody-binding sites and demonstrating possible restriction of bivalent binding of antibody to shaded site depending on orientation of binding. The axes of symmetry are labeled. Vectors A and B illustrate two possible orientations of antibody binding. Binding with the free arm pointing in direction A would allow binding of the second arm; binding with the free arm pointing in direction B might not.

bind to spannable sites such as site 1A in HRV-14 are strong neutralizers. We believe that antibodies which bind to a site with a symmetrically related partner that is close enough for bivalent binding may still not bind bivalently if they bind to this site with the wrong orientation. Figure 7 illustrates this effect of antibody orientation. If an antibody were to bind to PV-1 site 3A as shown, oriented so that its second arm was facing in the direction indicated by vector A, then it would likely bind its second arm to the same virion. If, on the other hand, the antibody were oriented so that its second arm was facing in direction B, it would be unlikely to be able to twist enough to find another site related by twofold symmetry and would eventually bind its second arm to another virion, causing virus aggregation. Some antibody target sites might be placed on the virus surface so as to force antibodies to bind in orientations that prevent bivalent binding.

FUTURE WORK

In summary we have described the use of neutralization curves to distinguish two types of neutralization, strong and weak. We propose that strong neutralization is caused by cross-linking adjacent protomers across a twofold axis of symmetry and that the effect of this cross-linking is to lock pentamers together. This model accommodates the observation that strongly neutralizing antibodies were observed only in the group targeted against sites separated by distances within the spanning limits of the IgG molecule, previously estimated to be roughly 9 to 15 nm (12). If our interpretation is correct, then the limits for spanning sites on a picorna-virus virion may be nearer to 7 to 12 nm.

The pentamer cross-linking model also accommodates the observation that many antibodies directed against strong sites are not strongly neutralizing. We predict that these antibodies do not anchor in an orientation favorable for reaching their partner site. This prediction could be tested directly by the methods of X-ray crystallography if it were possible to make high-quality crystals of virus-antibody complexes. Alternately, it may be possible to examine the orientation hypothesis indirectly by examining crystals of synthetic peptide representing a strong site complexed to Fab fragments of strongly and weakly neutralizing antibodies. Efforts to analyze such crystals are currently under way in collaboration with Rossmann's laboratory at Purdue University.

Acknowledgments. We thank Michael Rossmann and members of his laboratory, James Hogle and Dave Filman, and Marie Chow and Guy Page for stimulating discussions and important contributions leading up to this study. We also thank Ruth Rueckert for excellent technical assistance.

This work was supported by grant MV-33 from the American Cancer Society and Public Health Service grant AI24939 from the National Institutes of Health.

Literature Cited

1. **Blondel, B., R. Crainic, O. Fichot, G. DuFraisse, A. Candrea, D. Diamond, M. Girard, and F. Horaud.** 1986. Mutations conferring resistance to neutralization with monoclonal antibodies in type 1 poliovirus can be located outside or inside the antibody-binding site. *J. Virol.* **57**:81–90.

2. **Brioen, P., D. Dekegel, and A. Boeyé.** 1983. Neutralization of poliovirus by antibody-mediated polymerization. *Virology* **127**:463–468.

3. **Brioen, P., B. Rombaut, and A. Boeyé.** 1985. Hit-and-run neutralization of poliovirus. *J. Gen. Virol.* **66**:2495–2499.

4. **Brioen, P., A. A. M. Thomas, and A. Boeyé.** 1985. Lack of quantitative correlation between the neutralization of poliovirus and the antibody-mediated pI shift of the virion. *J. Gen. Virol.* **66**:609–613.

5. **Della-Porta, A. J., and E. G. Westaway.** 1977. A multi-hit model for the neutralization of animal viruses. *J. Gen. Virol.* **38**:1–19.

6. **Diamond, D. C., B. A. Jameson, J. Bonin, M. Kohara, S. Abe, H. Itoh, T. Komatsu, M. Arita, S. Kuge, A. Nomoto, A. D. M. E. Osterhaus, R. Crainic, and E. Wimmer.** 1985. Antigenic variation and resistance to neutralization in poliovirus type 1. *Science* **229**:1090–1093.

7. **Dimmock, N. J.** 1984. Mechanism of neutralization of animal viruses. *J. Gen. Virol.* **65**:1015–1022.

8. **Dulbecco, R., M. Vogt, and A. G. R. Strickland.** 1956. A study of the basic aspects of neutralization of two animal viruses, western equine encephalitis virus and poliomyelitis virus. *Virology* **2**:162–205.

9. **Emini, E. A., S.-Y. Kao, A. J. Lewis, R. Crainic, and E. Wimmer.** 1983. Functional basis of poliovirus neutralization determined with monospecific neutralizing antibodies. *J. Virol.* **46**:466–474.

10. **Emini, E. A., P. Ostapchuk, and E. Wimmer.** 1983. Bivalent attachment of antibody onto poliovirus leads to conformational alteration and neutralization. *J. Virol.* **48**:547–550.

11. **Hogle, J. M., M. Chow, and D. J. Filman.** 1985. Three-dimensional structure of poliovirus at 2.9 A resolution. *Science* **229**:1358–1365.

12. **Icenogle, J., H. Shiwen, G. Duke, S. Gilbert, R. Rueckert, and J. Anderegg.** 1983. Neutralization of poliovirus by a monoclonal antibody: kinetics and stoichiometry. *Virology* **127**:412–425.

13. **Icenogle, J. P., P. D. Minor, M. Ferguson, and J. M. Hogle.** 1986. Modulation of humoral response to a 12-amino-acid site on the poliovirus virion. *J. Virol.* **60**:297–301.

14. **Jones, T. A.** 1978. A graphics model building and refinement system for macromolecules. *J. Appl. Cryst.* **11**:268–272.

15. **Mandel, B.** 1976. Neutralization of poliovirus: a hypothesis to explain the mechanism and the one-hit character of the neutralization reaction. *Virology* **69**:500–510.

16. **Mandel, B.** 1979. Interaction of viruses with neutralizing antibodies, p. 37–121. *In* H. Fraenkel-Conrat and R. R. Wagner (ed.), *Comprehensive Virology*, vol. 15. Plenum Publishing Corp., New York.

17. **Minor, P. D., D. M. A. Evans, M. Ferguson, G. C. Schild, G. Westrop, and J. W. Almond.** 1985. Principal and subsidiary antigenic sites of VP1 involved in the neutralization of poliovirus type 3. *J. Gen. Virol.* **65:**1159–1165.

18. **Minor, P. D., M. Ferguson, D. M. A. Evans, J. W. Almond, and J. P. Icenogle.** 1986. Antigenic structure of polioviruses of serotypes 1, 2, and 3. *J. Gen. Virol.* **67:**1283–1291.

19. **Minor, P. D., M. Ferguson, A. Philips, D. I. Magrath, A. Huovilainen, and T. Hovi.** 1987. Conservation *in vivo* of protease cleavage sites in antigenic sites of poliovirus. *J. Gen. Virol.* **68:**1857–1865.

20. **Minor, P. D., G. C. Schild, J. Bootman, D. M. A. Evans, M. Ferguson, P. Reeve, M. Spitz, G. Stanway, A. J. Cann, R. Hauptmann, L. D. Clarke, R. C. Mountford, and J. W. Almond.** 1983. Location and primary structure of a major antigenic site for poliovirus neutralization. *Nature* (London) **301:**674–679.

21. **Page, G. S., A. G. Mosser, J. M. Hogle, D. J. Filman, R. R. Rueckert, and M. Chow.** 1988. Three-dimensional structure of poliovirus type 1 neutralizing determinants. *J. Virol.* **62:**1781–1794.

22. **Rossmann, M. G., E. Arnold, J. W. Erickson, E. A. Frankenberger, J. P. Griffith, H.-J. Hecht, J. E. Johnson, G. Kamer, M. Luo, A. G. Mosser, R. R. Rueckert, B. Sherry, and G. Vriend.** 1985. Structure of a human common cold virus and functional relationship to other picornaviruses. *Nature* (London) **317:**145–153.

23. **Sherry, B., A. G. Mosser, R. J. Colonno, and R. R. Rueckert.** 1986. Use of monoclonal antibodies to identify four neutralization immunogens on a common cold picornavirus. *J. Virol.* **57:**246–257.

24. **Taniguchi, K., and S. Urasawa.** 1987. Different virus-precipitating activities of neutralizing monoclonal antibodies that recognize distinct sites of poliovirus particles. *Arch. Virol.* **92:**27–40.

25. **Thomas, A. A. M., P. Brioen, and A. Boeyé.** 1985. A monoclonal antibody that neutralizes poliovirus by cross-linking virions. *J. Virol.* **54:**7–13.

26. **Thomas, A. A. M., R. Vrijsen, and A. Boeyé.** 1986. Relationship between poliovirus neutralization and aggregation. *J. Virol.* **59:**479–485.

27. **van der Werf, S., C. Wychowski, P. Bruneau, B. Blondel, R. Crainic, F. Horodniceanu, and M. Girard.** 1983. Localization of a poliovirus type 1 neutralization epitope in viral capsid polypeptide VP1. *Proc. Natl. Acad. Sci. USA* **80:**5080–5084.

28. **Wetz, K., P. Willingmann, H. Zeichhardt, and K.-O. Habermehl.** 1986. Neutralization of poliovirus by polyclonal antibodies requires binding of a single IgG molecule per virion. *Arch. Virol.* **91:**207–220.

29. **Wiegers, K.-J., and R. Dernick.** 1987. Binding site of neutralizing monoclonal antibodies obtained after *in vivo* priming with purified VP1 of poliovirus type 1 is located between amino acid residues 93 and 104 of VP1. *Virology* **157:**248–251.

Molecular Aspects of Picornavirus Infection and Detection
Edited by Bert L. Semler and Ellie Ehrenfeld
© 1989 American Society for Microbiology, Washington, DC 20006

Chapter 10

Molecular and Biochemical Aspects of Human Rhinovirus Attachment to Cellular Receptors

Richard J. Colonno, Gordon Abraham,
and Joanne E. Tomassini

BACKGROUND

Human rhinoviruses (HRVs) are members of the *Picornaviridae* and are the major causative agent of the common cold in man. To date 100 antigenically distinct HRV serotypes have been isolated from clinical specimens and shown to represent distinct serotypes by reciprocal cross-neutralization studies (12). Recent competition binding and immunological studies have demonstrated that 88 of these serotypes could be divided into major (78 serotypes) and minor (10 serotypes) receptor families (1, 8). Of particular importance was the finding that nearly 90% of the tested HRV serotypes and at least three of the coxsackievirus A serotypes utilize a single cellular receptor (8). Similar to other animal viruses, the attachment of HRVs to cellular receptors was found to be the determining factor in defining viral tropisms and was crucial for the initiation of HRV infection in susceptible cells.

Although the isolation of murine monoclonal antibodies capable of blocking the attachment of the picornaviruses has provided some insight into the nature of viral attachment, we know relatively little about the

Richard J. Colonno, Gordon Abraham, and Joanne E. Tomassini • Department of Virus and Cell Biology, Merck Sharp & Dohme Research Laboratories, West Point, Pennsylvania 19486.

molecular nature of cellular receptors involved in attachment or the virion attachment site itself. Progress in this area was the subject of a recent review (6).

ANTIBODY BLOCKADE OF VIRAL ATTACHMENT

Of the five antireceptor monoclonal antibodies shown to block the attachment of poliovirus, group B coxsackieviruses, or HRVs (4, 8, 10, 15, 17), the most useful in providing information regarding the nature of virion interaction with cellular receptors has been the monoclonal antibody (designated 1A6) which recognizes the cellular receptor for the major group of HRVs (8). The 1A6 antibody could block the attachment only of the major group of HRVs and coxsackievirus A serotypes, while having no effect on the binding of 16 other viruses tested (8). In addition, the binding affinity of the 1A6 antibody for the receptor protein was far superior to that determined for several HRV serotypes. Treatment of cell monolayers with only nanomolar concentrations of the 1A6 antibody provided complete protection against challenge by all major-group viruses. Previous studies have demonstrated that the affinity of this antibody was high enough to actually displace previously bound virions (8). To determine whether the size of the 1A6 antibody was a major determinant in its ability to block HRV attachment, Fab fragments were prepared from the 1A6 antibody and used in HeLa cell membrane-binding assays involving HRV-14 and HRV-15. The results obtained from these assays clearly demonstrated that the isolated Fab fragments were as effective at blocking attachment of radiolabeled HRVs on a molar basis as was intact antibody (data not shown). Further proof that the Fab fragments are as effective in receptor blockade as intact antibody was obtained in virion displacement studies in which the abilities of intact antibody and derived Fabs to displace previously bound HRV-14 virions were compared. The results (Fig. 1) show that both the Fab fragments and the intact antibody could displace equivalent amounts of HRV-14 from membranes. These results indicate that a bivalent attachment of the antibody to multiple receptor proteins is not required to block virus binding and that a molecule only 50 kilodaltons (kDa) in size can block viral attachment as effectively as an intact antibody having a molecular size of 150 kDa.

ISOLATION OF HRV RECEPTOR PROTEIN

The strong affinity of the 1A6 antibody for the cellular receptor enabled the successful isolation of a 90-kDa protein from HeLa cells by

Figure 1. HRV-14 displacement by antibody and Fab fragments. Duplicate tubes each containing 1 mg of HeLa R-19 cell membranes and [^{35}S]methionine-labeled HRV-14 (1.19 × 10^8 PFU/4 × 10^5 cpm) in McCoy 5A medium containing 1% fetal calf serum, 10 mM MgCl$_2$, and 20 mM HEPES (*N*-2-hydroxyethylpiperazine-*N'*-2-ethanesulfonic acid) were incubated for 1 h at 23°C in a final volume of 0.37 ml. After incubation, 0.5 ml of medium was added, and the membrane-virus com-

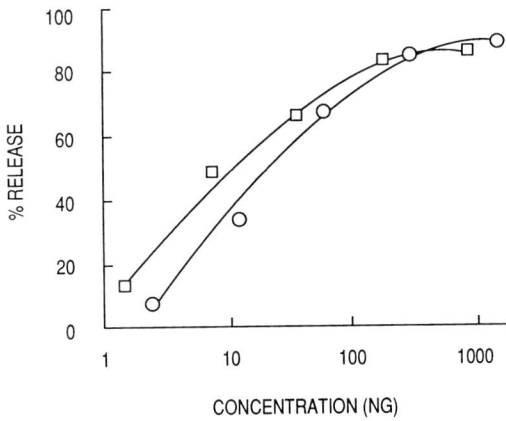

plexes were pelleted at 13,800 × *g* for 2 min. The resulting pellet was suspended and washed with 0.5 ml of medium. Each pellet was suspended in 0.37 ml of medium, and 45 μl was dispensed into new tubes and incubated at 25°C for 1 h with the indicated quantities of purified 1A6 antibody (□) or Fab fragments (○) in a total volume of 50 μl. Fab fragments were derived by a procedure previously described (7). Medium (0.5 ml) was then added, and the tubes were pelleted as described above. The amount of radiolabeled virus found in the supernatant (displaced) and pellets was determined by scintillation counting.

immunoaffinity chromatography (22). Confirmation that the 90-kDa protein was involved in HRV attachment was obtained by showing that rabbit polyclonal antiserum prepared against this protein had the same receptor specificity as the 1A6 antibody. The polyclonal antibody was also capable of immunoprecipitating the 90-kDa receptor protein and blocking the binding of 1A6 antibody to receptors on HeLa cell membranes (22). These studies clearly demonstrated that the 90-kDa receptor protein was involved in HRV attachment. The 90-kDa protein was determined to be an acidic glycoprotein having a pI of 4.2 (23). Digestion of the receptor protein with *N*-glycanase and endoglycosidases D, H, and F, followed by analysis of the digestion products on sodium dodecyl sulfate-polyacrylamide gel electrophoresis (SDS-PAGE) showed that hybrid, complex-type oligosaccharides were N-linked to the receptor protein (23). Sialic acid was found at the nonreducing ends of the carbohydrate by neuraminidase digestion of receptor protein. Deglycosylation of the 90-kDa protein by *N*-glycanase digestion resulted in a protein with an apparent molecular size of 60 kDa, suggesting that carbohydrates accounted for one-third of the molecular mass of the receptor protein as observed by SDS-PAGE (23). This cell surface protein is found only on cells derived from human or higher-primate sources (8) and undoubtedly determines the host range for this group of viruses.

DOES THE HRV RECEPTOR TURN OVER?

The normal cellular function of the major HRV group receptor remains unknown. Cell culture studies have demonstrated that a functional HRV receptor is not required for cell growth and division since the treatment of transformed and primary tissue culture cells with receptor antibody had no effect on cell division and growth over a 6-day period (9). Experiments were performed to determine whether attachment of 1A6 antibody could elicit a turnover of this protein. HeLa R-19 cell monolayers were treated with saturating quantities of 1A6 antibody, and unbound antibody was removed by washing. The confluent monolayers were then incubated at 37°C and assayed for their ability to bind either radiolabeled HRV-14 or 1A6 antibody over a period of 55 h. The results (Fig. 2) indicated that the receptor blockade remained in effect for up to 45 h before significant breakthrough could be detected. This experiment demonstrated that the HRV receptor is not regenerated rapidly, if at all, when cells are in a stationary phase. This conclusion is supported by the experimental finding that it is extremely difficult to metabolically label this protein with [^{35}S]methionine in confluent monolayers, compared with the relatively good incorporation which can be obtained when labeling is performed on actively dividing cells (data not shown). The ability of the 1A6 antibody to protect cells for an extended period of time is also testimonial to the stability of this monoclonal antibody in the presence of serum. A large proportion of the eventual breakdown in receptor blockade may actually be the result of overgrowth of the cell monolayer rather than receptor regeneration. However, the evolutionary retention of this protein and the finding that it is ubiquitous among human cells suggest a functional role in cells.

VIRION ATTACHMENT SITE

Until recently, knowledge concerning the structure and location of the HRV virion attachment site, which interacts with cell surface receptors, had been quite limited. Crystallographic studies on HRV-14, a serotype belonging to the major HRV group, and poliovirus have shed some light on this problem (13, 20). Among the many structural features determined for these two viruses was the discovery of a deep canyon present on the surface of the viral capsid surrounding each vertex of

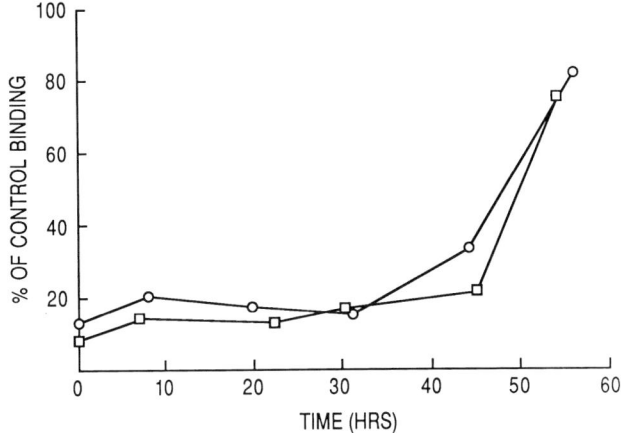

Figure 2. Availability of the HRV cellular receptor protein. Confluent HeLa cell monolayers in 48-well cluster plates were washed and incubated for 30 min at 34°C with 40 μg of 1A6 antibody in 0.2 ml of McCoy 5A medium supplemented with 10% fetal calf serum–10 mM $MgCl_2$. The antibody solution was then removed, and each well was washed twice with warm medium. After a final wash, 0.5 ml of fresh medium was placed into each well, and the cells were incubated at 37°C. At the times indicated, duplicate antibody-treated and untreated wells on individual plates were assayed for their ability to bind [^{35}S]methionine-labeled HRV-15 (□) or ^{125}I-labeled 1A6 antibody (○), as previously described (8). Briefly, supernatants were removed from monolayers after incubation and combined with two subsequent medium washes to determine unbound radioactivity. Cell monolayers were then incubated 10 min with 1% sodium dodecyl sulfate at 34°C, and the contents of each well were removed and combined with two subsequent medium washes to determine bound radioactivity.

fivefold symmetry on the icosahedral capsid (Fig. 3). The rim of the canyon measures up to 3.5 nm across and descends to 2.2 nm at the deepest regions of the canyon (8a, 20). Based on the nucleic acid sequence comparisons, similar canyon structures most likely exist within the capsid structures of other picornaviruses, such as coxsackievirus B (14). Immunological studies (21) have been used to map the four neutralization sites of HRV-14 to protein projections found on the surface of the viral capsid. These four sites are probably not involved in virion attachment since they represent hypervariable regions rather than the highly conserved sequences expected for the virion attachment site and because they fail to cross-neutralize other HRV serotypes (data not shown). However, neutralization by Fab fragments derived from the neutralizing monoclonal antibodies mapping to each of the four sites does inhibit virus binding (7). The precise mechanism(s) involved in this inhibition is unclear and may be the result of conformational alterations or steric

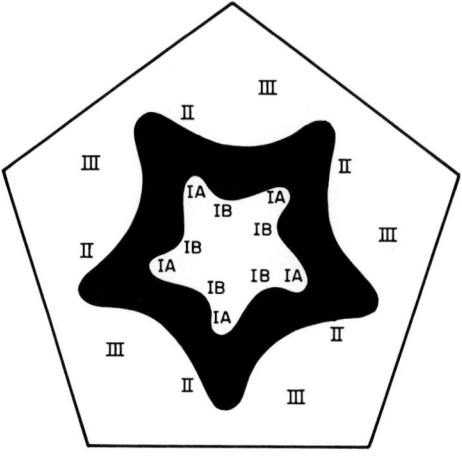

Figure 3. Schematic diagram of HRV canyon. The approximate contour of the viral canyon is shown in the context of a viral pentamer and was derived from the previously determined atomic structure of HRV-14 (20). The axis of fivefold symmetry is located in the center of the diagram, and the proximal location of four neutralization sites located on VP1 (IA and IB), VP2 (II), and VP3 (III) are as indicated.

hindrance. Since the virion attachment site is apparently inaccessible to immunoglobulins, Rossmann et al. (20) proposed that this deep canyon might contain or act as the virion attachment site. While some circumstantial evidence exists to support this hypothesis, direct evidence has only recently been obtained. In these studies (8a), 10 mutants were created by single amino acid substitutions or insertions at positions 103, 155, 220, and 223 of the structural protein VP1 of HRV-14, using mutagenesis of an infectious cDNA clone (16). The four amino acids are located in the deepest regions of the canyon and are all within 0.5 nm of each other. These amino acids were of interest because of their relative proximity and because their side chains invariably extended upward from the canyon floor. Only one of the constructions, in which the Pro-155 of VP1 was replaced with Tyr, failed to yield infectious virus and presumably represents a lethal mutation. Seven of the recovered mutants had small plaque phenotype and exhibited binding affinities significantly lower than wild-type virus. One mutant, in which Gly replaced Pro-155, showed a ninefold enhancement in binding affinity and changed its binding phenotype from a poor binding serotype to one of the strongest observed among HRVs. Despite an increased binding affinity, the Gly-155 mutant demonstrated retarded growth characteristics suggesting that this amino acid change may impede normal proteolytic processing or maturation (8a). The altered binding phenotypes displayed by eight of the nine viable mutants created by site-directed mutagenesis represent conclusive evidence that the canyon floor is involved in receptor interaction. Whether the observed binding affinities were the result of simple charge alterations or instead reflect gross anatomical changes in the floor of the canyon is not known. The region selected for amino acid substitutions is one that is highly conserved, not only among HRV serotypes but also among a number of other picornaviruses such as poliovirus and coxsackievirus B (13, 14, 20). It is highly unlikely that this particular region determines which cellular receptor protein a virus can utilize since these viruses clearly belong to different receptor families (6). Instead, it is more probable that this conserved region interacts with a common domain found on several receptor proteins and that it is the shape and contour of the canyon entrance and walls which define the accessibility of a receptor protein to the viral canyon and the stability of the virus-receptor interaction. Further studies will help to determine what other regions of the canyon are necessary for receptor interaction.

RHINOVIRUS BINDING IS A REVERSIBLE PROCESS

Previous studies on the binding of HRVs to cells have used HRV-2 as the prototype virus (18). The main conclusion from this work was that HRV-2 resembled other picornaviruses in its attachment to cellular receptors and subsequent processing (11). In general, most of the virions that become cell associated in physiological conditions elute spontaneously as noninfectious "A particles" which contain RNA but differ from complete viral particles in that they lack the smallest capsid protein, VP4 (11). Such A particles were believed to be intermediates in the capsid uncoating process during the initiation of viral infections. It followed that VP4 must have a particular role in the primary attachment of picornaviruses to cellular receptors as its removal from the capsid during spontaneous elution led to noninfectious A particles.

Studies were recently undertaken to determine whether the major HRV group of viruses behave in the same manner (2). The recent discovery that myristic acid is covalently bound to the N terminus of VP4 proteins (5, 19) enabled VP4 to be specifically radiolabeled and subsequently characterized for its role in receptor attachment. Two major group serotypes, HRV-14 and HRV-67, and one minor group virus, HRV-2, were selected for these studies. The HRVs were radiolabeled with either [^{35}S]methionine to label all four proteins or with [^{3}H]myristic acid which would only be incorporated into VP4. Viruses were allowed to attach to receptors on HeLa R-19 cell membranes and then were displaced by the 1A6 antibody or allowed to elute spontaneously at 20°C in phosphate-buffered saline (PBS). The results (Table 1) showed that under the conditions used, only 7 to 19% of the attached virions would spontaneously elute when incubated in PBS. This contrasted with up to

Table 1. Displacement of Radiolabeled Virus[a]

Virus	Radioactive precursor	Displacement (%)		Ratio, PFU/counts per minute	
		PBS	Antibody	Initial	Displaced
HRV-2	Myristate	19	17	ND[b]	
	Methionine	8	6	ND	
HRV-14	Myristate	7	61	ND	
	Methionine	7	82	47	31
HRV-67	Myristate	14	87	ND	
	Methionine	15	95	15	8

[a] Radiolabeled viruses were prepared as previously described (1, 19) and purified by equilibrium centrifugation in metrizamide density gradients (1). Metrizamide was removed from virus preparations by two centrifugations through Centicon-30 microconcentrators (Amicon Corp., Danvers, Mass.), using McCoy 5A medium as the diluent. After appropriate dilution in McCoy 5A medium containing 2.5% fetal calf serum, viruses were assayed in a HeLa cell membrane-binding assay as previously described (8) for 45 min at 20°C, and unbound virus was removed by centrifugation and washing with PBS. Membrane-virus complexes were then incubated for an additional 45 min with either PBS alone to measure spontaneous elution or with 1 μg of 1A6 antibody in a final volume of 50 μl. After incubation, membranes were pelleted, and the percentage of radiolabeled virus displaced was determined by scintillation counting. Relative infectivities of virions before binding and after displacement from membranes were determined by plaque assay (1), and PFU/counts per minute ratios were determined.
[b] ND, Not determined.

95% of the major group serotypes which were displaced by the 1A6 antibody. The specificity of the 1A6 antibody was again demonstrated by the failure of the antibody to displace the minor group serotype, HRV-2. Comparison of the displacement patterns obtained by using myristate-labeled virus versus methionine-labeled virus showed no significant difference in the proportion of virus displaced. These results indicated that, after displacement, VP4 elutes with displaced virions rather than remaining associated with membranes. The displaced [³H]myristate radioactivity comigrated with VP4 during electrophoresis and sedimented at the same rate as complete virus, indicating that VP4 remained an integral part of the viral particle after detachment from receptors (2). The finding that displaced particles had PFU/radioactivity ratios very similar to that of the initial virus (Table 1) demonstrated that the displaced virions were highly infectious and probably biochemically unchanged.

These results differ from those reported previously for HRV-2. The major group viruses used were able to bind reversibly to receptors at physiological temperatures and were biochemically and functionally complete after elution or displacement from receptors. No evidence was found for the existence of A particles. This result indicates that the interaction of the viral canyon with the cellular receptor protein is not sufficient to release the internal VP4 or the viral genome RNA. In addition, it is now apparent that the myristic acid component of the VP4 is not involved in the interaction between virus and cellular receptors.

Literature Cited

1. **Abraham, G., and R. J. Colonno.** 1984. Many rhinovirus serotypes share the same cellular receptor. *J. Virol.* **51**:340–345.
2. **Abraham, G., and R. J. Colonno.** 1988. Characterization of human rhinoviruses displaced by an anti-receptor monoclonal antibody. *J. Virol.* **62**:2300–2306.
3. **Callahan, P. L., S. Mizutani, and R. J. Colonno.** 1985. Molecular cloning and complete sequence determination of the RNA genome of human rhinovirus type 14. *Proc. Natl. Acad. Sci. USA* **82**:732–736.
4. **Campbell, B. A., and C. E. Cords.** 1983. Monoclonal antibodies that inhibit attachment of group B coxsackieviruses. *J. Virol.* **48**:561–564.
5. **Chow, M., J. F. E. Newman, D. Filman, J. M. Hogle, D. J. Rowlands, and F. Brown.** 1987. Myristylation of picornavirus capsid protein VP4 and its structural significance. *Nature* (London) **327**:482–486.
6. **Colonno, R. J.** 1987. Cell surface receptors for picornaviruses. *BioEssays* **5**:270–274.
7. **Colonno, R. J., P. L. Callahan, D. M. Leippe, R. R. Rueckert, and J. E. Tomassini.** 1989. Inhibition of rhinovirus attachment by neutralizing monoclonal antibodies and their Fab fragments. *J. Virol.* **63**:36–42.
8. **Colonno, R. J., P. L. Callahan, and W. J. Long.** 1986. Isolation of a monoclonal antibody that blocks attachment of the major group of human rhinoviruses. *J. Virol.* **57**:7–12.
8a. **Colonno, R. J., J. H. Condra, S. Mizutani, P. L. Callahan, M. E. Davies, and M. A. Murcko.** 1988. Evidence for the direct involvement of the rhinovirus canyon in receptor binding. *Proc. Natl. Acad. Sci. USA* **85**:5449–5453.
9. **Colonno, R. J., J. E. Tomassini, and P. L. Callahan.** 1987. Isolation and characterization of a monoclonal antibody which blocks attachment of human rhinoviruses, p. 93–102. *In* M. A. Brinton and R. R. Rueckert (ed.), *Positive Strand RNA Viruses.* Alan R. Liss, Inc., New York.
10. **Crowell, R. L., A. K. Field, W. A. Schleif, W. J. Long, R. J. Colonno, J. E. Mapoles, and E. A. Emini.** 1986. Monoclonal antibody that inhibits infection of HeLa and rhabdomyosarcoma cells by selected enteroviruses through receptor blockade. *J. Virol.* **57**:438–445.
11. **Crowell, R. L., and B. J. Landau.** 1983. Receptors in the initiation of picornavirus infections, p. 1–42. *In* H. Fraenkel-Conrat and R. R. Wagner (ed.), *Comprehensive Virology,* vol. 18. Plenum Publishing Corp., New York.
12. **Hamparian, V. V., R. J. Colonno, M. K. Cooney, E. C. Dick, J. M. Gwaltney, Jr., J. H. Hughes, W. S. Jordan, Jr., A. Z. Kapikian, W. J. Mogabgab, A. Monto, C. A. Phillips, R. R. Rueckert, J. H. Schieble, E. J. Stott, and D. A. J. Tyrrell.** 1987. A collaborative report: rhinoviruses—extension of the numbering system from 89 to 100. *Virology* **159**:191–192.
13. **Hogle, J. M., M. Chow, and D. J. Filman.** 1985. Three-dimensional structure of poliovirus at 2.9 Å resolution. *Science* **229**:1358–1365.
14. **Jenkins, O., J. D. Booth, P. D. Minor, and J. W. Almond.** 1987. The complete nucleotide sequence of coxsackievirus B4 and its comparison to other members of the picornaviridae. *J. Gen. Virol.* **68**:1835–1848.
15. **Minor, P. D., P. A. Pipkin, D. Hockley, G. C. Schild, and J. W. Almond.** 1984.

Monoclonal antibodies which block cellular receptors of poliovirus. *Virus Res.* **1**:203–212.

16. **Mizutani, S., and R. J. Colonno.** 1985. In vitro synthesis of an infectious RNA from cDNA clones of human rhinovirus type 14. *J. Virol.* **5**:628–632.

17. **Nobis, R., R. Zibirre, G. Meyer, J. Kuhne, G. Warnecke, and G. Koch.** 1985. Production of a monoclonal antibody against an epitope on HeLa cells that is the functional poliovirus binding site. *J. Gen. Virol.* **66**:2563–2569.

18. **Noble, J., and K. Lonberg-Holm.** 1973. Interactions of components of human rhinovirus type 2 with HeLa cells. *Virology* **51**:270–278.

19. **Paul, A. V., A. Shultz, S. E. Pincus, S. Oroszlan, and E. Wimmer.** 1987. Capsid protein VP4 of poliovirus is N-myristoylated. *Proc. Natl. Acad. Sci. USA* **84**:7827–7831.

20. **Rossmann, M. G., E. Arnold, J. W. Erickson, E. A. Frankenberger, J. P. Griffith, H.-J. Hecht, J. E. Johnson, G. Kamer, M. Luo, A. G. Mosser, R. R. Rueckert, B. Sherry, and G. Vriend.** 1985. Structure of a human common cold virus and functional relationship to other picornaviruses. *Nature* (London) **317**:145–153.

21. **Sherry, B., A. G. Mosser, R. J. Colonno, and R. R. Rueckert.** 1986. Use of monoclonal antibodies to identify four neutralization immunogens on a common cold picornavirus, human rhinovirus 14. *J. Virol.* **57**:246–257.

22. **Tomassini, J. E., and R. J. Colonno.** 1986. Isolation of a receptor protein involved in attachment of human rhinoviruses. *J. Virol.* **58**:290–295.

23. **Tomassini, J. E., T. R. Maxson, and R. J. Colonno.** 1989. Characterization of an acidic glycoprotein required for attachment of human rhinoviruses. *J. Biol. Chem.* **264**, in press.

Molecular Aspects of Picornavirus Infection and Detection
Edited by Bert L. Semler and Ellie Ehrenfeld
© 1989 American Society for Microbiology, Washington, DC 20006

Chapter 11

Towards a Molecular Vaccine for Foot-and-Mouth Disease

F. Brown

INTRODUCTION

Most present-day vaccines against virus diseases are either live attenuated or inactivated. The live attenuated vaccines have been derived from the virulent strains by growing them either in an unnatural host or in tissue culture cells until they no longer cause clinical disease but still grow well enough to evoke a protective immune response. Killed vaccines are produced by growing the virulent viruses in large amounts, either in animals or in tissue culture cells, and then inactivating them with a chemical agent under conditions which do not alter the antigenic structure of the proteins which elicit the immune response. With the exception of the genetically engineered hepatitis B vaccine, all the currently available vaccines are prepared in one or the other of these two ways. There is no doubt that they have been highly successful in controlling several diseases and, in the case of smallpox, actually eradicating the disease. Nevertheless, their empirical basis, the occasional side reaction (e.g., fever with some attenuated vaccines, anaphylactic shock with killed vaccines), and the ever-present problem of innocuity have encouraged several groups to investigate the possibility of producing vaccines in a more defined manner. This has been made possible by the information which has accumulated during the last quarter of a century on the molecular biology and structure of viruses. The impact of molecular biology on the under-

F. Brown • Department of Virology, Wellcome Biotech Ltd., Langley Court, Beckenham, Kent BR3 3BS, United Kingdom.

standing of biological processes in general has also been seen in virology, and the relatively simple structure of viruses compared with other self-replicating entities has allowed their functions to be studied readily at the molecular level. Consequently, there has been considerable progress in our understanding of the structures required to evoke a protective immune response. The ultimate goal in designing vaccines, however, will only be reached when the immune response itself is understood at the molecular level. Recent work by Berzofsky, Grey, and Rothbard, among others, is starting to provide this understanding. In this paper, the progress of these approaches with a synthetic vaccine against foot-and-mouth disease will be described.

THE PROBLEM OF VACCINATION AGAINST
FOOT-AND-MOUTH DISEASE

Foot-and-mouth disease is the most economically important disease of farm animals. Cattle, sheep, pigs, and goats are all affected, and the disease spreads rapidly. Losses in terms of production are estimated at approximately 25%. Moreover, indirect losses such as embargoes on trading can be devastating to those countries whose economy depends significantly on the export of farm animals and their products.

The disease occurs in many countries and is controlled either by slaughter or by vaccination. In countries such as Britain, Canada, the United States, Australia, New Zealand, and Japan, where the disease does not normally occur, the slaughter policy is applied. In those countries where the disease is endemic, vaccination is the accepted policy. Foot-and-mouth disease occurs worldwide, and the size of the problem means that more than 1 billion doses of vaccine are given each year. Properly applied, the current inactivated vaccines are effective, and the situation in Western Europe, where very few cases occur each year, is ample testimony to the effectiveness of the comprehensive vaccination programs which are undertaken there. The situation in other parts of the world is less encouraging, partly because of the more hostile terrain and weather conditions, making it difficult to preserve the potency of the vaccines until they are injected. Another serious problem, however, is the antigenic variation exhibited by the virus. Its occurrence as seven distinct serotypes, which are not cross-reactive, and a multiplicity of variants within each serotype adds greatly to the difficulties in providing effective vaccines. For this reason vaccine manufacturers keep a close surveillance on the antigenic relationship between field isolates and the viruses being used for preparing the vaccines.

CURRENT VACCINES: ATTRIBUTES AND DEFICIENCIES

The vaccines used at present are prepared by growing the virus in large amounts in a variety of cells. The most favored cells are either calf or pig kidney primary cells or the BHK-21 cell line, but substantial amounts of virus for vaccine production are still grown in bovine tongue epithelial fragments, a method, described by Frenkel in 1947 (8), which incidentally was the first application of large-scale culture techniques for growing a virus.

The virus was traditionally inactivated with formaldehyde, and indeed this reagent is still used by some manufacturers. However, the superior inactivation kinetics when the imines are used has resulted in the gradual change in most cases to the use of these reagents. The inactivated virus is then adsorbed onto aluminum hydroxide gel and inoculated with saponin as an adjuvant. In some instances an oil adjuvant is used instead of the aqueous adjuvant.

An assessment of the potency of the vaccines can be obtained from the level of the virus-neutralizing antibody in the serum 28 days after inoculation. If the serum diluted to about 1/45 will neutralize 100 50% infective doses of the homologous virus, the animal will be protected against intradermal challenge, directly into the tongue, with 10,000 50% infective doses of virus. However, regulatory bodies normally insist on a challenge test 28 days after inoculation of the vaccine.

The large-scale production of the vaccines has reached the stage where these stringent requirements are met with a low-priced product. It is thus a fair question to ask why efforts should be made to find alternative methods for producing them. The list of disadvantages with the current vaccines is not long, but some are important.

1. The virus must be produced in a high-containment facility to prevent its escape. The handling of the virus also means the imposition of restrictions on the movement of personnel working with the virus.
2. The innocuity of the product must be ensured. Several recent cases of the disease have been traced to contaminated vaccine.
3. The product must be inoculated at regular intervals to ensure protection. In some countries this can be as frequent as three times each year.
4. A cold chain is essential to maintain the potency of the product. This is frequently difficult to achieve.
5. Anaphylactic shock is an infrequent problem but nevertheless

occurs on a sufficient number of occasions to pose a problem in terms of litigation.

These problems are sufficient incentives to justify the search for alternative methods of immunization. The next section deals with the approaches which have been made during the past 3 decades as the methods of molecular biology have been applied to the problems.

THE MOLECULAR BASIS FOR VACCINATION

The general structural features of the virus which are required to ensure a protective immune response have now been determined. With the development of methods for studying nucleic acids and proteins at the most fundamental level, the problem of vaccination can now also be studied at the molecular level. Moreover, the exponential growth in our knowledge of the immune response, taking in the recent advances at the molecular level, should lead to a major change in our attitudes towards the problems of antigen presentation.

Structure of the Virus and Its Relevance to the Immune Response

In common with other members of the family of *Picornaviridae*, foot-and-mouth disease virus (FMDV) consists of one molecule of single-stranded RNA and 60 copies of each of four proteins, VP1 through VP4. There are two distinctive properties of FMDV which are particularly important when considering its immunogenic activity: (i) its extreme vulnerability to environments below pH 7, when it is disrupted into its infectious RNA, an aggregate of VP4, and a pentameric unit comprising VP1 to VP3 (the disrupted virus has very little immunogenic activity); and (ii) its vulnerability to proteolytic cleavage, resulting in a dramatic loss of infectivity due to impairment of cell attachment and a considerable decrease in immunogenicity with several strains (18). The only measurable physical alteration is the cleavage of protein VP1.

The disruption of the virus particle below pH 7 results in a dramatic fall in infectivity (>5 log), and the immunogenic activity is reduced by about 100-fold. Cleavage of VP1 by trypsin pointed to the major importance of this protein in the protective immune response and led to the anticipation that immunity could be achieved with this protein alone. Indeed, in 1973 Laporte and his colleagues (12) showed that the separated protein induced neutralizing antibody in pigs although there was no response to VP2 and VP3. The amount of VP1 required to elicit a protective level of response, however, was much greater than that found with intact virus particles. Thus, about 500 μg of VP1 was required, as two inoculations, to achieve protective levels of neutralizing antibody, compared with about 10 μg of virus particles as one inoculation.

The large difference in response could be ascribed to the difference in configuration of the VP1 when it was released from the constraints imposed by the architecture of the virus particle. A similar argument would account for the low activity of the 12S particle. However, there remains the possibility that the immunogenic site is a complex of the surface proteins which is disrupted when the virus is converted to 12S particles or VP1 is cleaved in situ with trypsin. Mapping of neutralizing monoclonal antibodies indicates that more than one site is involved (20), but the interpretation of these results awaits the three-dimensional structure of the particle.

LOCALIZATION OF IMMUNOGENIC EPITOPES ON VP1

The apparent immunodominance of VP1 focused attention on this protein and led to studies to determine the location of immunogenic epitopes. The approach used by Anderer (1) with tobacco mosaic virus and by Sela and his colleagues (11) with the RNA-containing MS2 bacteriophage was applied to the FMDV protein by Strohmaier and his colleagues (10, 17). They found that fragments of VP1 produced either by cleavage of the separated protein with cyanogen bromide or by hydrolysis of the protein in situ with proteolytic enzymes would elicit the formation of neutralizing antibody. When the amino acid sequence of VP1 became available, they were able to assign the positions of these fragments on the sequence and by a process of elimination predict those sequences which would have immunogenic activity (Fig. 1). In a second approach, Bittle and his colleagues (2) reasoned that since antigenic variation would be determined by amino acid sequence variation, comparison of sequences of viruses belonging to different serotypes would pinpoint potentially important sites. Three regions of sequence variability, at positions 41-60, 138-160, and 194-205, were found, the latter two of which correspond to those predicted by Strohmaier and his colleagues.

Bittle et al. (2) tested the immunogenic activity of 20-mer peptides encompassing the entire sequence and found that one inoculation of the 141-160 sequence elicited levels of neutralizing antibody which would protect guinea pigs against infection. The German group (15) also found that this region of VP1 would elicit neutralizing antibody.

Supportive evidence that this region contains the immunodominant site was provided by the observation that naturally occurring antigenic variants of a virus belonging to serotype A, subtype 12, differed only at positions 148 and 153 in their capsid proteins (16). These variants could be

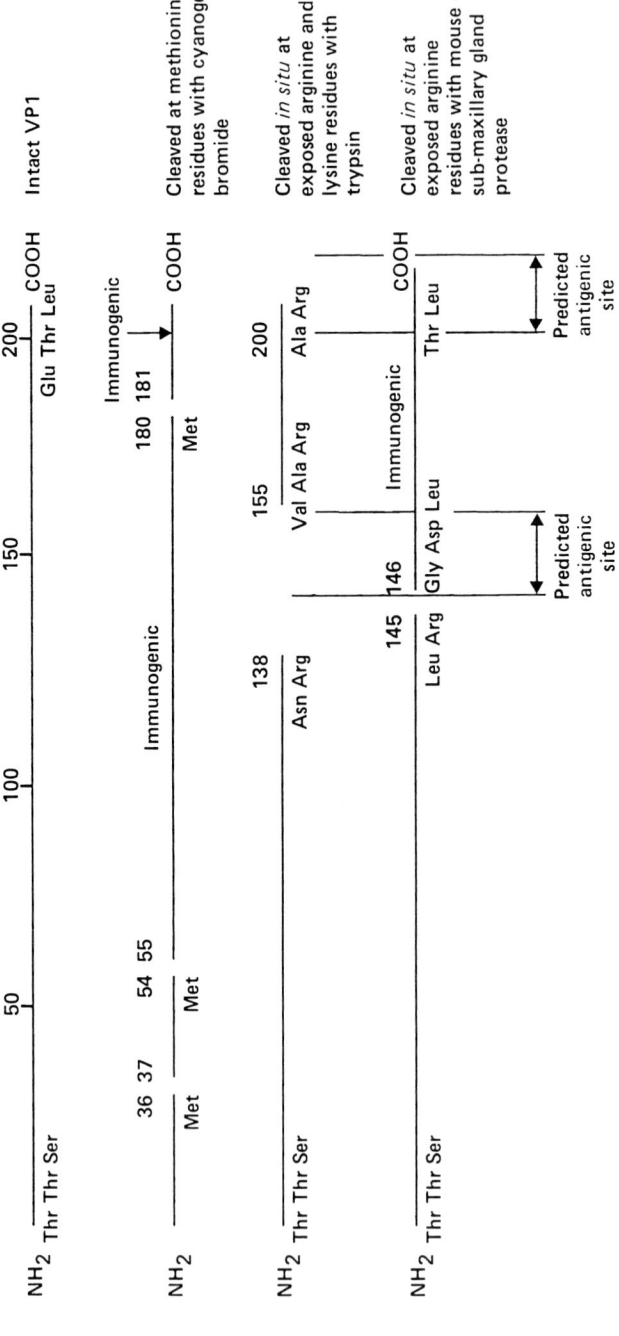

Figure 1. Location of immunogenic sites on VP1 of FMDV (adapted from reference 17).

distinguished readily by cross-neutralization tests. Moreover, antisera produced by injection of the corresponding 141-160 peptides distinguished between the variants in cross-neutralization tests in the same way as the antivirion sera.

These observations inevitably led to the conclusion that a vaccine could be produced with the 141-160 sequence as its basis. The next section reviews the results obtained with this peptide.

IMMUNE RESPONSE TO THE 141-160 PEPTIDE OF VP1

The first experiments with the peptide were made with it coupled to the highly antigenic protein, keyhole limpet hemocyanin. Neutralizing antibody responses which protected guinea pigs against infection with 1,000 50% infective doses of virus could be obtained with a single inoculation. When Freund incomplete adjuvant was used, as little as 30 μg of virus afforded protection, but the response was lower when aluminum hydroxide was used as adjuvant. The response in cattle and pigs was not as good as in guinea pigs, and in general a single inoculation did not afford protection. These results led us to conclude that the way in which the peptide was presented to the host could be crucial, so several different methods have been investigated.

Attachment to the N Terminus of β-Galactosidase

The ordered coupling of peptides to proteins has not been investigated in any detailed manner. The literature reveals that several methods have been used, ranging from the crude uncontrolled linking of the two with glutaraldehyde to the more specific linking via an added cysteine at one of the termini. We had found that the peptide alone was immunogenic in guinea pigs and that coupling it to different carrier protein molecules did not lead to an enhanced response (6). Indeed, in some instances we found that the response was lower. Consequently we have investigated whether a more defined attachment of the peptide to a carrier protein would lead to better antibody responses.

In our first experiments, the peptide sequence was fused to β-galactosidase at its N terminus by expressing the gene coding for the peptide ligated to the gene coding for β-galactosidase in *Escherichia coli* cells. This construction has the potential added advantage that the antigenic sites on β-galactosidase which are recognized by helper T cells are known. However, Winther and his colleagues (19) showed that this fusion protein was no more immunogenic for mice or guinea pigs than the chemically linked peptide. Broekhuijsen et al. (3) have also reported similar results.

Table 1. Antipeptide Antibody, Neutralizing Antibody, and Protective Immune Response of Guinea Pigs Inoculated with Fusion Proteins Containing One, Two, or Four Copies of the FMDV Peptide

Peptide sample	Dose (μg of peptide)	Antipeptide antibody ($\log_{10}$) at:		Neutralizing antibody ($\log_{10}$) at:		Protection[a] at 56 days
		28 days	56 days	28 days	56 days	
One copy	5	1.2	1.8	<0.6	<0.6	NP
Two copies	10	3.8	3.9	2.0	1.6	P
Four copies	20	3.3	3.5	1.5	1.6	P
Synthetic peptide 137–160 Cys	5			<0.6	<0.6	NP

[a] P, Protected; NP, not protected.

However, we have shown in joint experiments with the Dutch group (4) that proteins consisting of two or four copies of the peptide linked to β-galactosidase at the N terminus have much greater immunogenic activity than the single-copy construct (Table 1). Moreover, 40 μg of the peptide in the four-copy construct, given as a single inoculation, protected pigs against challenge infection with 60,000 50% infective doses of virus.

The reason for the much greater activity of the two- and four-copy constructs is not known. It could be the repetition of the antigenic determinant which is important, and a detailed structural study to compare the configuration of the peptide sequence on the one-, two-, and four-residue constructions could be rewarding.

Expression as Part of the Hepatitis B Virus Core Particle

The response to the peptide sequence when it forms part of the virus particle is very high. As little as 1 μg of virus, which contains 0.02 μg of the 141-160 sequence on the 60 copies of VP1, will provide protection against challenge infection. Consequently we have investigated whether multiple copies of the peptide presented on a particle of the same size as a picornavirus will evoke a similar high response (5). The core protein of hepatitis B virus self-assembles into particles which are 27 nm in diameter and contain several hundred copies of the molecule. Moreover, hybrid proteins in which foreign amino acid sequences are expressed at the N terminus of the hepatitis B virus core protein will also self-assemble into particles (P. E. Highfield, personal communication). On the basis of this information, the DNA coding for the FMDV peptide, six amino acids of the precore particle, and the entire core protein has been expressed (5). Although this construct poisoned the *E. coli* system we were using, it was found that it could be expressed in vaccinia virus. Moreover, the expressed hybrid protein assembled into particles which reacted with antisera prepared against hepatitis B core particles, the FMDV peptide, and, crucially, the FMDV particle. The particles were separated from the vaccinia virus and tested for their immunogenic activity in guinea pigs. One injection of as little as 2 μg of the particles, containing the equivalent of 0.2 μg of peptide, elicited very high neutralizing antibody levels which protected against challenge infection.

The excellent response to the peptide when it is expressed in this form may be explained in purely physical terms. However, other factors may be playing a role in the enhanced response, since Milich and McLachlan (14) have shown that the hepatitis B virus core protein can induce antibody responses via both T-cell-dependent and T-cell-independent pathways. Moreover, the core-specific helper-T cells can help B cells produce antibody against the envelope antigens of hepatitis B virus as well as the core proteins, even though these antigens are on different molecules. These immunological properties may account for the excellent response to the FMDV peptide when it is presented as part of the core particle. It is clear that the exploration of this method of presentation may provide important clues for the practical application of peptide vaccines.

IS THE 141-160 SEQUENCE ALONE SUFFICIENT FOR PROTECTION?

Although the 141-160 peptide region of VP1 elicits a neutralizing antibody response which protects guinea pigs against experimental infection, the response in pigs and cattle is lower despite the fact that the antibodies which are elicited are qualitatively similar in all three species.

To realize their full potential as vaccines, it is necessary for the peptides to combine with helper-T-cell receptors and Ia antigens in addition to antibody-binding sites. To study whether the restriction of the immune response in cattle could be overcome, we have investigated in preliminary experiments whether such nonresponsiveness can be overcome in a laboratory animal model (7). It has been shown recently that nonimmunogenic B-cell epitopes of the malaria circumsporozoite (9) and hepatitis B virus surface antigen (13) can be overcome either by conjugation with a ''natural'' T-cell epitope or even by copolymerization with a ''foreign'' T-cell epitope. We found that only two of six strains of congenic B10 mice, differing only at the locus of the *H-2* complex, responded to the free peptide in incomplete Freund adjuvant. These were the $H-2^k$ and $H-2^r$ strains. Among the nonresponding mice was the B10.D2 strain, which belongs to the $H-2^d$ haplotype and for which several T-helper-cell epitopes have been described. Three of the epitopes, one from ovalbumin (amino acids 323 to 339 [OVA]) and two from sperm whale myoglobin (amino acids 132 to 148 [SWMI] and amino acids 105 to 121 [SWMII]), were added to the C terminus of the 141-160 FMDV peptide. As a control, a fourth peptide consisting of the 141-160 peptide and the 161-177 sequence was used. Each peptide also had a nonnatural cysteine residue added to the C terminus for increased immunogenicity.

Neither the B10.D2 nor the BALB/c strain of mice (both $H-2^d$ haplotype) produced any antibody response to the 141-160 + 161-177 peptide, but each of the 141-160 + OVA, 141-160 + SWMI, and 141-160 + SWMII peptides elicited a good response (Fig. 2). Unexpectedly, the 141-160 + SWMII peptide did not elicit any neutralizing antibody in either strain after one injection, although the BALB/c mice produced neutralizing antibody after two injections. This suggests that T-helper-cell epitopes could control antibody production of specific B-cell clones.

These results provide further evidence that genetically controlled nonresponsiveness to the 141-160 peptide can be overcome by adding foreign T-cell epitopes and that functional (i.e., neutralizing) antibody can be produced. The significance of this result for the vaccination of cattle with the 141-160 peptide is that the poor response to the uncoupled sequence may be overcome by presenting it with a suitable T-helper-cell epitope.

THE POTENTIAL FOR A PEPTIDE VACCINE

The results presented here show that two important problems regarding the presentation of the FMDV peptide can be solved. The first was the relatively poor immune response to the peptide, whether uncoupled or coupled to a carrier protein, compared with the response to the virus particle. By presenting the peptide either as a repeat unit attached to the N terminus of β-galactosidase or as a single copy attached to the N terminus of the hepatitis B virus core protein, greatly enhanced responses are obtained. With the hepatitis B virus core protein construct, antibody responses approaching those obtained with the virus particle have been demonstrated.

The second problem was the genetic restriction of the immune response in different species. In preliminary experiments in mice belonging to different haplotypes, we have shown that the failure of $H-2^d$ mice

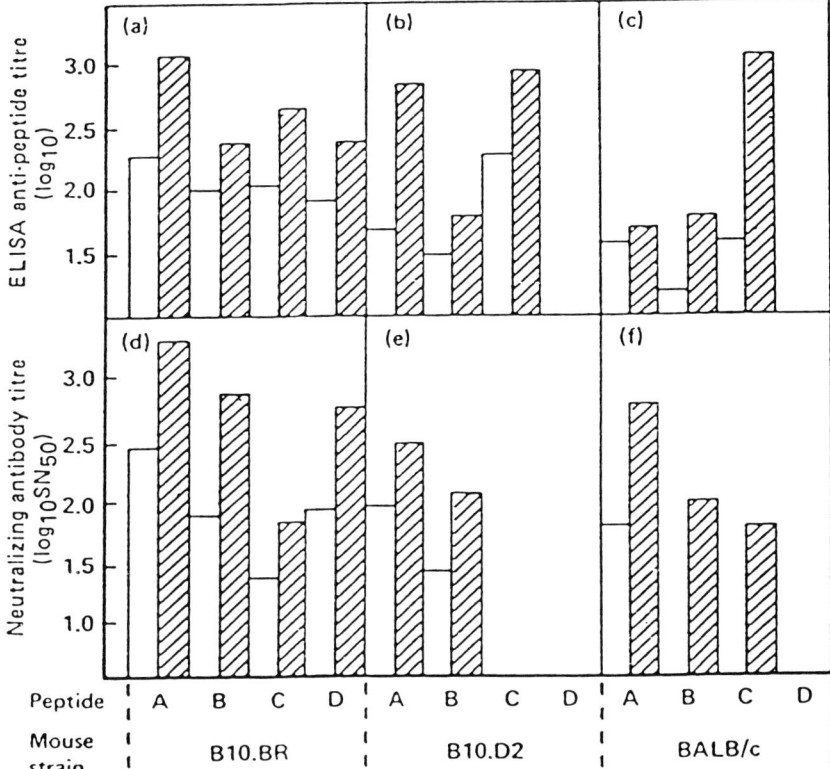

Figure 2. Antipeptide and neutralizing antibody responses of mice belonging to the $H\text{-}2^k$ (B10.BR) and $H\text{-}2^d$ (B10.D2 and BALB/c) haplotypes to the 141-160 peptide of FMDV linked at its C terminus to (A) OVA, (B) SWMI, (C) SWMII, or (D) 161-177 from FMDV VP1 and inoculated with Freund incomplete adjuvant. The open columns give the values for sera collected 28 days after a single inoculation, and the hatched columns give the values 28 days after a second inoculation 63 days after the first. ELISA, Enzyme-linked immunosorbent assay.

to respond to the FMDV peptide can be overcome by covalently linking it to foreign T-cell epitopes from ovalbumin and sperm whale myoglobin. These results are important in considerations of the immune response to peptide vaccines in general.

Literature Cited

1. **Anderer, F. A.** 1963. Versuche zur Bestimmung der serologisch terminaten Gruppen des Tobakmosaikvirus. *Z. Naturforsch. Teil B* **188**:1010–1014.
2. **Bittle, J. L., R. A. Houghten, H. Alexander, T. M. Shinnick, J. G. Sutcliffe, R. A. Lerner, D. J. Rowlands, and F. Brown.** 1982. Protection against foot-and-mouth disease by

immunization with a chemically synthesized peptide predicted from the viral nucleotide sequence. *Nature* (London) **298**:30–33.

3. **Broekhuijsen, M. P., T. Blom, M. Kottenhagen, P. H. Pouwels, R. H. Meloen, S. J. Barteling, and B. E. Enger-Valk.** 1986. Synthesis of fusion proteins containing antigenic determinants of foot-and-mouth disease virus. *Vaccine* **4**:119–124.

4. **Broekhuijsen, M. P., J. M. M. Van Rijn, A. J. M. Blom, P. H. Pouwels, B. E. Enger-Valk, F. Brown, and M. J. Francis.** 1987. Fusion proteins with multiple copies of the major antigenic determinant of foot-and-mouth disease virus protect both the natural host and laboratory animals. *J. Gen. Virol.* **68**:3137–3143.

5. **Clarke, B. E., S. E. Newton, A. R. Carroll, M. J. Francis, G. Appleyard, A. D. Syred, P. E. Highfield, D. J. Rowlands, and F. Brown.** 1987. Improved immunogenicity of a peptide epitope after fusion to hepatitis B core protein. *Nature* (London) **330**:381–384.

6. **Francis, M. J., C. M. Fry, D. J. Rowlands, J. L. Bittle, R. A. Houghten, R. A. Lerner, and F. Brown.** 1987. Immune response to uncoupled peptides of foot-and-mouth disease virus. *Immunology* **61**:1–6.

7. **Francis, M. J., G. Z. Hastings, A. D. Syred, B. McGinn, F. Brown, and D. J. Rowlands.** 1987. Non-responsiveness to a foot-and-mouth disease virus peptide overcome by addition of foreign helper T-cell determinants. *Nature* (London) **330**:168–170.

8. **Frenkel, H. S.** 1947. La culture du virus de la fièvre aphteuse sur l'epithelium de la langue des bovides. *Bull. Off. Int. Epizoot.* **82**:155–162.

9. **Good, M. F., W. K. Maloy, M. N. Lunde, H. Margalit, J. L. Cornette, G. L. Smith, B. Moss, L. H. Miller, and J. A. Berzofsky.** 1987. Construction of synthetic immunogen: use of new T-helper epitope on malaria circumsporozoite protein. *Science* **235**:1059–1062.

10. **Kaaden, O. R., K. H. Adam, and K. Strohmaier.** 1977. Induction of neutralizing antibodies and immunity in vaccinated guinea-pigs by cyanogen bromide peptides of VP3 in foot-and-mouth disease virus. *J. Gen. Virol.* **34**:397–400.

11. **Langebeheim, H., R. Arnon, and M. Sela.** 1976. Antiviral effect on MS2 coliphage obtained with a synthetic antigen. *Proc. Natl. Acad. Sci. USA* **73**:4636–4640.

12. **Laporte, J., J. Grosclaude, J. Wantyghem, S. Bernard, and P. Rouze.** 1973. Neutralisation en culture cellulaire du pouvoir infectieux du virus de la fièvre aphteuse par des serums provenant de porcs immunisés à l'aide d'une protéine virale purifiée. *C. R. Acad. Sci.* **276**:3399–3401.

13. **Leclerc, C., G. Przewlocki, M. Schutze, and L. Chedid.** 1987. A synthetic vaccine constructed by copolymerization of B and T cell determinants. *Eur. J. Immunol.* **17**:269–273.

14. **Milich, D. R., and A. McLachlan.** 1986. The nucleocapsid of hepatitis B virus in both a T-cell-independent and a T-cell-dependent antigen. *Science* **234**:1398–1401.

15. **Pfaff, E., M. Mussgay, H. O. Bohm, G. E. Schulze, and H. Schaller.** 1982. Antibodies against a preselected peptide recognize and neutralize foot-and-mouth disease virus. *EMBO J.* **1**:869–874.

16. **Rowlands, D. J., D. V. Sanger, and F. Brown.** 1972. Stabilizing the immunising antigen of foot-and-mouth disease virus by fixation with formaldehyde. *Arch. Gesamte Virusforsch.* **39**:274–283.

17. **Strohmaier, K., R. Franze, and K.-H. Adam.** 1982. Localisation and characterisation of the antigenic portion of the foot-and-mouth disease virus protein. *J. Gen. Virol.* **59**:295–306.

18. **Wild, T. F., J. N. Burroughs, and F. Brown.** 1969. Surface structure of foot-and-mouth disease virus. *J. Gen. Virol.* **4**:313–320.

19. **Winther, M. D., G. Allen, R. H. Bomford, and F. Brown.** 1986. Bacterially expressed antigenic peptide from foot-and-mouth disease virus capsid elicits variable immunologic responses in animals. *J. Immunol.* **136:**1835–1840.

20. **Xie, Q.-C., D. McCahon, J. R. Crowther, G. J. Belsham, and K. C. McCullough.** 1987. Neutralization of foot-and-mouth disease can be mediated through any of at least three separate antigenic sites. *J. Gen. Virol.* **68:**1637–1647.

Molecular Aspects of Picornavirus Infection and Detection
Edited by Bert L. Semler and Ellie Ehrenfeld
© 1989 American Society for Microbiology, Washington, DC 20006

Chapter 12

Antigenic Structure of Hepatitis A Virus

Stanley M. Lemon and Li-Hua Ping

Introduction

Hepatitis A virus (HAV) is a hepatotropic picornavirus with a host range restricted to certain primates (20). It is endemic throughout much of the world, but particularly prevalent in regions with poor sanitation. Although infection with the virus is usually asymptomatic in very young children, most adult infections are accompanied by jaundice and overt hepatitis. HAV-related illness may be prolonged and is occasionally fatal. At present, passive immunization with pooled human serum globulin represents the only available means of prevention of hepatitis A. Intervention with immune globulin requires timely recognition of exposure, however, and provides protection for only several months. Since the successful adaptation of HAV to growth in cell culture (34), extensive efforts have been expended on the development of both inactivated and attenuated whole-virus vaccines (2, 33). It remains uncertain, however, whether virus can be propagated in vitro in yields sufficient for economical production of an inactivated vaccine. Furthermore, although HAV may become attenuated during in vitro passage, no specific markers of attenuation have yet been described, and an acceptably attenuated variant of HAV which retains sufficient infectivity for humans has not yet been identified. For these reasons, synthetic peptide vaccines or polypeptide immunogens expressed from recombinant DNA represent potentially attractive alternative approaches to HAV immunization.

Stanley M. Lemon and Li-Hua Ping • Division of Infectious Diseases, Department of Medicine, University of North Carolina at Chapel Hill, Chapel Hill, North Carolina 27599-7030.

Current understanding of the antigenic structure of HAV is rudimentary. Attempts to characterize the immunogenic sites of HAV at the molecular level have been impeded by the slow and generally noncytopathic growth of HAV in cell culture and by difficulties in obtaining quantities of virus sufficient for crystallographic studies. Nonetheless, progress has been made. This review will focus on the antigenic structure of HAV, a firm knowledge of which will be essential for the future success of novel approaches to HAV vaccine development.

PHYSICAL AND BIOLOGICAL PROPERTIES OF HAV

The HAV virion is roughly spherical, approximately 27 nm in diameter, and has multiple features that support its classification among the picornaviruses (8). The virus may be propagated in a variety of primate cell types in vitro, but generally grows slowly, remains noncytopathic, and is largely cell associated (3, 11, 34). More rapidly replicating variants displaying a cytopathic effect in vitro have been described recently, however (1, 10, 41).

Infectious particles sediment at about 160S and band isopycnically in CsCl at 1.32 to 1.33 g/ml (26, 36). Dense particles banding at 1.40 to 1.44 g/ml, typical of picornaviruses, and chloroform-resistant, infectious light particles banding at approximately 1.27 g/ml have also been described (26). Virion RNA (35S) is plus sense, single stranded, 7,478 nucleotides in length, and polyadenylated at the 3' end (7, 9, 30). It is linked to a small peptide, VPg, at the 5' end (42). Partial cDNA transcripts of various HAV strains have been molecularly cloned, and the complete nucleotide sequences of three virus strains have been reported (7, 30, 32, 40). A complete genomic cDNA construct and RNA transcribed from it have recently been shown to be infectious in vitro (6).

The nucleotide sequence of HAV suggests that the genomic organization of the virus resembles that of other picornaviruses. Consistent with these data, at least three major structural (capsid) polypeptides have been identified: VP1 (300 amino acids in length), VP2 (222 amino acids), and VP3 (246 amino acids) (17, 43). A fourth, considerably smaller polypeptide (VP4, 23 amino acids) is suggested by the genomic sequence, but has not been conclusively identified as present in the virion. If VP4 of HAV, like VP4 of many other picornaviruses, is N-terminus myristylated at a postulated consensus myristylation site (4), it may be only 17 amino acids in length and may be preceded in the polyprotein by a short leader peptide. Coincident with these unique structural features, HAV appears to be exceptionally stable at high temperatures. Held at 80°C for 10 min in the presence of 1 M $MgCl_2$, the infectious titer of HAV drops by only 1.5 $\log_{10}$ (37).

HAV is antigenically distinct from other known picornaviruses, and HAV capsid proteins show essentially no sequence homology with poliovirus or other picornaviruses at either the nucleotide or amino acid level (7, 27, 30, 32). The amino acid sequences of the capsid proteins of HAV have been aligned with those of other picornaviruses, however, based on similarities in amino acid sequences, the crystallographic structures of poliovirus, rhinovirus, and mengovirus, and available hydrogen-bonding maps (A. Palmenberg, personal communication). These alignments suggest the absence in HAV of portions of immunogenic loop structures in VP1 (residues 216 to 232 in type 1 poliovirus) and VP2 (residues 130 to 181 in poliovirus). Assuming these alignments to be correct, these structural differences may explain in part why HAV is more thermostable than poliovirus. Because this analysis also suggests the absence in HAV of amino acid residues comprising neutralization immunogenic sites in poliovirus (15, 29) and rhinovirus (35), it raises the possibility that HAV may have a less complex antigenic structure than either of these other picornaviruses.

ASSAY METHODS FOR NEUTRALIZING ANTIBODY TO HAV

Because most cell culture-adapted isolates of HAV are noncytopathic, conventional plaque assays have limited application to the study of HAV. Accordingly, an indirect plaque assay has been developed (radioimmunofocus assay) in which foci of viral replication ("radioimmunofoci") are detected by autoradiography following removal of the agarose overlay, acetone fixation of the cell sheet, and staining with ^{125}I-labeled antibody (24). This technique remains the only quantal assay available for all cell culture-adapted strains of HAV. Neutralizing antibody to HAV may be measured by inhibition of radioimmunofocus formation following exposure of virus to antibody (21). This assay is approximately 100-fold more sensitive than competition radioimmunoassays, which are widely used for detection of antibody to the virus (38). As serum neutralizing antibody titers measured by this method are in the range of 1:10 to 1:40 2 months after administration of protective doses of pooled human immune globulin, it is reasonable to assume that such titers represent protective antibody levels in humans. Other methods have been developed for measurement of neutralizing antibody, including inhibition of fluorescent focus formation and antigen production in cells inoculated with virus-antibody mixtures. More recently, the selection of rapidly growing HAV variants yielding a demonstrable cytopathic effect (1, 10,

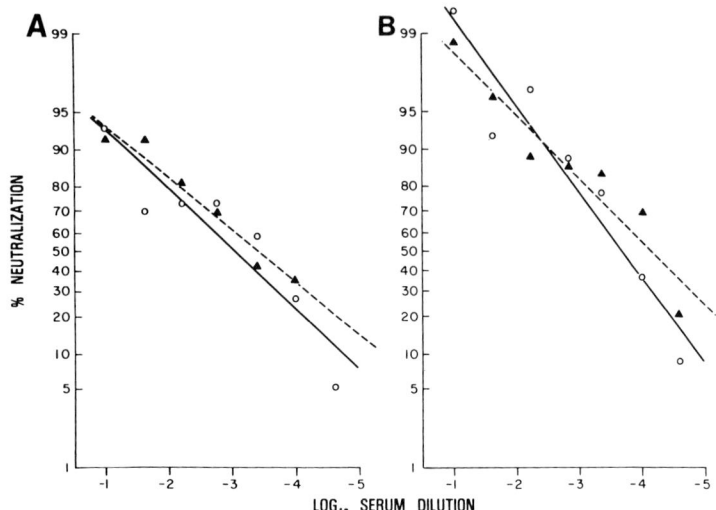

Figure 1. Cross-neutralization of HM175 (A) and PA21 (B) strains of HAV with 10-fold dilutions of postinfection sera collected from primates infected with HM175 (○) or PA21 (▲) virus. Each point represents the mean of neutralization values determined in replicate cultures by the radioimmunofocus inhibition method. Best-fit lines were obtained by linear regression analysis of the logistic function of the percent viral neutralization obtained with HM175 (—) and PA33 (– – –) immune sera. The slopes of these best-fit lines were not statistically different ($P > 0.05$). Reprinted from reference 22.

41) offers the promise of conventional plaque-reduction assays for detection of neutralizing antibody.

ANTIGENIC CONSERVATION AMONG HAV STRAINS

Despite a longstanding impression that significant antigenic variation does not exist among different strains of HAV, relatively few virus strains have been studied in detail. Only two isolates of HAV have been compared by classical polyclonal antibody cross-neutralization methods (Fig. 1) (22). This approach suggested no significant differences in the neutralization antigen(s) of the HM175 and PA21 virus strains, although HM175 virus was recovered from a human in Australia and PA21 virus from a naturally infected New World owl monkey (*Aotus trivirgatus*) in Panama. In support of these findings, a high level of cross-protection was documented in owl monkeys serially challenged with HM175 and PA33 (epizootiologically related to PA21) viruses by intravenous inoculation (25). Nonetheless, despite the absence of significant antigenic variation between the PA21 and HM175 strains of HAV, it has become evident recently that there are marked differences in the RNA sequence in the P1 region of these viruses (25). cDNA probes prepared from the capsid-encoding region of HM175 virus do not hybridize under conditions of normal stringency with RNA from PA21 virus, leading us to suspect that PA21 may be a distantly related virus endogenous to the owl monkey and genetically distinct from human HAV. These data suggest that the biological constraints acting to conserve the antigenic structure of HAV, although not yet defined, are exceptionally strong.

AMINO ACID SEQUENCE VARIABILITY AMONG CAPSID POLYPEPTIDES

The amino acid sequences of the capsid polypeptides of HAV are well conserved among those strains of HAV for which data are available (7, 27, 30, 32). Each of the three major capsid polypeptides from several different strains demonstrates >97% amino acid sequence homology (see J. Ticehurst, J. I. Cohen, S. M. Finestone, R. H. Purcell, R. W. Jansen, and S. M. Lemon, this volume). The very close relatedness of most HAV strains to each other stands in marked contrast to their distance from other picornaviruses phylogenetically (as suggested by differences in nucleotide sequences of viral RNA) and is surprising given the greater variability evident among other picornaviruses (Palmenberg, personal communication).

ANTIGENICITY AND IMMUNOGENICITY OF HAV CAPSID POLYPEPTIDES

Considerable evidence suggests that the immunodominant neutralization sites of HAV are assembled and not sequential immunogens (14). Western blots of denatured HAV capsid proteins separated by sodium dodecyl sulfate-polyacrylamide gel electrophoresis do not stain with neutralizing monoclonal antibodies (17). Moreover, as with most other picornaviruses, purified HAV capsid polypeptides are weak immunogens with respect to their abilities to elicit antibodies reactive with the native capsid. Nonetheless, low levels of antibodies capable of neutralizing virus and immunoprecipitating whole virus were produced in rats immunized with multiple doses of individual purified capsid polypeptides (VP1, VP2, and VP3) (16). Antibody raised to VP1 demonstrated the strongest response when tested in competitive inhibition assays against polyclonal anti-HAV.

Although there is little relatedness evident in the amino acid sequences of the VP1s of type 1 poliovirus and HAV, the surface probability profiles of the N-terminal halves of these two polypeptides (aligned on the basis of minimal similarities in amino acid sequences) reveal common features which suggest that certain higher-ordered structures may be conserved (13, 30). Specifically, this type of analysis suggest that the immunogenic "C3 loop" of poliovirus (VP1 residues 89 to 103) (15, 29, 44) may be conserved as a surface structure in HAV. Furthermore, a synthetic peptide representing residues 11 to 25 in the HAV VP1 sequence, selected because it corresponded in the surface probability profile to an immunogenic region in the poliovirus VP1 sequence, has been reported to elicit a very low titer of neutralizing antibody in rabbits (13). In poliovirus, the corresponding VP1 residues are not present on the surface of the virion (15), although they may be externalized during capsid structural changes accompanying attachment of virus to cellular receptors. These HAV VP1 residues are therefore unlikely to constitute a major immunogenic site. Other laboratories have examined the immunogenicity of synthetic peptides corresponding to regions in the capsid proteins of HAV, but none have reported favorable results to date.

    With respect to proteins expressed from recombinant DNA, the N-terminal 60 amino acids of HAV VP1 have been reported to be antigenic when expressed as a fusion protein with *Escherichia coli* β-galactosidase (31). However, it is not clear whether the antigenic activity obtained was that of the native virus or of denatured VP1. Similarly, expression of the entire VP1 polypeptide along with the carboxy-terminal 46 amino acids of VP3 as a TrpE fusion protein in *E. coli* has resulted in an immunogenic 88-kilodalton protein capable of eliciting antibodies reactive with denatured VP1 but not intact virus (19). This fusion protein was not capable of inducing a neutralizing antibody response in immunized rabbits, although it could prime for a neutralizing antibody response to normally subimmunogenic doses of intact virus and therefore probably contains relevant T-cell epitopes.

MONOCLONAL ANTIBODY MAPPING OF HAV NEUTRALIZATION EPITOPES

    Murine monoclonal antibodies to HAV have been developed in a number of laboratories (12, 17, 28). We have evaluated 13 different monoclonal antibodies obtained from five different sources (Table 1). Each of the monoclonal antibodies listed in Table 1 is capable of neutralizing HM175 strain HAV, even though most were raised against epidemiologically unrelated strains of virus. However, the magnitude of the reduction in titer of HAV following exposure of virus to high antibody concentrations varies among the different monoclonal antibodies, with

Table 1. Monoclonal Antibodies to HAV[a]

Monoclonal antibody	Immunizing virus strain	Source or reference
K2-4F2	HM790	(28)
K3-4C8	HM790	(28)
K3-2F2	HM790	(28)
B5-B3	KMW-1	R. Tedder
6A5	CR326	(17)
1B9	CR326	(17)
2D2	CR326	(17)
3E1	CR326	(17)
10.09	CF53	D. Crevat and E. Deloince
813	CF53	D. Crevat and E. Deloince
AG3	S84-1	C. Li
AD2	S84-1	C. Li
AE8	S84-1	C. Li

[a] All antibodies neutralized strain HM175.

B5-B3 generally yielding the lowest reductions in titer of infectious HM175 virus. The significance of this variation in antibody activity, which is independent of antibody titer, is not clear.

We examined eight of these monoclonal antibodies (K2-4F2, K3-4C8, K3-2F2, B5-B3, 6A5, 1B9, 3E1, and 2D2) for their ability to bind to several different strains of HAV (HM175, PA21, HAS15, and LCDC) and found that the epitopes recognized by each of these monoclonal antibodies are present on each of these virus strains (E. Cox and S. M. Lemon, unpublished data). Since the monoclonal antibodies were raised against strains of virus other than those used as antigens in these tests (Table 1), the relevant epitopes appear to be shared among many if not all HAV strains. Furthermore, the binding profiles of several of these antibodies to the HM175 and PA21 strains of HAV show little difference, despite the sequence differences evident in the P1-region RNA of these strains (25). Recent data suggest that the epitope on HM175 virus that is recognized by K3-4C8 may be sensitive to trypsin while the K2-4F2 epitope is not (38a).

Each monoclonal antibody listed in Table 1 is capable of significantly competing with radiolabeled, postconvalescent polyclonal human antibody for attachment to purified virus captured by polyclonal antibody on a solid-phase support (17, 28, 39). Individual monoclonal antibodies, tested at concentrations in substantial excess over that of the radiolabeled antibody, inhibit the binding of the polyclonal antibody by 40 to 80% (Fig. 2). Furthermore, several combinations of two monoclonal antibodies (B5-B3 plus K2-4F2, or B5-B3 plus K3-4C8) demonstrate potent additive

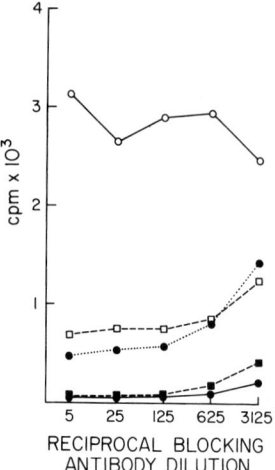

Figure 2. Competition radioimmunoassay evaluating monoclonal antibodies and polyclonal sera for their ability to block the attachment of radiolabeled polyclonal human convalescent immunoglobulin G to virus. The target antigen was gradient-purified HM175 strain HAV bound to polyclonal human convalescent-phase antibody coating wells of a polyvinyl chloride microtiter plate. Symbols: Competing antibodies included (○—○) preimmune and (●—●) postconvalescent polyclonal sera, (□– – –□) K2-4F2, (●····●) B5-B3, and (■– – –■) a 1:1 mixture of K2-4F2 and B5-B3. The mixture of monoclonal antibodies provides blocking activity comparable to polyclonal convalescent serum at low dilutions. Reprinted from reference 39.

activity in competing with convalescent polyclonal antibody for attachment to HM175, GR8, or PA21 strains of HAV (39; Cox and Lemon, unpublished data) (Fig. 2). This observation suggests that the relevant epitopes are immunodominant in humans.

Finally, we examined the ability of monoclonal antibodies to compete with each other for attachment to purified virions (32a, 39) (Fig. 3). With the exception of B5-B3, each antibody competes strongly with K3-4C8 (or K2-4F2) for attachment to HAV. The other antibodies may be ranked, however, according to the extent to which they compete with B5-B3 for attachment to the virion. 3E1 strongly competes with B5-B3 for attachment, while at the other extreme K3-4C8 and K3-2F2 reproducibly enhance the binding of B5-B3 to the virus. This latter observation suggests that the binding of K3-4C8 or K3-2F2 results in conformational changes in the capsid protein(s) which either render the B5-B3 epitope more accessible to antibody or favorably influence amino acid residues forming the B5-B3 epitope to assume the conformation necessary for antibody binding. These positive interactions between the K3-4C8, K3-2F2, and B5-B3 epitopes suggest that they are closely spaced on the virion surface.

As K3-4C8, K3-2F2, and 813 do not competitively inhibit attachment of B5-B3 under the assay conditions (Fig. 3), nonspecific steric interactions do not appear to prevent the simultaneous attachment of different antibodies to HAV, provided they recognize spatially distinct epitopes. On the other hand, 1B9, 2D2, and 3E1 each strongly compete with both

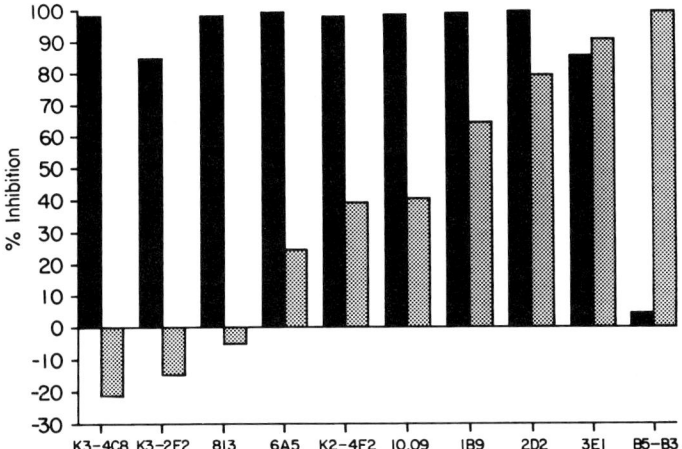

Figure 3. Competition between monoclonal antibodies for attachment to HAV fixed to a solid-phase support as in Fig. 2. Dilutions of 10 different murine monoclonal antibodies were added to the virus-coated wells prior to the addition of much smaller quantities of ^{125}I-labeled monoclonal antibody K3-4C8 (solid columns) or B5-B3 (stippled columns). The quantity of radiolabeled antibody bound to virus was compared with the quantity bound in the absence of any competing antibody. Reprinted from reference 32b.

K3-4C8 and B5-B3 for attachment to the virus. Although several interpretations are possible, these data suggest that the monoclonal antibodies are directed against a cluster of closely spaced epitopes, presumably present in 60 copies on the surface of the virion.

NEUTRALIZATION ESCAPE MUTANTS OF HAV

Virus mutants resistant to monoclonal antibody-mediated neutralization have proven helpful in identifying immunogenic sites on other picornaviruses (15, 29, 35), but the application of this approach to characterization of HAV immunogenic sites has been hampered by slow and inefficient virus replication in vitro. Moreover, as discussed above, most HAV isolates are noncytopathic and highly cell associated and demonstrate a substantial nonneutralizable fraction in vitro. Most clonally isolated virus variants recovered from radioimmunofocus assay agarose overlays following stringent monoclonal neutralization reactions demonstrate continuing neutralization susceptibility. This fact reflects the substantial persistent virus fraction that is characteristic of HAV stocks, even following chloroform or detergent treatment (23). Because the persistent fraction exceeds the spontaneous mutation rate, the selection

Table 2. HM175 Mutants Selected for Resistance to K2-4F2[a]

Variant	VP3 amino acid 70	K2-4F2 neutralization resistance[b]	K2-4F2 binding
HM175	Asp	1.43 ± 0.26	+
S18	His	−0.12	−
S20	Asp	0.88	+
S23	His	0.03	−
S27	His	−0.06	−
S28	His	0.02	−
S30	His	−0.09	−

[a] All variants bound K3-4C8 antibody.
[b] Log_{10} reduction in titer of infectious virus (mean ± standard deviation).

of HAV neutralization-escape mutants has required a process of repeated neutralization and passage in the presence of antibody, followed by clonal isolation from agarose overlays (39). The neutralization resistance of such mutants is associated with reduced binding of antibody to the virion, as measured by solid-phase immunoaffinity cDNA-RNA hybridization or radioimmunoassay (32a, 39).

A number of monoclonal neutralization-escape mutants of HM175 strain HAV have been isolated (39). Mutant S30 was initially selected by neutralization and passage in the presence of monoclonal antibody K2-4F2, and even after passage in the absence of antibody it showed no decrease in titer when incubated with this antibody (Table 2). The entire capsid-encoding region of S30 RNA was sequenced by primer extension (32a) and compared with the sequence of parental HM175 virus obtained from cloned cDNA (18). The only mutation identified in the capsid-encoding region of the S30 virus was a G-to-C substitution at base 1677, predicting a change in the Asp-70 residue of VP3 to His. This substitution was also present in four other mutants selected against K2-4F2 (mutants S18, S23, S27, and S28) but absent in mutant S20, which had reverted to neutralization susceptibility during final large-volume cell culture passage (Table 2) (32a). Because of the method of mutant selection (39), these five virus variants may represent sibling clones originally derived from a common neutralization-resistant virus. Asp-70 of VP3 was unchanged in several other mutants (S1, S32, S33, S34, and S57) which had been selected for resistance to other antibodies.

Mutant S32 was selected for resistance to antibody B5-B3, but demonstrated only partial resistance to this antibody following amplification in the absence of antibody. The entire P1 region of S32 RNA was also sequenced. The only mutation identified was C-to-T substitution at base

Figure 4. Amino acid residues recognized to contribute to neutralization immunogenic sites involving VP3 of HAV, poliovirus (sites 3b and 4) (29), and rhinovirus (neutralization immunogenic site III) (35). The identified residues are sites of mutations in monoclonal antibody escape mutants. Amino acid sequences are aligned according to Palmenberg (personal communication).

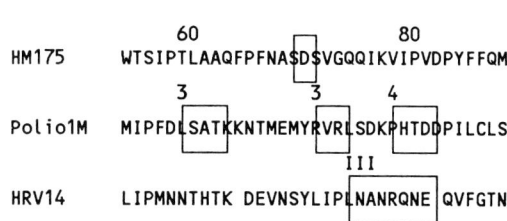

position 2512 which predicts a change in the Ser-102 residue of VP1 to Leu. However, since S32 demonstrated only partial neutralization resistance following amplification in the absence of antibody, the role played by Ser-102 in the antigenic structure of the virus is unclear.

Asp-70 of VP3 aligns closely with recognized neutralization immunogenic sites involving VP3 of poliovirus (3b or 4) (15, 29) and rhinovirus (NIm-III) (35) (Fig. 4). These data thus confirm that there is conservation of function as well as structure among these very distantly related viruses. Site 3b and the neutralization immunogenic site III are discontinuous (conformational) immunogens in poliovirus and rhinovirus, involving residues within VP3 and near the carboxy terminus of VP1. The site identified in VP3 of HAV may be analogous to these immunogenic sites and therefore may also involve VP1. This hypothesis is consistent with observations, based on the cross-linking of Fab fragments of intact virus, that epitopes recognized by monoclonal antibodies 2D2 and 6A5 involve residues of VP1 (17), and with studies showing that immunization with either purified VP1 or VP3 may induce very low levels of neutralizing antibody to HAV (16).

Mutant S30 demonstrates broad resistance to the monoclonal antibodies listed in Table 1 (AG3, AE8, and AD2 have not yet been tested) (Table 3). While the extent of neutralization resistance is variable, it is greatest against K2-4F2 and other monoclonal antibodies behaving in a fashion similar to K2-4F2 in the solid-phase competition studies (Fig. 3). Radioimmunoassays in which the mutant virus was bound to a solid-phase support through convalescent-phase polyclonal antiserum and subsequently probed with radioiodinated monoclonal antibody indicate that S30 has lost the capacity to bind K2-4F2. This is also true for its sibling mutants S18, S23, S27, and S28, but not the revertant S20 virus (Ping et al., submitted). The binding of K3-4C8 and B5-B3 to mutant S30 does not appear to be altered, however, despite the fact that S30

Table 3. Resistance of Mutant S30 to Monoclonal Antibodies

Monoclonal antibody	Neutralization resistance[a]	Antibody binding
K3-4C8	0.25/1.1	+
813	−0.09/0.91	
6A5	−0.03/1.5	
K2-4F2	−0.09/1.33	−
10.09	−0.16/1.33	
1B9	−0.08/0.8	
2D2	0.19/1.8	
3E1	0.31/1.1	
B5-B3	0.13/0.72	+

[a] Log_{10} reduction in titer with S30/log_{10} reduction with HM175.

demonstrates at least partial resistance to neutralization by these antibodies.

Although incomplete at present, these studies indicate that Asp-70 of VP3 contributes to a major, immunodominant antigenic site on the surface of the native HAV particle. Virus selected for resistance to monoclonal antibody K2-4F2 and having a single amino acid substitution at this residue is at least partially resistant to neutralization by every other monoclonal antibody tested (32b, 39), suggesting that this site represents part of a single functional immunogenic domain recognized by each of these monoclonal antibodies. An alternative interpretation is that the mutation at Asp-70 of VP3 renders the virus resistant to neutralization by monoclonal antibodies binding at sites distant on the capsid surface from the VP3 residue. It would be surprising if this latter interpretation were correct, however, given the clustered locations on the virion surface of mutations identified in functionally grouped escape mutants of poliovirus and rhinovirus (15, 35). Nonetheless, the finding that amino acid substitutions at both VP3-70 and VP1-102 are capable of conferring at least partial resistance to antibody B5-B3 challenges the assumption that mutations act only by inducing local changes in the capsid structure. If the proposed alignments of the HAV capsid polypeptides with those of other picornaviruses are correct (Palmenberg, personal communication), and assuming the structure of HAV resembles that of other picornaviruses with known crystal structures, these two residues are most likely located at sites distant from each other on the surface of the virus. VP1-102 would be near the fivefold axis of symmetry, while VP3-70 would be near the threefold axis (M. Rossman, personal communication). Thus, these data suggest that mutations may have at least weak effects distant from the primary antibody-binding site, or that the structure of HAV cannot be adequately predicted from existing structures of other picornaviruses.

IMPLICATIONS FOR VACCINE DEVELOPMENT

Since conventional inactivated or attenuated vaccines are not yet available for prevention of hepatitis A, recombinant or synthetic peptide approaches to HAV vaccine development are attractive alternatives which should be pursued. Enthusiasm for such approaches should be bolstered by the fact that relatively low levels of serum neutralizing antibody (about 1:40) correlate with substantial protection from symptomatic hepatitis A in humans (38), and by the data presented above which suggest that the immunogenic neutralization epitopes of HAV may be limited in number. Recent advances in enhancing the immunogenicity of synthetic peptide foot-and-mouth disease virus vaccines by the inclusion of appropriate T-cell determinants (5) lend further support to this argument and suggest the ultimate success of such vaccine development strategies. Because the immunodominant immunogen(s) of HAV appears to be assembled and conformationally dependent, however, the major challenge facing these approaches to vaccine development will be the development of synthetic immunogens mimicking the conformation of the relevant peptide sequences in their native state. Possible solutions to this problem may ultimately be suggested by the mapping of additional mutations present in HAV neutralization escape mutants, combined with crystallographic studies of the HAV capsid.

Acknowledgments. We are grateful to I. Gust, R. Tedder, J. Hughes, E. Emini, D. Crevat, and C. Li for their generous gifts of monoclonal antibodies to HAV, and to A. Palmenberg, M. Rossman, and J. Johnston for providing data in advance of publication.

This work was supported in part by contract DAMD17-85C-5272 from the U.S. Army Medical Research and Development Command, by Public Health Service grant R01-AI22279 from the National Institute of Allergy and Infectious Diseases, and by Technical Services Agreement V24/181/7 with the World Health Organization Programme for Accelerated Vaccine Development.

Literature Cited

1. **Anderson, D. A.** 1987. Cytopathology, plaque assay, and heat inactivation of hepatitis A virus strain HM175. *J. Med. Virol.* **22**:35–44.
2. **Binn, L. N., W. H. Bancroft, S. M. Lemon, R. H. Marchwicki, J. W. LeDuc, C. J. Trahan, E. C. Staley, and C. M. Keenan.** 1986. Preparation of a prototype inactivated hepatitis A virus vaccine from infected cell cultures. *J. Infect. Dis.* **153**:749–756.
3. **Binn, L. N., S. M. Lemon, R. H. Marchwicki, R. R. Redfield, N. L. Gates, and W. H. Bancroft.** 1984. Primary isolation and serial passage of hepatitis A virus strains in primate cell cultures. *J. Clin. Microbiol.* **20**:28–33.
4. **Chow, M., J. F. E. Newman, D. Filman, J. M. Hogle, D. J. Rowlands, and F. Brown.**

1987. Myristylation of picornavirus capsid protein VP4 and its structural significance. *Nature* (London) **327**:482–486.

5. **Clarke, B. E., S. E. Newton, A. R. Carroll, M. J. Francis, G. Appleyard, A. D. Syred, P. E. Highfield, D. J. Rowlands, and F. Brown.** 1987. Improved immunogenicity of a peptide epitope after fusion to hepatitis B core protein. *Nature* (London) **330**:381–384.

6. **Cohen, J. I., J. R. Ticehurst, S. M. Feinstone, B. Rosenblum, and R. H. Purcell.** 1987. Hepatitis A virus cDNA and its RNA transcripts are infectious in cell culture. *J. Virol.* **61**:3035–3039.

7. **Cohen, J. I., J. R. Ticehurst, R. H. Purcell, A. Buckler-White, and B. M. Baroudy.** 1987. Complete nucleotide sequence of wild-type hepatitis A virus: comparison with different strains of hepatitis A virus and other picornaviruses. *J. Virol.* **61**:50–59.

8. **Coulepis, A. G., S. A. Locarnini, E. G. Westaway, G. A. Tannock, and I. D. Gust.** 1982. Biophysical and biochemical characterization of hepatitis A virus. *Intervirology* **18**:107–127.

9. **Coulepis, A. G., G. A. Tannock, S. A. Locarnini, and I. D. Gust.** 1981. Evidence that the genome of hepatitis A virus consists of single-stranded RNA. *J. Virol.* **37**:473–477.

10. **Cromeans, T., M. D. Sobsey, and H. A. Fields.** 1987. Development of a plaque assay for a cytopathic, rapidly replicating isolate of hepatitis A virus. *J. Med. Virol.* **22**:45–56.

11. **Daemer, R. J., S. M. Feinstone, I. D. Gust, and R. H. Purcell.** 1981. Propagation of human hepatitis A virus in African green monkey kidney cell culture: primary isolation and serial passage. *Infect. Immun.* **32**:388–393.

12. **Dawson, G. J., R. H. Decker, D. K. Norton, W. H. Bryce, R. O. Whittington, I. I. Tribby, and I. K. Mushahwar.** 1984. Monoclonal antibodies to hepatitis A virus. *J. Med. Virol.* **12**:1–8.

13. **Emini, E. A., J. V. Hughes, D. S. Perlow, and J. Boger.** 1985. Induction of hepatitis A virus-neutralizing antibody by a virus-specific synthetic peptide. *J. Virol.* **55**:836–839.

14. **Gerlich, W. H., and G. G. Frösner.** 1983. Topology and immunoreactivity of capsid proteins in hepatitis A virus. *Med. Microbiol. Immunol.* **172**:101–106.

15. **Hogle, J. M., M. Chow, and D. J. Filman.** 1985. Three-dimensional structure of poliovirus at 2.9 Å resolution. *Science* **229**:1358–1365.

16. **Hughes, J. V., C. Bennett, L. Stanton, D. L. Linemeyer, and S. W. Mitra.** 1985. Hepatitis-A virus structural proteins: sequencing and ability to induce virus-neutralizing antibody responses, p. 255–259. *In* R. A. Lerner, R. M. Chanock, and F. Brown (ed.), *Molecular and Chemical Basis of Resistance to Parasitic, Bacterial and Viral Vaccines.* Cold Spring Harbor Laboratory, Cold Spring Harbor, N.Y.

17. **Hughes, J. V., L. W. Stanton, J. E. Tomassini, W. J. Long, and E. M. Scolnick.** 1984. Neutralizing monoclonal antibodies to hepatitis A virus: partial localization of a neutralizing antigenic site. *J. Virol.* **52**:465–473.

18. **Jansen, R. W., J. E. Newbold, and S. M. Lemon.** 1988. Complete nucleotide sequence of a cell culture-adapted variant of hepatitis A virus: comparison with wild-type virus with restricted capacity for *in vitro* replication. *Virology* **163**:299–307.

19. **Johnston, J. A., S. A. Harmon, L. N. Binn, O. C. Richards, E. Ehrenfeld, and D. F. Summers.** 1988. Antigenic and immunogenic properties of a hepatitis A virus capsid protein expressed in Escherichia coli. *J. Infect. Dis.* **157**:1203–1211.

20. **Lemon, S. M.** 1985. Type A viral hepatitis: new developments in an old disease. *N. Engl. J. Med.* **313**:1059–1067.

21. **Lemon, S. M., and L. N. Binn.** 1983. Serum neutralizing antibody response to hepatitis A virus. *J. Infect. Dis.* **148**:1033–1039.

22. **Lemon, S. M., and L. N. Binn.** 1983. Antigenic relatedness of two strains of hepatitis A virus determined by cross-neutralization. *Infect. Immun.* **42**:418–420.

23. **Lemon, S. M., and L. N. Binn.** 1985. Incomplete neutralization of hepatitis A virus *in vitro* due to lipid-associated virions. *J. Gen. Virol.* **66:**2501–2505.

24. **Lemon, S. M., L. N. Binn, and R. H. Marchwicki.** 1983. Radioimmunofocus assay for quantitation of hepatitis A virus in cell cultures. *J. Clin. Microbiol.* **17:**834–839.

25. **Lemon, S. M., S.-F. Chao, R. W. Jansen, L. N. Binn, and J. W. LeDuc.** 1987. Genomic heterogeneity among human and nonhuman strains of hepatitis A virus. *J. Virol.* **61:**735–742.

26. **Lemon, S. M., R. W. Jansen, and J. E. Newbold.** 1985. Infectious hepatitis A virus particles produced in cell culture consist of three distinct types with different buoyant densities in CsCl. *J. Virol.* **54:**78–85.

27. **Linemeyer, D. L., J. G. Menke, A. Martin-Gallardo, J. V. Hughes, A. Young, and S. W. Mitra.** 1985. Molecular cloning and partial sequencing of hepatitis A viral cDNA. *J. Virol.* **54:**247–255.

28. **MacGregor, A., M. Kornitschuk, J. G. R. Hurrell, N. I. Lehmann, A. G. Coulepis, S. A. Locarnini, and I. D. Gust.** 1983. Monoclonal antibodies against hepatitis A virus. *J. Clin. Microbiol.* **18:**1237–1243.

29. **Minor, P. D., M. Ferguson, D. M. A. Evans, J. W. Almond, and J. P. Icenogle.** 1986. Antigenic structure of polioviruses of serotypes 1, 2, and 3. *J. Gen. Virol.* **67:**1283–1291.

30. **Najarian, R., D. Caput, W. Gee, S. J. Potter, A. Renard, J. Merryweather, G. Van Nest, and D. Dina.** 1985. Primary structure and gene organization of human hepatitis A virus. *Proc. Natl. Acad. Sci. USA* **82:**2627–2631.

31. **Ostermayr, R., K. von der Helm, V. Gauss-Müller, E. L. Winnacker, and F. Deinhardt.** 1987. Expression of hepatitis A virus cDNA in *Escherichia coli*: antigenic VP1 recombinant protein. *J. Virol.* **61:**3645–3647.

32. **Paul, A. V., H. Tada, K. von der Helm, T. Wissel, R. Kiehn, E. Wimmer, and F. Deinhardt.** 1987. The entire nucleotide sequence of the genome of human hepatitis A virus (isolate MBB). *Virus Res.* **8:**153–171.

32a. **Ping, L.-H., R. W. Jansen, J. T. Stapleton, J. I. Cohen, and S. M. Lemon.** 1988. Identification of an immunodominant antigenic site involving capsid protein VP3 of hepatitis A virus. *Proc. Natl. Acad. Sci. USA* **85:**8281–8285.

32b. **Ping, L.-H., R. W. Jansen, J. T. Stapleton, P. C. Murphy, J. I. Cohen, and S. M. Lemon.** 1988. An immunodominant neutralization site residing on capsid protein VP3 of human hepatitis virus. *UCLA Symp. Mol. Cell. Biol. N. Ser.* **84:**527–537.

33. **Provost, P. J., R. P. Bishop, R. J. Gerety, M. R. Hilleman, W. J. McAleer, E. M. Scolnick, and C. E. Stevens.** 1986. New findings in live, attenuated hepatitis A vaccine development. *J. Med. Virol.* **20:**165–175.

34. **Provost, P. J., and M. R. Hilleman.** 1979. Propagation of human hepatitis A virus in cell culture *in vitro*. *Proc. Soc. Exp. Biol. Med.* **160:**213–221.

35. **Rossmann, M. G., E. Arnold, J. W. Erickson, E. A. Frankenberger, J. P. Griffith, H.-J. Hecht, J. E. Johnson, G. Kamer, M. Luo, A. G. Mosser, R. R. Rueckert, B. Sherry, and G. Vriend.** 1985. Structure of a human common cold virus and functional relationship to other picornaviruses. *Nature* (London) **317:**145–153.

36. **Siegl, G., C. G. Frösner, V. Gauss-Müller, J.-D. Tratschin, and F. Deinhardt.** 1981. The physicochemical properties of infectious hepatitis A virions. *J. Gen. Virol.* **57:**331–341.

37. **Siegl, G., M. Weitz, and G. Kronauer.** 1984. Stability of hepatitis A virus. *Intervirology* **22:**218–226.

38. **Stapleton, J. T., R. W. Jansen, and S. M. Lemon.** 1985. Neutralizing antibody to hepatitis A virus in immune serum globulin and in the sera of human recipients of immune serum globulin. *Gastroenterology* **89:**637–642.

38a. **Stapleton, J. T., and D. K. Lange.** 1988. Trypsin sensitivity of epitopes within the hepatitis A virus immunodominant neutralization site, p. 235–240. *In* H. Ginsburg, F. Broun, R. A. Lerner, and R. Channock (ed.), *Vaccines 88.* Cold Spring Harbor Laboratory, Cold Spring Harbor, N.Y.

39. **Stapleton, J. T., and S. M. Lemon.** 1987. Neutralization escape mutants define a dominant immunogenic neutralization site on hepatitis A virus. *J. Virol.* **61:**491–498.

40. **Ticehurst, J. R., V. R. Racaniello, B. M. Baroudy, D. Baltimore, R. H. Purcell, and S. M. Feinstone.** 1983. Molecular cloning and characterization of hepatitis A virus cDNA. *Proc. Natl. Acad. Sci. USA* **80:**5885–5889.

41. **Venuti, A., C. Di Russo, N. del Grosso, A.-M. Patti, F. Ruggeri, P. R. De Stasio, M. G. Martiniello, P. Pagnotti, A. M. Degener, M. Midulla, A. Pana, and R. Perez-Bercoff.** 1985. Isolation and molecular cloning of a fast-growing strain of human hepatitis A virus from its double-stranded replicative form. *J. Virol.* **56:**579–588.

42. **Weitz, M., B. M. Baroudy, W. L. Maloy, J. R. Ticehurst, and R. H. Purcell.** 1986. Detection of a genome-linked protein (VPg) of hepatitis A virus and its comparison with other picornaviral VPgs. *J. Virol.* **60:**124–130.

43. **Wheeler, C. M., B. H. Robertson, G. Van Next, D. Dina, D. W. Bradley, and H. A. Fields.** 1986. Structure of the hepatitis A virion: peptide mapping of the capsid region. *J. Virol.* **58:**307–313.

44. **Wychowski, C., S. van der Werf, O. Siffert, R. Crainic, P. Bruneau, and M. Girard.** 1983. A poliovirus type 1 neutralization epitope is located within amino acid residues 93 to 104 of viral capsid polypeptide VP1. *EMBO J.* **2:**2019–2024.

Part III

GENETIC DETERMINANTS OF VIRAL DISEASE AND APPLICATIONS TO DIAGNOSIS

Molecular Aspects of Picornavirus Infection and Detection
Edited by Bert L. Semler and Ellie Ehrenfeld
© 1989 American Society for Microbiology, Washington, DC 20006

Chapter 13

Sequence Alignments of Picornaviral Capsid Proteins

Ann C. Palmenberg

INTRODUCTION

In 1975, Howard Temin, David Baltimore, and Renato Dulbecco were awarded the Nobel Prize in Medicine for tumor virus work leading to the discovery of reverse transcriptase. Picornavirologists should be keenly aware of this fact because the revolutionary advances in our field brought about by recombinant engineering are based squarely upon the fortunate proclivity of this enzyme to copy RNA templates into cDNA. The once fragile RNA genomes, notorious for degrading during experiments, are now easily converted into stable cDNAs which can be manipulated or amplified. The enzyme has also made possible such simplified approaches to RNA sequencing that it is not uncommon for a laboratory to determine 2,000 to 3,000 nucleotides from a new viral strain within a week of its isolation. To judge from recent picornaviral literature, it is obvious that a lot of reverse transcriptase has indeed been utilized for just this purpose.

As of early 1988, complete genomes or genomic fragments from at least 70 distinct viral strains, including representative members from all major and minor classification groups, had been cloned and sequenced. More than 250,000 picornavirus-specific nucleotides have already been published, in a tally which does not include an additional 100,000 bases from unpublished, in press, or redundant determinations. A partial list of strains and references is presented in Table 1.

Ann C. Palmenberg • Institute for Molecular Virology and Department of Veterinary Science, University of Wisconsin, Madison, Wisconsin 53706.

Table 1. Origin of Sequences

Virus	Serotype/strain	Alignment abbreviation[a]	Reference
FMDV	A 10/61	FMDA10	4, 5, 7
	A 12/199ab	FMDA12	47, 60
	A 22/550/USSR/65	FMDA22	60
	Asia 1	FMDAsia1	—[b]
	C 1	FMDC1	3, 59
	C 3/India1/Brazil/71	FMDC3	9, 33
	O 1/BFS/Britain/68	FMDO1Bfs	9
	O 1/Kaufbeuren	FMDO1K	20
	1 South African Terr.	FMDSat1	—[b]
	2 South African Terr.	FMDSat2	—[b]
	3 South African Terr.	FMDSat3	—[b]
EMC virus	Rueckert	EMCR	40
	D (diabetogenic)	EMCD	—[c]
	B (nondiabetogenic)	EMCB	—[c]
Mengovirus	M (medium plaque)	Mengo	32, —[d]
TME virus	BeAn 8386	TMEBeAn	44
	GDVII	TMEGd7	43
	DA	TMEDa	37
HAV	Los Angeles (Chiron)	HALA	34, —[e]
	CR326	HACr326	31
	Mbb	HAMbb	42
	Hm175	HA175	1, 11
Poliovirus	1 Mahoney	Polio1M	28, 45
	1 LSc/2ab (Sabin)	Polio1S	36, 57
	2 Lansing	Polio2La	29
	2 P712/Ch/2ab (Sabin)	Polio2S	57
	3 Leon/37	Polio3Le	56
	3 23127 (Finland)	Polio3F	24
	3 Leon/12a$_1$b (Sabin)	Polio3S	57
Coxsackie-virus	A21	CoxA21	—[f]
	B1	CoxB1	26
	B3 Nancy	CoxB3	30, 58
	B4 J.V.B. Benschoten	CoxB4	27
BEV	Vg/5/27	BEV	17
Human rhinovirus	1a	Rhino1a	—[g]
	1b	Rhino1b	25
	2	Rhino2	53
	14	Rhino14	6, 55
	39	Rhino39	—[g]
	49	Rhino49	—[g]
	89	Rhino89	14

[a] See Fig. 2.
[b] Berwyn Clarke, personal communication.
[c] Y.-S. Bae, H.-Y. Eun, and J.-W. Yoon, personal communication.
[d] Unpublished corrections.
[e] J. Ticehurst and R. Ralston, personal communication.
[f] P. Hughes and G. Stanway, personal communication.
[g] G. Cordova, E. Doran, and B. Korant, unpublished data.

In view of this huge pile of available information and the ever-present seductive temptation to determine "just one more isolate," it is sometimes useful to stress the point that generation of sequence data, by itself, is not a particularly enlightening scientific endeavor. Sequences become valuable only when they are used to extract information about the viruses from which they originate. Without critical evaluation and perspective, any given sequence could just as well be a random collection of nucleotides (or amino acids).

Though it is correct to point out that the final answer to nearly every imaginable virological question is probably cryptically encoded somewhere within each RNA genome, no one presently knows how to decipher most of this information. The subtle nuances of base or amino acid arrangements which determine tertiary and quaternary structure of nucleic acids and proteins lie frustratingly beyond today's technology. Yet, despite our ignorance, there is a wealth of information that can and should be picked from existing sequences. Open reading frames, base composition, codon frequency, enzyme recognition sites, and the occasional gross morphological feature (e.g., stable RNA secondary structures) can readily be identified within individual segments and give powerful hints as to the coded function.

Another important tool is the comparative examination of multiple sequences, since generally it is believed that regions with strong sequence similarities probably also share analogous genetic or biochemical functions. This is especially true if the relevant genomes are closely related by phylogeny. In these cases, specific identities or differences (e.g., natural or engineered mutations) can accurately pinpoint the genetic purpose of individual bases or amino acids (e.g., epitope mapping or enzyme active sites). More typically, the commonalities are somewhat less precise, covering widely spaced regions with minimum discernible conservation of residues. Though more difficult to recognize, these segments too can provide a useful basis for identifying extended portions with analogous function. Both similarities and variations can be positively informative, as the alternating patterns of conserved and nonconserved sequences locate and distinguish those regions responsive to differential selective pressures. Properly identified, these relationships may intimate legitimate biological or phylogenic homologies characteristic of whole proteins, genomes, or even entire virus groups.

The obvious key to accurate comparisons lies in the ability to correctly detect and match equivalent regions within multiple sequences. With closely related sequences (>60% identity), the optimum alignment is usually unambiguous. However, correlations among more distantly related fragments (<40% identity) become progressively more difficult.

Computers sometimes help in this regard, as they can rapidly evaluate multiple arrangements for the best solution. Popular comparative programs such as those of Needleman and Wunsch (35) or Sellers (52) rely on matrix searching algorithms that maximize the total number of matches between pairs of sequences by the careful insertion of gaps. The selection of screening parameters strongly influences the outcome, and the programs are fully capable of ignoring regions of excellent similarity if they can produce an alignment with equal or better quality in some other way. The cataloging of analogous regions among distantly related sequences therefore depends to a great extent upon the parameters used to detect these segments.

Until recently these uncertainties precluded formation of reliable capsid protein alignments among the various genera of picornaviruses. Of course, the P1 proteins from viruses like rhino-, polio-, and coxsackievirus could be reasonably compared among themselves (14, 27, 57), but these strains share >45% amino acid identity throughout their polyproteins and the alignment guideposts are fairly obvious. Equivalent P1 region comparisons between poliovirus and hepatitis A virus (HAV) (<15 to 17% amino acid identity), coxsackievirus and foot-and-mouth disease virus (FMDV) (20 to 22% identity), or mengovirus and rhinovirus (<25% identity), for example, are not as apparent. Unlike the P2 and P3 region proteins, which share a minimum of 25% amino acid identity among representatives of the four genera (unpublished data), the viral capsid proteins (especially VP1) are encoded in highly variable regions of the RNA. The wonderfully broad spectrum of properties (antigenicity, receptor recognition, buoyant density, pH stability, drug sensitivity, and thermal lability) which make each strain unique is reflected in this mutability and camouflages the essential core of sequence sameness that logically must be held among all icosahedral viruses. It wasn't until the crystallographic structures of several picornaviruses had been solved that the unity among these capsid proteins became discernible enough to allow formation of comprehensive alignments.

HOW THE ALIGNMENTS ARE FORMED

Resolution of the three-dimensional atomic structures of polio1M (23), rhino14 (48), and mengovirus (32) (see Table 1 for strain abbreviations) revealed that these T=1 (pseudo T=3) animal viruses have tertiary and quaternary organization very similar to that of the T=3 plant viruses. The VP1, VP2, and VP3 proteins share as common structural motif a remarkable well-conserved wedge-shaped, eight-stranded antiparallel β-barrel configuration. They differ from each other, and from the plant capsid proteins, primarily in the size and shape of the loops connecting particular β strands and in the lengths and orientations of their terminal extensions (see Fig. 1). Stripped of these decorations, the β-barrel segments of VP1, VP2, and VP3 of polio1M, rhino14, and mengovirus are readily superimposable and provide unambiguous, fixed structural points upon which reliable amino acid alignments can be built.

Michael Rossmann, Edward Arnold, and Ming Luo at Purdue University have constructed high-resolution hydrogen bonding diagrams from the refined coordinates of rhino14 and mengovirus (personal communication). The diagrams show the precise interactions of residues within the structures and define the elements which begin and end each strand of the β sheets (or α helices). They also help in determining the exactly equivalent amino acids within the other corresponding viral structure. Although some small segments remain unresolved due to crystallographic disorder, reasonably complete sequence alignments for each of VP1, VP2, and VP3 were established for rhino14 and mengovirus by superimposition of their respective structures (32).

Since entero- and rhinoviruses share strong sequence identity, it was then possible to add seven strains of poliovirus, four strains of coxsackievirus (three type B and one type A), six different rhinoviruses, and bovine enterovirus (BEV) to this alignment by reiterative comparison with rhino14. Six additional cardioviruses (three encephalomyocarditis [EMC] viruses and three Theiler's murine encephalomyocarditis [TME] viruses) were added by similar comparisons with mengovirus and with each other (>60% amino acid identity). Throughout these progressively more elaborate procedures, care was taken not to disturb the relative orientations of the key, formative sequences (mengovirus/rhino14). Each addition was aided by extensive computer searches to optimize the total matched residues for each possible pairwise comparison and for the compilation as a whole.

The University of Wisconsin Genetics Computer Group (UW GCG) program GAP, based on the method of Needleman and Wunsch (15, 35), was modified to use the peptide scoring comparison table of Roger Staden (54). This table originates with the relatedness odds matrix of M. Dayhoff (13) and reflects frequency, structural, biochemical, and genetic parameters for all possible pairwise amino acid substitutions. The program compared two sequences at a time and aligned them by the insertion of gaps to maximize the highest-value path through the matrix. The gap weight penalty was 0.5, and the gap length penalty was 0.03. Higher (1.0 gap weight and 0.1 gap length penalties) and lower (0.1 gap weight and 0.03 gap length penalties) stringencies were sometimes used for comparison or to optimize specific regions. The final overall alignments for

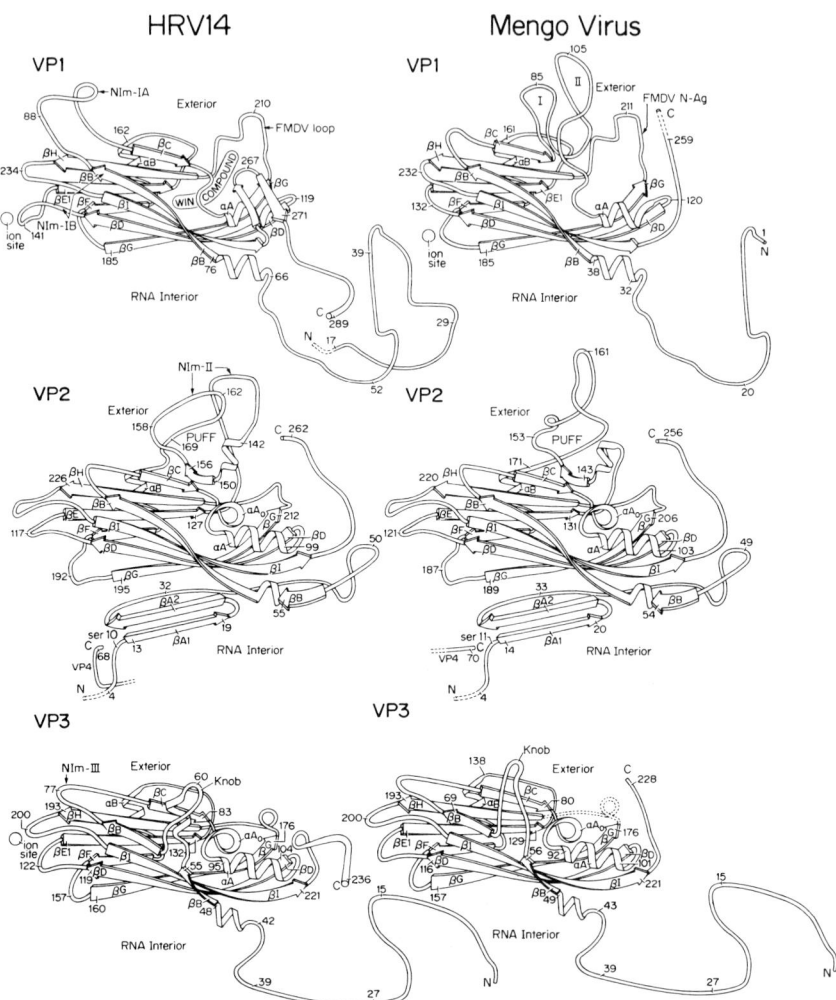

Figure 1. Ribbon drawings of VP1, VP2, and VP3 from rhino14 and mengovirus. This representation of the three-dimensional crystal structures is reproduced from Fig. 6 in reference 32 with the kind permission of Michael Rossmann (copyright 1987 by the AAAS). Amino acid numbering, secondary structural nomenclature, and location of the neutralization immunogenic sites (NIm) are shown. Segments indicated with dashed lines are crystallographically disordered. The α-helix occurring at the amino end of β-B in all diagrams is designated as α-Z in the text and sequence alignments.

entero-, rhino-, and cardioviruses do not vary significantly within these ranges.

Regrettably, precise addition of FMDV and HAV capsid sequences could not be satisfactorily achieved by sequential comparisons of individual proteins. Results varied too widely with each pairwise search. Ideally, one requires hydrogen bonding profiles for these viruses too, so that accurate superimposition with the rhinovirus and mengovirus maps would force unambiguous results. For the present, however, these data are unavailable. Therefore, alternative, though admittedly less accurate, methods were employed to engineer the best possible fit with the information at hand.

Because of the large assortment of already compiled sequences, it seemed probable that buried somewhere within the rhino-/entero-/cardiovirus alignments were enough total comparative data to reasonably match and identify equivalent regions in FMDV and HAV. The UW GCG PROFILE program series was adapted for this purpose. These programs read an existing sequence alignment and convert the data for each column into a table of position-specific symbol comparison values and gap weights (21, 22). New sequences can then be compared against the completed matrix for the best overall fit. Profile searches emphasize and seek out similarities within conserved regions, while tolerating diversity in variable regions. To prevent undue bias, each viral serotype in the starting alignments was accorded a single "vote" for the profile calculations. Serotypes represented by more than one sequence (e.g., polio1S and polio1M) proportioned their vote equally. Thus, during comparisons with FMDV, for example, no one serotype could unfairly influence the balloting just because its member strains were more frequently represented.

In this manner, capsid alignments of all seven FMDV serotypes relative to the entero-, rhino-, and cardioviruses were formed. Subsequent crystallographic modeling of the FMDA12 three-dimensional structure on the basis of these predictions was consistent with surface epitope maps, suggesting that this sequence fit may actually be quite good (31a). It is hoped the same may eventually be said for the HAV alignments, too. These sequences were fit using revised profiles, which also included the newly added FMDV data to provide the broadest possible search base. The resulting VP2 and VP3 HAV alignments are dependable, but VP1 fit may still be subject to some select local shifting as comparative methods improve. The present alignments for VP4, VP2, VP3, and VP1, formed with representative members of each picornaviral genus, are presented in Fig. 2A, B, C, and D, respectively. The VP4 data are based primarily on

Continued on p. 229

(A)

```
                    Block 1                                                    Block 2                               Block 3
FMDA10    GAgqSspatGSQnqsgn...tGSiINnyyMqqYqNsmstqlgdntisggsn ( 48)   EgstDttsthtntqnndwFsklassaFtglfgalLa   ....DPpr..tyg.qfsnlFsg.aVnaFsnml.PlLa ( 85)
FMDA12    GAgqSspatGSQnqsgn...tGSiINnyyMqqYqNsmdtqlgdnAisggsn ( 48)   EgstDttsthtntqnndwFsklassaFtglfgalLa   ....DPpr..tyg.qfsnlLsg.aVnaFsnmi.PlLa ( 85)
FMDC1     GAgqSspatGsQnqsgn...tGSiINnyyMqqYqNsmdtqlgdnAisggsn ( 48)   EgstDttsthtntqnndwFsklassaFsglfgalLa   ....DPpr..tyg.qfsnlLsg.aVnaFsnmi.PlLa ( 85)
FMDO1K    GAgqSspatGsQnqsgn...tGSiINnyyMqqYqNsmdtqlgdnAisggsn ( 48)   EgstDttsthtntqnndwFsklassaFsglfgalLa   ....DPpk..tyg.qfsnlLsg.aVnaFsnml.PlLa ( 85)
FMDSat3   GAgqSspctGsQnqsgn...tGSiINnyyMqqYqNsmdtqlgdnAisggsn ( 48)   EgstDttsthtntqnndwFsklaqsaIsglfgalLa   ....DPpk..tyg.qfsnlLsg.aVnaFsnml.PlLa ( 85)
EMCR      GnstSsDknnsssegne...GviINnfysNqYqNsid...lSanA..ags. ( 42)                                          ....gDapq..tng.qlsnilgg.aanaFatma.PlLl ( 70)
EMCB      GnstSsDkmnsssdgne...GviINnfysNqYqNsid...lSanA.tgs.  ( 42)                                          ....gDapq..nng.qlssiLgg.aanaFatma.PILl ( 70)
EMCD      GnstSsDkmnsssdgne...GviINnfysNqYqNsid...lSanA.tgs.  ( 42)                                          ....gDapq..nng.qlsniLgg.aanaFatma.PILl ( 70)
Mengo     GnstSsDknnssegne...GviINnfysNqYqNsid...lSanA.tgs.   ( 42)                                          ....gDapq..mng.qlsniLgg.aanaFatma.PlLl ( 70)
TMEBeAn   GnssSsDksnsQssgne...GviINnfysNqYqNsid...lSasg.gna.  ( 42)                                          ....gDapq..tng.qlsniLgg.aanaFatma.PlLm ( 71)
TMEGd7    GnasSsDksnsQssgne...GviINnfysNqYqNsid...lSasg.gna.  ( 42)                                          ....gDapq..nng.qlssiLgg.aanaFatma.PILm ( 71)
TMEDa     GnasSsDksnsQssgne...GviINnfysNqYqNsid...lSasg.gna.  ( 42)                                          ....gDapq..nng.qlssiLgg.aanaFatma.PILl ( 71)
HALA      GifqtvgsgldHils...                                  ( 15)                                          ...............La ( 17)
HACr326   GifqtvgsgldHils...                                  ( 15)                                          ...............La ( 17)
HAHm175   GifqtvgsgldHils...                                  ( 15)                                          ...............La ( 17)
HAMbb     GifqtvgsgldHils...                                  ( 15)                                          ...............La ( 17)
Polio1M   GAQVSSQkvGaHensnraygGStINYttINYYrDsA...SnaA...skq   ( 43)   DFsqDPs.............                    KFTE.PIkDVLiktaPmLn ( 68)
Polio1S   GAQVSSQkvGaHensnraygGStINYttINYYrDsA...SnaA...skq   ( 43)   DFsqDPs.............                    KFTE.PIkDVLiktsPmLn ( 68)
Polio2La  GAQVSsQkvGaHensnraygGStINYttINYYrDsA...SnaA...skq   ( 43)   DFaqDPs.............                    KFTE.PIkDVLiktaPtLn ( 68)
Polio2S   GAQVSsQkvGaHensnraygGStINYttINYYrDsA...SnaA...skq   ( 43)   DFaqDPs.............                    KFTE.PIkDVLiktaPmLn ( 68)
Polio3Le  GAQVSSQkvGaHensnraygGStINYttINYYkDsA...SnaA...skq   ( 43)   DYsqDPs.............                    KFTE.PLkDVLiktaPaLn ( 68)
Polio3F   GAQVSSQkvGaHensnraygGStINYttINYYkDsA...SnaA...skq   ( 43)   DYsqDPs.............                    KFTE.PLkDVLiktaPaLn ( 68)
Polio3S   GAQVSSQkvGaHensnraygGStINYttINYYkDsA...SnaA...skq   ( 43)   DYsqDPs.............                    KFTE.PLkDVLiktaPaLn ( 68)
CoxA21    GAQVStQktGaHenqnvaanSStINYttINYYkDsA...SnsA..trq    ( 43)   DLsqDPs.............                    KFTE.PVkDLMiktaPaLn ( 68)
CoxB1     GAQVStQktGaHetglnasgnSiIHYtnINYYkDaA...SnsA...nrq   ( 43)   DFtqDPg.............                    KFTE.PVkDLMiktaPaLn ( 68)
CoxB3     GAQVStQktGaHetglnasgnSiIHYtnINYYkDaA...SnsA...nrq   ( 43)   DFtqDPg.............                    KFTE.PVkDIMiksmPaLn ( 68)
CoxB4     GAQVStQktGaHetslsasgnSiIHYtnINYYkDaA...SnsA...nrq   ( 43)   DFtqDPs.............                    KFTE.PVkDVMiksIPaLn ( 68)
BEV       GAQLSrNtadsHtgtyatgGStINYnnINYYshaA...Saaq..nkq     ( 43)   EFtQDPs.............                    KFTQ.PIaDVIketavpLk ( 68)
Rhino1a   GAQVSrQnvGtHstqnsvnGSsLNYfnINYFkDaA...SsgA...srl    ( 43)   DFsqDPs.............                    KFTD.PVkDVLekgiPtLq ( 68)
Rhino1b   GAQVSrQnvGtHstqnsvnGSsLNYfnINYFkDaA...SsgA...srl    ( 43)   DFsqDPs.............                    KFTD.PVkDVLekgiPtLq ( 68)
Rhino2    GAQVSrQnvGtHstqnsvnGSsLNYfnINYFkDaA...SngA...skl    ( 43)   DFsqDPs.............                    KFTD.PVkDVLekgiPtLq ( 68)
Rhino14   GAQVStQksGsHenqniltnGSnqtFtvINYYkDaA...Stss...agg   ( 43)   sLsmDPs.............                    KFTE.PVkDLMlkgaPaLn ( 68)
Rhino89   GAQVSrQnvGtHstqnsvnGSsLNYfnINYFkDaA...SsgA...srl    ( 43)   DFsqDPs.............                    KFTD.PVkDVLekgiPtLq ( 68)
Consensus GAQVS-Q--G-H-------GS-INY--INYY-D-A----S--A-----            DF--DP-------------                     -KFTE-PV-DVL----P-L-
Structure                                                            aa.........            aa                   aaaa
```

Figure 2. Amino acid alignments for picornaviral capsid proteins. Formation of alignments and abbreviations are described in the text and Table 1. The parenthetic values serve as reference points for the numbering of residues within unaligned sequences. (A) VP4; (B) VP2; (C) VP3; (D) VP1. (Figure 2 continues on following 10 pages.)

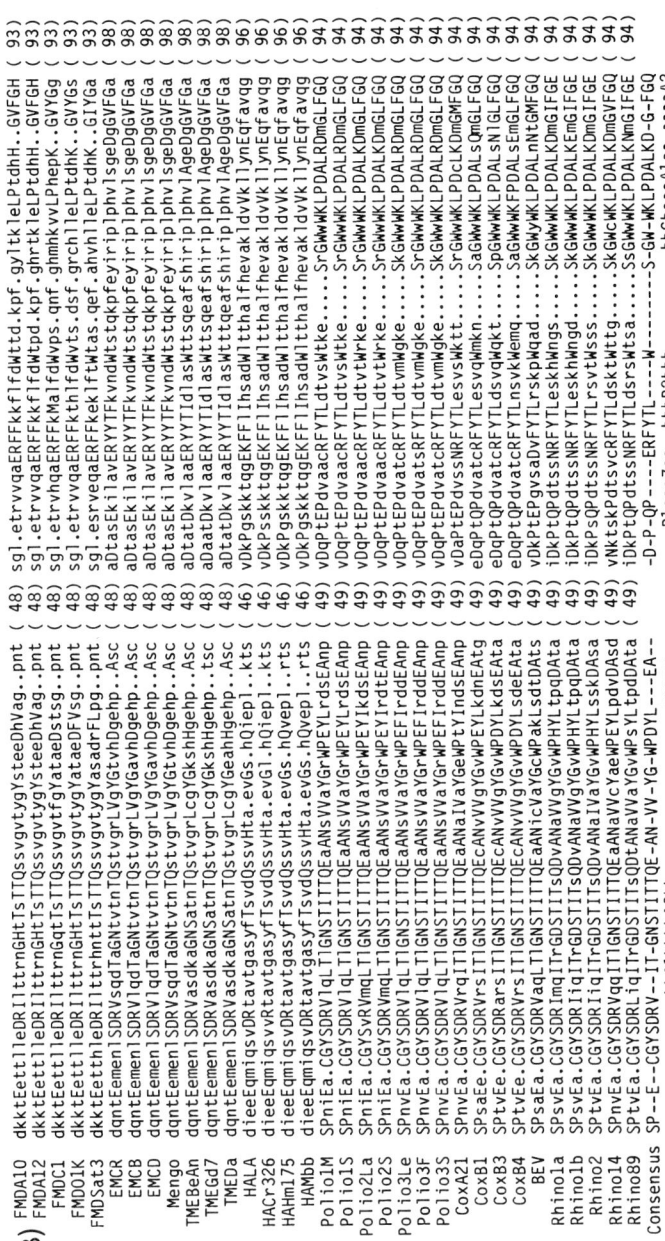

Figure 2. *Continued.*

```
FMDA10    lVdsyaymRnGwdVEVcavgnqFNgcLLVamVPEwk.............. (130)  ................afd (133)
FMDA12    lVdsyaymRnGwdVEVsavgnqFNgcLLVamVPEwk.............. (130)  ................tfd (133)
FMDC1     lVksyaymRnGwdVEVtavgnqFNGcLqaalVPEmgdi............ (132)  ..................s (133)
FMDO1K    ltdsyaymRnGwdVEVtavgnqFNGcLLVamVPElysi............ (132)  ..................q (133)
FMDSat3   mVesHayvRnGwdVQVsatstqFNgGtLLVamVPElhsl........... (132)  ..................d (133)
EMCR      aLrrHYLvKtGwrVQVQCNAsqFHaGgLLVfmaPEyp............. (135)  .tldafam.dnrwsk........dnlpngtrtqtnkkgpfamd (168)
EMCB      aLrrHYLvKtGwrVQVQCNAsqFHaGsLLVfmaPEyp............. (135)  .tldafam.dnrwsk........dnlpngtktqtnrkgpfamd (168)
EMCD      aLrrHYLvKtGwrVQVQCNAsqFHaGsLLVfmaPEyp............. (135)  .tldafam.dnrwsk........dnlpngtktqtnrkgpfamd (168)
Mengo     tLrrHYLvKtGwrVQVQCNAsqFHaGsLLVfmaPEyp............. (135)  .tldvfam.dnrwsk........dnlpngtrtqtnrkgpfamd (168)
TMEBeAn   tLrrHYLcKtGwrVQVQCNAsqFHaGsLLVfmaPEfytgkgtktgtme.. (146)  .psdpftm.dtewrsp......qgaptgyryds.rtgffatn (179)
TMEGd7    tLrrHYLcKtGwrVQVQCNAsqFHaGsLLVfmaPEfytgktksgtme... (146)  .psdpftm.dttwrsp......qsaptgyrydr.qagffamn (179)
TMEDa     tLrrHYLcKtGwrVQVQCNAsqFHaGsLLVfmaPEfytgktktgdme... (146)  .ptdpftm.dttwrap......qgaptgyryds.rtgffamm (179)
HALA      lLrYHtyaRfGieIQVQiNptpFQQGglicamVPg............... (131)  ................dqs (134)
HACr326   lLrYHtyaRfGieIQVQiNptpFQQGglicamVPs............... (131)  ................dqs (134)
HAHm175   lLrYHtyaRfGieIQVQiNptpFQQGglicamVPg............... (131)  ................dqs (134)
HAMbb     lLrYHtyaRfGieIQVQiNptpFQQGglicamVPg............... (131)  ................dqs (134)
Polio1M   NMyYHYLgRsGytVHVQCNAskFHQGaLgVfaVPEmcLagdsntttmhts (144)  yqnanpgekggtfgtftpdnnqtsparrfcpvdyll......gn (183)
Polio1S   NMyYHYLgRsGytVHVQCNAskFHQGaLgVfaVPEmcLagdsntttmhts (144)  yqnanpgekggtfgtftpddnqtsparrfcpvdylf......gn (183)
Polio2La  NMfYHYLgRaGytVHVQCNAskFHQGaLgVfaVPEmcLagds.tthmftk (143)  yenanpgekGgefkgsftldtnatnparnfcpvdylf......gs (182)
Polio2S   NMfYHYLgRsGytVHVQCNAfkFHQGaLgVfaIPEycLagds.tthmftk (143)  yenanpgekGgefkgsftldtnatnparnfcpvdylf......gs (182)
Polio3Le  NMyYHYLgRsGytVHVQCNAskFHQGaLgVfaIPEfcLagdsd.taryts (143)  yananpgerGgkfysqfnkdnavtspkrefcpvdyll......gc (182)
Polio3F   NMyYHYLgRsGytVHVQCNAskFHQGsLgVfaIPEfcLagdsd.taryts (143)  yananpgekGgkfyaqfnkdtavtspkrefcpvdyll......gc (182)
Polio3S   NMyYHYLgRsGytVHVQCNAskFHQGaLgVfaIPEycLagdsd.karyts (143)  yananpgerGgkfysqfnkdnavtspkrefcpvdyll......gc (182)
CoxB1     NMyYHYLgRsGytIHVQCNAskFHQGaLgVflIPEfvMacntesktsyvs (144)  yinanpgeGgefntnypnsntdvsegrqfaaldyll......gs (183)
CoxB3     NMqYHYLgRtGytdHVQCNAskFHQGcLLVcvVPEaeMgcsn..lnntp (141)  ...kfaelsGgdnarmftdtevgtsndkkvqtavwna......gm (177)
CoxB4     NMqYHYLgRsGytIHVQCNAskFHQGcLLVvcVPEaeMgcat..ldntp (141)  ..ssaellGgdtakefadkpvasgsnklvqrvvyna......gn (177)
BEV       NMqYHYLgRsGytIHVQCNAskFHQGcLLVvcVPEaeMgctn..aenap (141)  ...aydlcGgetaksf.eqnaatg.ktavqtavcna......gm (175)
Rhino1a   NaqFHYlyRgGwaVHVQCNAtkFHQGtLlVlaIPEhqlatqeq....pa (139)  fdrtmpgseGgtfqep............fwl......ed (160)
Rhino1b   NMyYHFLgRsGytVHVQCNAskFHQGtLLVamIPEhqLasakh.gsvtag (143)  yklthpgeaGrdvsqe.rdaslr.qps..ddswlnf......d (176)
Rhino2    NMyYHFLgRsGytVHVQCNAskFHQGtLLVamIPEhqLasakn.gsvtag (143)  ynlthpgeaGrvgqq.rdanlr.qps..ddswlnf......d (176)
Rhino14   NMfYHYLgRsGýtIHVQCNAskFHQGtLIValIPEhqlasalh.gnvnvg (143)  ynythpgetGrevkae.trlnpdlqpt..eeywlnf......d (177)
Rhino89   NMfYHHsLgRsGytIHVQCNsskFHQGlLIVaaIPEhqLasats.gnvsvg (143)  ynhthpgeGrevvps.rtssdnkrps..ddswlnf......d (177)
Consensus NM-YHYL-R-G.-VHVQCNA-KFHQG-LLV-VPE--L-            --------G--------  aPaappp   pp....ppppppp..  aaaBla......a
Structure aaaabbbbbDbbbbbb     bbbEbbbb                     p.pppppppa       b
```

Figure 2. Continued.

220

```
FMDA10    trekyqltLFPHQFIspRTNmtAhItvPYLgvnryDqyk..KHkpWtLVV (181)   MvLSPLtvsntaapqIkYyanIAPtyvhvaGe.....lpskE (218)
FMDA12    treeyqltLFPHQFIspRTNmtAhItvPYLgvnryDqyk..KHkpWtLVI (181)   MvLSPLtvsntaatqIkYyanIAPtyvhvaGe.....lpskv (218)
FMDC1     drekyqltLFPHQFINpRTNmtAhItvPYYgvnryDqyk..KHkpWtLVV (181)   MvVaPLtntagaqqIkYyanIAPtnvhvaGe.....lpskE (218)
FMDO1K    kreLyqltLFPHQFINpRTNttAhItvPFYgvmryDqVa..vHkpWtLVV (181)   MvVaPLtvntegapqIkYyanIAPtnvhvaGe.....rpskE (218)
FMDSat3   krdVsqltLFPHQyINpRTNttAhItvPYVgvnrhDqVq..mHkaWtLVV (181)   evMaPLttsnmqqdnVeVyanIAPtnvyraGe.....rpskQ (218)
EMCR      hqnFwqwtLYPHQFLNLRTNttvdLevPYVNiaptsswt..qHasWtLVI (216)   avVaPLtystgastsLdITasIqPVrpvFnGlRhetl..srQ (256)
EMCB      hqnFwqwtLYPHQFLNLRTNttvdLevPYVNiaptsswt..qHasWtLVI (216)   avVaPLtystgastsLdITasIqPVrpvFnGlRhetl..srQ (256)
EMCD      hqnFwqwtLYPHQFLNLRTNttvdLevPYVNiaptsswt..qHasWtLVI (216)   avVaPLtystgastsLdITasIqPVrpvFnGlRhetl..srQ (256)
Mengo     hqnFwqwtLYPHQFLNLRTNttvdLevPYVNiaptsswt..qHasWtLVI (216)   avVaPLtystgastsLdITasIqPVrpvFnGlRhevl..srQ (256)
TMEBeAn   hqnqwqwtVYPHQILNLRTNttvdLevPYVNvapssswt..qHaNWtLVV (227)   avLsPLqyatgsspdVqITasLqPVnpvFnGlRhetv..iaQ (267)
TMEGd7    hqnqwqwtVYPHQILNLRTNttvdLevPYVNvapssswt..qHaNWtLVV (227)   avLsPLqyatgsspdVqITasLqPVnpvFnGlRhetv..iaQ (267)
TMEDa     hqnqwqwtVYPHQILNLRTNttvdLevPYVNiaptsswt..qHaNWtLVV (227)   avFsPLqyasgsssdVqITasIqPVnpvFnGlRhetv..iaQ (267)
HALA      ygsIasltVYPHgLLNcniNNvvrIkVPFIytrgayhfkdpqypvWeLtI (184)   rvwseLnigtgtsaytsLnVlarftdlELhGltplst....Q (222)
HACr326   ygsIasltVYPHgLLNcniNNvvrIkVPFIytrgayhfkdpqypvWeLtI (184)   rvwseLnigtgtsaytsLnVlarftdlELhGltplst....Q (222)
HAHm175   ygsIasltVYPHgLLNcniNNvvrIkVPFIytrgayhfkdpqypvWeLtI (184)   rvwseLnigtgsaytsLnVlarftdlELhGltplst....Q (222)
HAMbb     ygsIasltVYPHgLLNcniNNvvrIkVPFIytrgayhfkdpqypvWeLtI (184)   rvwseLnigtgtsaytsLnVlarftdlELhGltplst....Q (222)
Polio1M   GtlLGNafVFPHQIINLRTNNcATLVLPYVNsLsIDsMv..KHnNWgIaI (231)   LPLaPLnfasesspeIPITLtIAPMccEFnGlRnitlpr1.Q (272)
Polio1S   GtlLGNafVFPHQIINLRTNNcATLVLPYVNsLsIDsMv..KHnNWgIaI (231)   LPLaPLnfasesspeIPITLtIAPMccEFnGlRnitlpr1.Q (272)
Polio2La  GvlaGNafVYPHQIINLRTNNcATLVLPYVNsLsIDsMt..KHnNWgIaI (230)   LPLaPLdfatessteIPITLtIAPMccEFnGlRnitvprt.Q (271)
Polio2S   GvlVGNafVYPHQIINLRTNNcATLVLPYVNsLsIDsMt..KHnNWgIaI (230)   LPLaPLdfvtessteIPITLtIAPMccEFnGlRnitvprt.Q (271)
Polio3Le  GvlLGNafVFPHQIINLRTNNsATLVLPYVNaLaIDsMv..KHnNWgIaI (230)   LPLsPLdfaqdssveIPITVtIAPMcsEFnGlRnvtapkf.Q (271)
Polio3F   GvlIGNafVFPHQIINLRTNNsATLVLPYVNaLsIDsMv..KHnNWgIaI (230)   LPLsPLdfaqdssveIPITVtIAPMcsEFnGlRnvtapk1.Q (271)
Polio3S   GvlLGNafVFPHQIINLRTNNsATLVLPYVNaLaIDsMv..KHnNWgIaI (230)   LPLsPLdfaqdssveIPITVtIAPMcsEFnGlRnvtapkf.Q (271)
CoxA21    GvlaGNafVFPHQIINLRTNNsATIVVPYINsLvIDcMa..KHnNWgIVI (231)   LPLaPLafaatsspqVPITVtIAPMctEFnGlRnitipvh.Q (272)
CoxB1     GvgVGNltIFPHQwINLRTNNsATIVMPYINsVpMDnMy..RHnN1tLMI (225)   IPFvPLnysegsspyVPITVtIAPMcaEYnGlRlass....Q (263)
CoxB3     GvgVGNltIFPHQwINLRTNNsATIVMPYtNsVpMDnMf..RHnNvtLMV (225)   IPFvPLdycpgsttyVPITVtIAPMcaEYnGlRlagh....Q (263)
CoxB4     GvgVGNltIFPHQwINLRTNNsATIVMPYINsVpMDnMF..RHnNftLMI (223)   IPFaPLdyvtgassyIPITVtVAPMsaEYnGlRlagh....Q (261)
BEV       GtsLGNslIYPHQwINLRTNNsATLILPYVNaIpIDsai..RHsNWtLaI (208)   IPVaPLkyaaetfpIVPITVtIAPMetEYnGlRraiasn..Q (248)
Rhino1a   GtlLGNx1IFPHQFINLRsNNsATLIVPYVNaVpMDsM1..RHnNWcLvI (224)   IPIsPLrsettssniVPITVsIsPMcaxFsGaRakni...kQ (263)
Rhino1b   GtlLGN11IFPHQFINLRsNNsATLIVPYVNaVpMDsM1..RHnNWsLvI (224)   IPIsPLrsettssnirPITVsIsPMcaEFsGaRaknv...rQ (263)
Rhino2    GtlLGNitIFPHQFINLRsNNsATIIaPYVNaVpMDsM1..RHnNWsLVI (225)   IPIcPLe.tssaintIPITIsIsPMcaEFSGaRakr....Q (261)
Rhino14   GtlLGN11IFPHQFINLRTNNtATIVIPYINsVpIDsMt..RHnNvsLMV (222)   IPIaPLtvptgatpsLPITVtIAPMctEFsGiRsksivp..Q (262)
Rhino89   GtlLGNipIYPHQyINLRTNNsATLILPYVNaVpMDsM1..RHnNWsLVI (225)   IPIcPLqvapggtqsIPITVsIsPMfsEFSGpRskvvfsttQ (267)
Consensus G--LGN--IFPHQFINLRTNN-ATIVPYVN-V-MD-M--RH-NW-LVI           IPL-PL--------IPITV-IAPM--EF-G-R-------Q
Structure a aaaB2aaabbFbbb   bbbG1bb   bG2   bbb                   bbHbbbb   bbbbbbbIbbbbbbbb
```

Figure 2. *Continued.*

(C)

```
          ← block 1 →                                    (pos)   ← block 2 →                                        (pos)
FMDA10    giFPVacadGyggLVTtDpktadpvYgkvynpPktnyPGrftNLLDVAEa  ( 50)  cpTFLrF........ddgkpyVvtraddtr1...LakFdVs1a.akhMsN  ( 88)
FMDA12    giFPVacsdGyggLVTtDpktadpvYgkeynpPktnyPrrftNLLDVAEa  ( 50)  cpTFLcF........ddgkpyVvtrtddtr1...LakFdVs1a.akhMsN  ( 88)
FMDC1     giFPVacsdGygnmVTtDpktadpAYgkvynpPrtaLPGrftNLLDVAEa  ( 50)  cpTFLmF........envpyVstrtdgqr1...LakFdVs1a.akhMsN  ( 87)
FMDO1K    giFPVacsdGyggLVTtDpktadpvYgkvfnpPrnqLPGrftNLLDVAEa  ( 50)  cpTFLrF........egvpyVtktdsdrv...LaqFdMs1a.akqMsN  ( 88)
FMDSat3   giiPVacndGyggFqntDpktadpiYglvsnaPtaFPGkftNLLDVAEa   ( 50)  cpTFLdF........gtpyVktgtivdei...LahIdLalg.tnqFkN  ( 87)
EMCR      spIPVtirehagtwyst1pdstvpiYgktpvaPsnyMvGeykDFLEIAQI  ( 50)  .pTFIgnk.i....p.navpyIeasn.tavktqpLatyqVt1s.cscLaN  ( 91)
EMCB      spIPVtirehagtwyst1pdstvpiYgktpvaPanyMvGeykDFLEIAQI  ( 50)  .pTFIgnk.i....p.navpyIeasn.tavktqpLatyqVt1s.cscLaN  ( 91)
EMCD      spIPVtirehagtwyst1pdstvpiYgktpvaPanyMvGeykDFLEIAQI  ( 50)  .pTFIgnk.i....p.navpyIeasn.tavktqpLatyqVt1s.cscLaN  ( 91)
Mengo     spIPVtirehagtwyst1pdstvpiYgktpvaPanyMvGeykDFLEIAQI  ( 50)  .pTFIgnk.m....p.navpyIeasn.tavktqpLavyqVt1s.cscLaN  ( 91)
TMEBeAn   spIPVtvrehkgcFystNpdttvpiYgktisTPsdyMcGefsDLLELcKL  ( 50)  .pTFLgnpntn..n.krypyfsatn.svpat.sMvdyqVa1s.cscMaN  ( 92)
TMEGd7    spIPVtvrehqgcFystNpdttvpiYgktisTPsdyMcGefsDLLELcKL  ( 50)  .pTFLgnpstd...n.krypyfsatn.svpat.slvdyqVa1s.csctaN  ( 92)
TMEDa     spIPVtvrehqgcFystNpdttvpiYgktisTPndyMcGefsDLLELcKL  ( 50)  .pTFLgnpns....mnkrypyfsatn.syptt.slvdyqVa1s.cscMcN  ( 92)
HALA      spIaVtvrehkgcFystNpdttvpiYgktisTPndyMcGefsDLLELcKL  ( 50)  ftTwtsIptlaaqfpfnasdsVgqqikvipvd.pyFfqmtntnpdqk.ci  ( 98)
HACr326   mmrnefrvsttenvVnlsNyEdarAkmsFaldqedwksdpsqgggikith  ( 50)  ftTwtsIptlaaqfpfnasdsVgqqikvipvd.pyFfqmtntnpdqk.ci  ( 98)
HAHm175   mmrnefrvsttenvVnlsNyEdarAkmsFaldqedwksdpsqgggikith  ( 50)  ftTwtsIptlaaqfpfnasdsVgqqikvipvd.pyFfqmtntnpdqk.ci  ( 98)
HAMbb     mmrnefrvsttenvVnlsNyEdarAkmsFaldqedwksdpsqgggikith  ( 50)  ftTwtsIptlaaqfpfnasdsVgqqikvipvd.pyFfqmtntnpdqk.ci  ( 98)
Polio1M   .GLPVmntPGSnQYLTaDNfQSPcALPeFDvTPpIdIPGeVkNMMELAEI  ( 49)  .DTMIpFDlsatkkntmemYrVrLsdkphtdd.pILcLsLspasdprLsH  ( 97)
Polio1S   .GLPVmmtPGSnQYLTaDNfQSPcALPeFDvTPpIdIPGeVkNMMELAEI  ( 49)  .DTMIpFDlsakkkntmemYrVrLsdkphtdd.pILcLsLspasdprLsH  ( 97)
Polio2La  .GLPVlntPGSnQYLTaDNyQSPcAiPeFDvTPpIdIPGeVrNMMELAEI  ( 49)  .DTMIpLNltnqrkntmdmYrVeLndaahsdt.pILcLsLspasdprLaH  ( 97)
Polio2S   .GLPVlntPGSnQYLTaDNyQSPcAiPeFDvTPpIdIPGeVrNMMELAEI  ( 49)  .DTMIpLNltsqrrntmdmYrVeLsdtahsdt.pILcLsLspasdprLaH  ( 97)
Polio3Le  .GLPVlntPGSnQYLTsDNhQSPcAiPeFDvTPpIdIPGeVkNVMELAEI  ( 49)  .DTMIpLNlestkrntmdmYrVtLsdsad1sq.pILcLsLspasdprLsH  ( 97)
Polio3F   .GLPVlntPGSnQYLTsDNhQSPcAiPeFDvTPpIdIPGeVkNVMELAEI  ( 49)  .DTMIpLNlentkrntmdmYrVtLsdsad1sq.pILcLsLspaadprLsH  ( 97)
Polio3S   .GLPVlntPGSnQYLTsDNhQSPcAiPeFDvTPpIdIPGeVkNMMELAEI  ( 49)  .DTMIpLNlestkrntmdmYrVtLsdsad1sq.pILcLsLspafdprLsH  ( 97)
CoxA21    .GLPtmntPGSnQfLTsDDfQSPcALPnFDvTPpIhIPGeVkNMMELAEI  ( 49)  .DTLIpMNavdgkvntmemYqIpLnd..n1skapIFcLsLspasdkrLsH  ( 96)
CoxB1     .GLPVmttPGStQfLTsDDfQSPsAmpQfDvTPeMqIPGrvkNLMEIAEV  ( 49)  .DsVpVVNntdnnvngIkaYqIpVsnsdnrr.qVFgFLqpgannvLnR  ( 97)
CoxB3     .GLPtmntPGScQfLTsDDfQSPsAmpQYDvTPeMrIPGeVkNLMEIAEV  ( 49)  .DsVpVQnvgeknsmeaYqIpVrsnegsgt.qVFgFLqpgyssvFsR  ( 97)
CoxB4     .GLPtmltPGStQfLTsDDfQSPsAmpQfDvTPeMiPGqVrNLMEIAEV   ( 49)  .DsVpVINnIkan1mtmeaYrVqVrstdemgg.q1FgFLqpgassvLqR  ( 97)
BEV       .GLPtcmltPGStQfLTsDDfQSPsAmQfDvTPeMiPGqYrNLMEIAEV   ( 49)  .EsILeaknreg.vegverYvIpVsvqda1d.q1YaLrLe1ggsgpLss  ( 96)
Rhino1a   .GLPVyitPGSgQfMTtDdmQSPcALPwYHpTkeIsPGeVkNLIEMcQV   ( 49)  .DTLIpVNnvgnnvgnvsmtVqLgnqmdmaq.eVFaikVdit.stpLat  ( 96)
Rhino1b   .GLPVyitPGSgQfMTtDDmQSPcALPwYHpTkeIsPGeVkNLIEMcQV   ( 49)  .DTLIpVNnvgtnvgnismtVqLgnqmdmaq.eVFaikVdit.sqpLat  ( 96)
Rhino2    .GLPVfitPGSgQfLTtDDfQSPcALPwYHpTkeIsPGeVkNLVEIcQV   ( 49)  .DsLVpINntdtyinsenmYsVvLqssinapd.kIFsIrtdva.sqpLat  ( 96)
Rhino14   .GLPtttPGSgQfLTtDDrQSPsALPnYEpTrIhIPGqVkNLLEIiQV    ( 49)  .DTLIpMNnnthtk.devnsYlIpLnanrqne..qVFgtnLfig.dgvFkt  ( 94)
Rhino89   .GLPVmltPGSgQfLTtDDtQSPsAFPyFHpTkeIfIPGqvrNLIEMcQV  ( 49)  .DTLIpVNnotqenvrsvnmtVdLrtqvdlak.eVFsIpVdia.sqpLat  ( 96)
Consensus -GLPV--PGS-QfLT-DD-QSP-ALP-FD-TP-I-IPG-V-NLLEIAEV         -DT-I--N--------Y-V-L--------IF-F-L------L-N
                                                 aaaaZaaa                                                  bbCbbbaaa.Alaa
Structure bbbbbb                                                         bb1kkkkkkk.kkkkkkbB2
```

Figure 2. *Continued.*

222

```
FMDAIO    TyLsgIaqYYTQysGtInLhFMFtGstdskaRYMVAYiPlGve..tpPdt (136)  peEAahciHaeWD.tGLnSkftFsIPyVSaaDYayIasd.....taetTn (180)
FMDAI2    TyLsgIaqYYTQysGtInLhFMFtGstdskaRYMVAYiPPGve..tpPet (136)  pegAahciHaeWD.tGLnSkftFsIPyVSaaDYayIasd.....taetTn (180)
FMDC1     TyLagLaqYYTQytGtInLhFMFtGptdakaRYMVAYvPPGmd...aPdn (134)  peEAahciHaeWD.tGLnSkftFsIPyISaaDYayIash.....eaetTc (178)
FMDO1K    TFLagLaqYYTQysGtInLhFMFtGptdakaRYMVAYaPPGme...pPkt (135)  peaAahciHaeWD.tGLnSkftFsIPyLSaaDYayIasg.....vaetTn (179)
FMDSat3   TyLagLaqYYaQysGSInLhFMYtGptqskaRFMVAYiPPaph..rsPdt (135)  pekAahcyHseWD.tGLnSkftFtVPyMSaaDFayTycd.....epeqas (179)
EMCR      TFLaaLsrnFaQyrGSLvytFVFtGtammkgKFLIAYtPPGag...kPts (138)  RdQAMqaTyaIWDL.GLnSsysFtVPfISptHFRmvgtd.....qvniTn (182)
EMCB      TFLaaLsrnFaQyrGSLvytFVFtGtammkgKFLIAYtPPGag...kPts (138)  RdQAMqaTyaIWDL.GLnSsysFtVPfISptHFRmvgtd.....qvniTn (182)
EMCD      TFLaaLsrnFaQyrGSLvytFVFtGtammkgKFLIAYtPPGag...kPts (138)  RdQAMqaTyaIWDL.GLnSsysFtVPfISptHFRmvgtd.....qvniTn (182)
Mengo     TFLaaLsrnFaQyrGSLvytFVFtGtammkgKFLIAYtPPGag...kPts (138)  RdQAMqaTyaIWDL.GLnSsysFtVPfISptHFRmvgtd.....qaniTn (182)
TMEBeAn   sMLaaVarnFnQyrGSLnFlFVFtGaamvkgKFLIAYtPPGag...kPtt (139)  RdQAMqaTyaIWDL.GLnSsfnFtaPfISptHYRqTsyt.....sptiTs (183)
TMEGd7    sMLaaVarnFnQyrGSLnFlFVFtGaamvkgKFrIAYtPPGag...kPtt (139)  RdQAMqaTyaIWDL.GLnSsfnFtaPfISptHYRqTsyt.....sptiTs (183)
TMEDa     sMLaaVarnFnQyrGSLnFlFVFtGaamvkgKFlIAYtPPGag...kPtt (139)  RdQAMqaTyaIWDL.GLnSsfvFtaPfISptHYRqTsyt.....satias (183)
HALA      TaLasIcqmFcFwrGdLvFdFqvfptkyhsgRLLFcvPgnelidvtgit (148)  lkQAttapcaVmDItGVQStlrFvPWISdtpYRvnrytksahqkgeyTa (198)
HACr326   TaLasIcqmFcFwrGdLvFdFqvfptkyhsgRLLFcvPgnelidvtgit (148)  lkQAttapcaVmDItGVQStlrFvPWISdtpYRvnrytksahqkgeyTa (198)
HAhm175   TaLasIcqmFcFwrGdLvFdFqvfptkyhsgRLLFcFvPgnelidvsgit (148)  lkQAttapcaVmDItGVQStlrFvPWISdtpYRvnrytksahqkgeyTa (198)
HAMbb     TaLasIcqmFcFwrGdLvFdFqvfptkyhsgRLLFcFvrgnelidvsgit (148)  lkQAttapcaVmDItGVQStlrFvPWISdtpYRvnrytksahqkgeyTa (198)
Polio1M   TMLGeIlnYYTHWaGSLkFtFLFcGfmmatgKLLVsYaPPGad...aPks (144)  RkEAMLGTHVIWDl.GLQSsctMvVPWISnttYRqTi......ddsfTe (186)
Polio1S   TMLGeIlnYYTHWaGSLkFtFLFcGsmmatgKLLVsYaPPGad...pPkk (144)  RkEAMLGTHVIWDl.GLQSsctMvVPWISnttYRqTi......ddsfTe (186)
Polio2La  TMLGeIlnYYTHWaGSLkFtFLFcGsmmatgKLLVsYaPPGae...aPks (144)  RkEAMLGTHVIWDl.GLQSsctMvVPWISnttYRqTi......ndsfTe (186)
Polio2S   TMLGeIlnYYTHWaGSLkFtFLFcGsmmatgKLLVsYaPPGae...aPks (144)  RkEAMLGTHVIWDl.GLQSsctMvVPWISnttYRqTt......ndsfTe (186)
Polio3Le  TMLGeVlnYYTHWaGSLkFtFLFcGsmmatgKiLVAYaPPGaq...pPts (144)  RkEAMLGTHVIWDl.GLQSsctMvVPWISnvtYRqTt......qdsfTe (186)
Polio3F   TMLGeVlnYYTHWaGSLkFtFLFcGsmmatgKLLVAYaPPGaq...pPts (144)  RkEAMLGTHVIWDl.GLQSsctMvVPWISnvtYRqTt......qdsfTe (186)
Polio3S   TMLGeVlnYYTHWaGSLkFtFLFcGsmmatgKiLVAYaPPGaq...pPts (144)  RkEAMLGTHVIWDl.GLQSsctMvVPWISnvtYRqTt......qdsfTe (186)
CoxA21    TMLGeIlnYYTHWtGSIrFtFLFcGsmmatgKLLLsYsPPGak...pPtn (143)  RkDAMLGTHIIWDl.GLQSscsMvaPWISntvYRrca......rdfTe (185)
CoxB1     TLLGeIlnYYTHWsGSIkLtFMFcGsamatgKFLLAYsPPGag...vPkn (144)  RrDAMLGTHVIWDV.GLQSscvLcVPWISqtHYRyvv.......edeyTa (186)
CoxB3     TLLGeIasYYTHWtGSLrFsFMFcGtanttIKLLLAYsPPGid...aPtk (143)  RvDAMLGTHVVWDV.GLQSscvLcIPWISqtHYRyva......sdecTa (186)
CoxB4     TLLGeIlnYYTHWsGSLkLtFVFcGsamatgKFLLAYsPPGag...aPds (144)  RkNAMLGTHVIWDV.GLQSsMvVPWISasHYRyvv.......ddkyTa (186)
BEV       sLLGtLakhYTQwsGSVeItcMFtGtfmttgKvLLAYtPPGgd...mPrn (133)  ReEAMLGTHVIWDF.GLQSsitLvIPWISasHFRgvsnd..dvInyqyqa (180)
Rhino1a   TLIGeIasYYTHWtGSLrFsFMFcGtanttIKLLLAYtPPGid...ePtl (143)  RkDAMLGTHVVWDV.GLQStisLvVPWVSasHFRlTa......dnkysm (185)
Rhino1b   TLIGeIssYFTHWtGSLrFsFMFcGtanttvKLLLAYtPPGia...ePtt (143)  RkDAMLGTHVWDV.GLQStisMvVPWISasHYRnTs.......pgrsTs (185)
Rhino2    TLIGeIssYFTHWtGSLrFsFMFcGtanttvKLLLAYtPPGia...ePtt (143)  RkDAMLGTHVWDV.GLQStisMvVPWISasHYRnTs.......pgrsTs (185)
Rhino14   TLIGeIvqYYTHWsGSLrFsLMYtGpalssaKLILAYtPPGar...gPqd (140)  RrEAMLGTHVVWDI.GLQStivMtIPWtSgvQFRyTd......pdtyTs (183)
Rhino89   TLIGeLasYYTHWtGSLrFsFMFcGsasstIKLLIAYtPPGvg...kPks (143)  RrEAMLGTHLWDV.GLQStasLvVPWVSasHFRfTt.......pdtyss (185)

Consensus TLLG-I--YYTHW-GSL-F-FMF-G------KLLLAY-PPG------P--        R--EAMLGTHVIWDV-GLQS-----VPWIS--HYR-T------------T-
Structure aaaA2aaaabbbbbbbbbbbbbbb  bbbEbbbb         a            aaBaaaabbbFbbb     bbGlbbb       bG2
```

Figure 2. *Continued.*

223

```
FMDA10    vqGwVcvyqiTH.....gkaendtLLvsaSAgkDFeLRLpiDprtq....  (221) .....vttQ (246)
FMDA12    vqGwVciyqiTH.....gkaeddtLVvsaSAgkDFeLRLpiDprsq....  (221) .....vttQ (246)
FMDC1     vqGwVclfqiTH.....gkadadaLVvLaSAgkDFeLRLpvDarqq....  (219) .....vttQ (246)
FMD01K    vqGwVclfqiTH.....gkadgdaLVvLaSAgkDFeLRLpvDarae....  (220) .....vttQ (246)
FMDSat3   aqGwWtlyqiTd.....thdpdsaVLvsVAgaDFeLRLpiNpatq....   (220) .....alaQ (238)
EMCR      adGwWtvwqlTpLtyPpgcptsakILtMVSAgkDFsLKMpispapwspq.  (231) .....alaQ (238)
EMCB      vdGwWtvwqlTpLtyPpgcptsakILtMVSAgkDFsLKMpispapwspq.  (231) .....ampQ (238)
EMCD      vdGwWtvwqlTpLtyPpgcptsakILtMVSAgkDFsLKMpispapwspq.  (231) .....ampQ (238)
Mengo     vdGwWtvwqlTpLtyPsgtptnsdILtLVSAgdDFtLRMpisptkwvpq.  (232) .....alpQ (238)
TMEBeAn   vdGwWtvwklTpLtyPsgtptnsdILtLVSAgdDFtLRMpisptkwvpq.  (232) .....alpQ (238)
TMEGd7    vdGwWtvwqlTpLtyPsgtpthsdILtLVSAgdDFtLRMpisptkwvpq.  (232) .....ampQ (238)
TMEDa     vdGwWtvwqlTpLtyPsgapvnsdILtLVSAgdDFtLRMpisptkwapq.  (232) .....ampQ (238)
HALA      i.GKLivycynRLtsPsnvashvrVnvyLSAinlecFa.plyhamd....  (242) .....alpQ (238)
HACr326   i.GKLivycynRLtsPsnvashvrVnvyLSAinlecFa.plyhamd....  (242) .....alpQ (238)
HAHm175   i.GKLivycynRLtsPsnvashvrVnvyLSAinlecFa.plyhamd....  (242) liartQ (240)
HAMbb     i.GKLivycynRLtsPsnvashvrVnvyLSAinlecFa.plyhamd....  (242) .....nfyQ (238)
Polio1M   g.GyIsvFYQTRIVVPlstpremdILgFVSAcnDFsVRLlRDtthieQk.  (234) .....nffQ (238)
Polio1S   g.GyIsvFYQTRIVVPlstpremdILgFVSAcnDFsVRLmRDtthieQk.  (234) .....nfyQ (238)
Polio2La  g.GyIsmFYQTRVVVPlstprkmdILgFVSAcnDFsVRLlRDtthisQe.  (234) .....ampQ (238)
Polio2S   g.GyIsmFYQTRVVVPlstprkmdILgFVSAcnDFsVRLlRDtthisQe.  (234) .....ampQ (238)
Polio3Le  g.GyIsmFYQTRIVVPlstpksmsMLgFVSAcnDFsVRLlRDtthisQs.  (234) .....alpQ (238)
Polio3F   g.GyIsmFYQTRIVVPlstpkamdMLgFVSAcnDFsVRLlRDtthisQa.  (234) .....ampQ (238)
Polio3S   g.GyIsmFYQTRIVVPlstpksmsMLgFVSAcnDFsVRLlRDtthisQs.  (234) .....alpQ (238)
CoxA21    g.GFItcFYQTRIVVPastptsmFMLgFVSAcpDFsVRLlRDtshisQsk  (234) liartQ (240)
CoxB1     a.GyVtcwYQTNIIVPadvqstcdILcFVSAcnDFsVRMlKDtpfirQd.  (234) .....nfyQ (238)
CoxB3     g.GFItcwYQTNIVVPadaqsscyIMcFVSAcnDFsVRLlKDtpfisQe.  (234) .....nffQ (238)
CoxB4     s.GFIscwYQTNVIVPaeaqkscyIMcFVSAcnDFsVRMlRDtqfikQt.  (234) .....nfyQ (238)
BEV       a.GhVtiwYQTNMVIPpgfpntagIImMlaAqpNFsFRIqKDredmtQt.  (238) .....ailQ (242)
Rhino1a   a.GyItcwYQTxLVVPpstpqtadMLcFVSAckDFcLRMaRDtdlhiQsg  (234) p..ieQ (238)
Rhino1b   a.GyItcwYQTNLVVPpntpqtadMLcFVSAckDFcLRMaRDtdlhiQsg  (234) p..ieQ (238)
Rhino2    .GyItcwYQTRLVIPpqtpptarLLcFVSgckDFcLRMaRDtnlhlQsg   (233) a..iaQ (237)
Rhino4    a.GFLscwYQTsLILPpettqqvyLLsFISAcpDFkLRLmKDtqtisQtv  (231) a...ltE (236)
Rhino89   a.GyItcwYQTNFVVPdstpdnakMvcMVSAckDFcLRLaRDtnlhtQeg  (234) v..ltQ (238)
Consensus --G-I---YQT-LVVP------------IL-FVSA--DF-LRL-RD------Q-------Q
Structure   bbbbbbHbbbbb            bbbbbbbIbbbbbbbb
```

Figure 2. *Continued.*

224

```
(D) FMDA10    .............................................  .tttgesadpvttvEnyggdtqv...qrRhhtdV.gFImd.....rfV ( 40)
    FMDA12    .............................................  .ttatgesadpvttvEnyggetqv..qrRhhtdV.sFImd.....rfV ( 40)
    FMDA22    .............................................  .tttgesadpvttvEnyggetqv..qrRqhtdV.tFImd.....rfV ( 40)
    FMDAsia1  .............................................  .tttgesadpvttvEnyggetqt..arRlhtdV.aFlld.....rfV ( 40)
    FMDC1     .............................................  .tttgesadpvttvEnyggetqv..qrRhhtdV.aFVld.....rfV ( 40)
    FMDC3     .............................................  .tttgesadpvttvEnyggetqi..qrRhhtdV.aFVld.....rfV ( 40)
    FMDO1Bfs  .............................................  .ttsagesadpvttvEnyggetqi..qrRqhtdV.sFImd.....rfV ( 40)
    FMDO1K    .............................................  .ttsagesadpvttvEnyggetqi..qrRqhtdV.sFImd.....rfV ( 40)
    FMDSat1   .............................................  .ttsageGadpvttdasahggntrtt..sRaHtdV.tFLld.....rft ( 40)
    FMDSat2   .............................................  .ttsageGaevvttnptthggkvttp..sRvHtdV.aFLld.....rst ( 40)
    FMDSat3   .............................................  .ttsageGadvvtthggkvsvp..rRqHtnV.eFLld.....rft ( 40)
    EMCR      .............................................  gVenaEkGvtentnatadfva.qpvy].penQtkVafFynRsspIgaft ( 47)
    EMCB      .............................................  gVenaErGvtedtdatadfva.qpvy].penQtkVafFydRsspIgaft ( 47)
    EMCD      .............................................  gVenaErGvtedtdatadfva.qpvy].penQtkVafFydRsspIgafa ( 47)
    Mengo     .............................................  gVdnaEkGvtentdatadfva.qpvy].penQtkVafFydRsspIgafa ( 47)
    TMEBeAn   .............................................  gVdnaEkGvsnddasvdfva.epvkl.penQrVafFydRavpIgmlr ( 47)
    TMEGd7    .............................................  gldnaEkGkvsnddasvdfva.epvkl.penQtrVafFydRavpIgmlr ( 47)
    TMEDA     .............................................  gsdnaEkGkvsnddasvdfva.epvkl.penQtrVafFydRavpIgmlr ( 47)
    HALA      vgddsggfsttvsteqnvpdpqvgittmrdlkgkanrgkmdvsgvqapvg ( 50)  alttiE.dpvlakkvPEtfpelkpgesrhtsdHmslykFMgRshfLctft ( 99)
    HACr-326  vgddsggfsttvsteqnvpdpqvgittmkdlkgkanrgkmdvsgvqapvg ( 50)  alttiE.dpalakkvPEtfpelkpgesrhtsdHmslykFMgRshfLctft ( 99)
    HAHm175   vgddsggfsttvsteqnvpdpqvgittmkdlkgkanrgkmdvsgvqapvg ( 50)  alttiE.dpvlakkvpPEtfpelkpgesrhtsdHmslykFMgRshfLctft ( 99)
    HAMbb     vgddsggfsttvsteqnvpdpqvgittmkdlkgkanrgkmdvsgvqapvg ( 50)  alttiE.dpvlakkvPEtfpelkpgesrhtsdHmslykFMgRshfLctft ( 99)
    Polio1M   ....glgqmlesmidntvretvgaatsrdalpnteasgpthskeip ( 42)  aLtavtEGatnplvpsDtvqtrhvvqh.rsRsEsslesFFaRgacVtimt ( 91)
    Polio1S   ....glgqmlesmidntvretvgaatsrdalpnteasgpahskeip ( 42)  aLtavtEGatnplvpsDtvqtrhvvqh.rsRsEsslesFFaRgacVait ( 91)
    Polio2La  ....glgdliegvvegvtrnaltpltpann]pdtqssgpahsketp ( 42)  aLtavtEGatnplvpsDtvqtrhviqk.rtRsEstVesFFaRgacVaiie ( 91)
    Polio2S   ....glgdmiegavegitknalvpptstns]pghkpsgpahskeip ( 42)  aLtavtEGatnplvpsDtvqtrhviqr.rtRsEstVesFFaRgacVaiie ( 91)
    Polio3Le  ....gledlisevagga]..tlslpkqqds]pdtkasgpahskevp ( 42)  aLtavtEGatnplapsDtvqtrhvvqr.rsRsEstlesFFaRgacVaiie ( 89)
    Polio3F   ....gvddlitevaqna]..alslpkpqsn]pdtkasgpahskevp ( 40)  tLtavtEGatnplvpsDtvqtrhviqq.rsRsEstlesFFaRgacVaiie ( 89)
    Polio3S   ....gledlisevagga]..tlslpkqqds]pdtkasgpahskevp ( 40)  aLtavtEGatnplapsDtvqtrhvvqr.rsRsEstlesFFaRgacVaiie ( 89)
    CoxA21    ....gledlidtaiknalrv.....sqplrpsqlkpqngvnsqevp ( 37)  aLtavtEGasgqapsDvvetrhviny.ktRsEscLesFFaRaacVtils ( 86)
    CoxB1     .........gpveesveramvr.......vadtvsskptnsesip ( 29)  aLtaaEtGhtsqvvpsDtmqtrhvkny.hsRsEsslenFLcRsacVyyat ( 78)
    CoxB3     .........gpvedaitaaigr.......vadtvgtgpnnseaip ( 29)  aLtaatGhtsqvvpgDtmqtrhvkny.hsRsEstIenFLcRsacVyfte ( 78)
    CoxB4     .........gpteesveramgr.......vadtiarngvnseqip ( 29)  aLtavEtrhtsqvpsDtmqtrhvhny.hsRsEsslenFLcRsacViyik ( 78)
    BEV       ....ndpgkmlkdaidkqvagaiv.....agttsthsvatdstp ( 36)  aLqaaEtGatstardesmietrivpt.hgiHtsVesFFgRsslVgmpl ( 85)
    Rhino1a   .........npvenyidev]nev]v.....vpnikeshttsnsap ( 32)  lLdaaEtGhtsnvqpeDaietryvits.qtRdEmslesFLgRsgcVhisr ( 81)
    Rhino1b   .........npvenyidev]nev]v.....vpnikeshttsnsap ( 32)  lLdaaEtGhtsnvqpeDaietryvmts.qtRdEmslesFLgRsgcVhisr ( 81)
    Rhino2    .........npvenyidev]nev]v.....vpninssnpttsnsap ( 32)  aLdaaEtGhtssvqpeDvvietryyqts.qtRdEmsLesFLgRsgcIhesk ( 81)
    Rhino14   ....glgdeleevivektkqt.....vasi.ssgpkhtqkvp ( 32)  iLtantGatmpvlpsDsietrtymh.fngsEtdVecFLgRaacVhvte ( 81)
    Rhino39   .........npvenyidgv]nev]v.....vpnireshpttsnaap ( 32)  aLdaaEtGhtsslqpeDtietryyqts.htRdEmsVesFLgRsgcIhist ( 81)
    Rhino49   .........npvenyidev]nev]v.....vpninsshpttsnsap ( 32)  aLdaaEtGhtssvqpeDietryyqts.qtkDEmsLesFLgRsgcIhesk ( 81)
    Rhino89   .........npvenyidsv]nev]v.....vpniqpstsvshaap ( 32)  aLdaaEtGhtssvqpeDmietryyitd.qtRdEts1esFLgRsgcIamie ( 81)

    Consensus -----------------------------------------------    -L---E-G-------D------R-E---V--FL-R---V----
    Structure                           aaaaa                    aaaaa   aaazZaa                   bbbbBBb
```

Figure 2. *Continued.*

225

```
FMDA10    kinslspt...hvidlmqthkhgI    ( 61)    ...........vgallRaa..TyyfsDlEIvVr....   ( 81)
FMDA12    kiksinpt...hvidlmqthqhgL    ( 61)    ...........vgallRaa..TyyfsDlEIvVr....   ( 81)
FMDA22    kiqnlnpt...hvidlmqthqhgL    ( 61)    ...........vgallRaa..TyyfsDlEIvVr....   ( 81)
FMDAsial  kftpkn.t..qtldlmqipshtL     ( 60)    ...........vgallRsa..TyyfsDlEIaLv....   ( 80)
FMDC1     kvtvsgnq...htldvmqahkdnI    ( 61)    ...........vgallRaa..TyyfsDlEIaVt....   ( 81)
FMDC3     kvhvsgnq...htldvmqvhkdsI    ( 61)    ...........vgallRaa..TyyfsDlEIaVt....   ( 81)
FMD01Bfs  kvtpqnqi...nildlmqvpshtL    ( 61)    ...........vgallRas..TyyfsDlEIaVk....   ( 81)
FMD01K    kvtpqnqi...nildlmqipshtL    ( 61)    ...........vgallRas..TyyfsDlEIaVk....   ( 81)
FMDSat1   lvgktndn..klvfdllstkeksL    ( 62)    ...........vgallRaa..TyyfsDlEVadcvg..   ( 84)
FMDSat2   hvhtdtta...fvvdlmdtkekaL    ( 61)    ...........vgaiRsa...TyyfcDlEVa.cvg..   ( 82)
FMDSat3   hvgkvnes...rtislmdtkehtL    ( 61)    ...........vgaisgsa..TyyfcDlEVaIl....   ( 81)

EMCR      vksgslesgfapfsngtcpnsviltpgpqfdpaydqlrpqrl.tei.wgn ( 95)  g.nee.tskvfplkskqdysFclfs.pFvVykcDlEVTLsphtsgn.... (138)
EMCB      vksgslesgtpfsnqtcpnsviltpgpqfdpaydqlrpqrl.tei.wgn ( 95)  g.nee.tskvfplkskqdysFclfs.pFvVykcDlEVTLsphtsgn.... (138)
EMCD      vksgslesgfapfsneqtcpnsviltpgpqfdpaydqlrpqrl.tei.wgn ( 95) g.nee.tskvfplkskqdysFclfs.pFvVykcDlEVTLsphtsgn.... (138)
Mengo     vksgslesgfapfsnkacpnsviltpgpqfdpaydqlrpqrl.tei.wgn ( 95)  g.nee.tsevfplktkqdysFclfs.pFvVykcDlEVTLsphtsga.... (138)
TMEBeAn   pgqnmet.tfnyqendyrlncllltplpsfcpdsssg.pqktkapvqwrw ( 95)  vrsggvnganfplmtkqdyaFlcfs.pFTYkcDlEVTVsalgmtr....  (140)
TMEGd7    pgqnmet.tfsyqendfrlncllltplpsycpdsssg.pvrtkapvqwrw ( 95)  vrsgganganfplmtkqdyaFlcfs.pFTYkcDlEVTVsamgagt....  (140)
TMEDa     pgqnies.tfvyqendlrlncllltplpsfcpdstsg.pvrtkapvqwrw ( 95)  vrsggtt..nfplmtkqdyaFlcfs.pFTYkcDlEVTVsalgtdt....  (138)

HALA      fnsnnkeytfpitlsstsnpphgLp  (124)    ......stLrwffnlFqLyRgplDLTliitg.atd....  (152)
HACr326   fnsnnkeytfpitlsstsnpphgLp  (124)    ......stLrwffnlFqLyRgplDLTliitg.atd....  (152)
HAHm175   fnsnnkeytfpitlsstsnpphgLp  (124)    ......stLrwffnlFqLyRgplDLTliitg.atd....  (152)
HAMbb     fnsnnkeytfpitlsstsnpphgLp  (124)    ......stLrwffnlFqLyRgplDLTliitg.atd....  (152)

Polio1M   vdnpast....tnkdklfavwklty  (112)    kdtvqLrrKleFFTYsRFDmEFTFvvtanftetnn  (147)
Polio1S   vdnsast....knkdklftvwklty  (112)    kdtvqLrrKleFFTYsRFDmEFTFvvtanftetnn  (147)
Polio2La  vdndapt....kraslfsvwklty   (112)    kdtvqLrrKleFFTYsRFDmEFTFvvtsnytdann  (147)
Polio2S   vdndapt....krasrlfsvwklty  (112)    kdtvqLrrKleFFTYsRFDmEFTFvvtsnyidann  (147)
Polio3Le  vdneqpt....traqklfamvrity  (110)    kdtvqLrrKleFFTYsRFDmEFTFvvtanftnann  (145)
Polio3F   vdneqpa....tnvqklfatwrity  (110)    kdtvqLrrKleFFTYsRFDmEFTFvvtanftnsnn  (145)
Polio3S   vdneqpt....traqklfamvrity  (110)    kdtvqLrrKleFFTYsRFDIEMTFvvtanftnann  (145)
CoxA21    ltnssks....geekkfniwnlty   (107)    tdtvqLrrKleFFTYsRFDIEMTFvftenypstas  (142)
CoxB1     ynnnse.....kgyaewvint      ( 94)    rqvaqLrRkleFFTYvRFDlELTFvitsaqepsta  (129)
CoxB3     yensga.....kryaewvltp      ( 94)    rqaaqLrrKleFFTYvRFDlELTFvitstqqpstt  (129)
CoxB4     yssaesnnl..kryaewvInt      ( 97)    rqvaqLrrKmemFTYiRcDmELTFvitshqemsta  (132)
BEV       latgtsit......hwrldf       ( 99)    refvqLrakmswFTYvRFDvEFTIiatsstgqnvt  (134)
Rhino1a   ikvdytdy...ngqdinftkwkitl  (103)    qemaqIrrKfelFTYvRFDsEVTLvpciagrgddi  (138)
Rhino1b   ikvdyndy...ngvnkrfttwkitl  (103)    qemaqIrrKfelFTYvRFDsEITLvpcisalsqdi  (138)
Rhino2    levltny....nkenftvwaln     (101)    qemaqIrrKfelFTYtRFDsEITLvpcisalsqdi  (138)
Rhino14   iqnkdatgidnh.reaklfndwkinl (106)    sslvqLrrKlelFTYvRFDsEyTIlatasqpdsan  (141)
Rhino39   itmkkeny....nehnfvdwkitl   (101)    qemaVrrKfemFTYvRFDsEITLvpciagrgedi   (136)
Rhino49   levltny....nennfvwnln      (101)    qemaqIrrKfelFTYtRFDsEITLvpcislskdi   (136)
Rhino89   fntssdkt..ehdkigkgfktwkvsl (105)    qemaqIrrKyelFTYtRFDsEITIvtaaaaqnds   (140)

Consensus ------------aa.aaaa-bCb-------------------------------------L--K--FTY-R-D-EITF---------
Structure bbb                                           aaaaAaaaabbbbbbbbbDbbbbbbb
```

Figure 2. *Continued.*

226

Figure 2 (continued) — multiple sequence alignment.

First alignment block:

```
FMDA10    .....hdgnltwVPnGaPe.........aalsnt.sNptaYnkap.  (111)
FMDA12    .....hdgnltwVPnGaPe.........aalsnt.gNptaYnkap.  (111)
FMDA22    .....hdgnltwVPnGaPe.........aalsnt.gNptaYlkap.  (111)
FMDAsia1  .....htpqkttwVPnGaPe........taldnq.tNptaYhkqp.  (110)
FMDC1     .....htgkltwVPnGaPv.........saldnt.tNptaYhkgp.  (111)
FMDC3     .....htgkltwVPnGaPv.........saldnt.aNptaYhkgp.  (111)
FMDO1Bfs  .....herdltwVPnGaPe.........kaldnt.tNptaYhkap.  (111)
FMDO1K    .....hegdltwVPnGaPe.........kaldnt.tNptaYhkap.  (111)
FMDSat1   .....tnawvgwtPnGsPv.........ltevg.dNpvVFsrrg.   (113)
FMDSat2   .....khkrvfwqPnGaPr.........ttqlg.dNpmVFshnn.   (111)
FMDSat3   .....gtawaawVPnGrPh.........tgrve.dNpvVhskgs.   (110)
EMCR      .....hgllvrwcPtGtPtk.pttqvlhevsslseg.rtpqVVsagpg (179)
EMCB      .....hglwrwcPtGtPtk.pttqvlhevsslseg.rtpqVVsagpg (179)
EMCD      .....hglwrwcPtGtPak.pttqvlhevsslseg.rtpqVVsagpg (179)
Mengo     .....hgllvrwcPtGtPtk.pttqvlhevsslseg.rtpqVVsagpg (179)
TMEBeAn   .....vasvlrwaPtGaPad.vtdqligytpslget.rNphMwIvgga (181)
TMEGd7    .....vssvlrwaPtGaPad.vtdqligytpslget.rNphMwIvgsg (181)
TMEDa     .....vasvlrwaPtGaPad.vtdqligytpslget.rNphMwIvgag (179)
HALA      .....vdgmawftPvGlavdtpwvekesalsidyktalgavrFntrrt (195)
HACr326   .....vdgmawftPvGlavdtpwvekesalsidyktalgavrFntrrt (195)
HAhm175   .....vdgmawftPvGlavdtpwvekesalsidyktalgavrFntrrt (195)
HAMbb     .....vdgmawftPvGlavdtpwvekesalsidyktalgavrFntrrt (195)
Polio1M   ghaln..qvyqimyVPpGaPvpe.....kwddytwqtssNpsIFyty. (187)
Polio1S   ghaln..qvyqimyVPpGaPipg.....kwddytwqtssNpsVFyty. (187)
Polio2La  ghaln..qvyqimyIPpGaPipg.....kwndytwqtssNpsIFyty. (187)
Polio2S   ghaln..qvyqimyIPpGaPipg.....kwndytwqtssNpsVFyty. (187)
Polio3Le  ghaln..qvyqimyVPpGaPtpk.....swddytwqtssNpsIFyty. (185)
Polio3F   ghaln..qvyqimyIPpGaPtpk.....swddytwqtssNpsIFyty. (185)
Polio3S   ghaln..qvyqimyIPpGaPtpk.....swddytwqtssNpsIFyty. (185)
CoxA21    gevrn..qcdqimyIPpGaPrps.....swddytwqsssNpsIFymy. (182)
CoxB1     tsvdapvqtqqimyVPpGgPvpt.....kvtdyawqtstNpsVFwte. (171)
CoxB3     qnqdaqilthqimyVPpGgPvpd.....kvdsyvvwtstNpsVFwte. (171)
CoxB4     tnsdvpvqthqimyVPpGgrvpd.....svndyvwqtstNpsIFwte. (174)
BEV       teqht..tyqvmyVPpGaPvps.....nqdsfqwqsqcNpsVFadt.  (173)
Rhino1a   gh.....ivmqymyVPpGaPips.....krndfswqsgtNmsIFwqh. (175)
Rhino1b   gh.....vvmqymyVPpGaPipk.....trndfswqsgtNmsIFwqh. (175)
Rhino2    gh.....itmqymyVPpGaPvpn.....srddyawqsgtNasVFwqh. (183)
Rhino14   yssn...lvvqamyVPpGaPnpk.....ewddytwqsasNpsVFFkv. (180)
Rhino39   gh.....ivmqymyVPpGaPvpk.....krddytwqsgtNasIFwqh. (173)
Rhino49   gh.....itmqymyVPpGaPvpk.....srddyawqsgtNasIFwqh. (173)
Rhino89   gh.....ivlqfmyVPpGaPvpe.....krddytwqsgtNasVFwqe. (177)
Consensus -------------VP-G-P----------N--VF----
Structure    bbbbEbbb       aaaBaa       bbbFbbb
```

Second alignment block:

```
FMDA10    ...ftRLaLPYtaphrvlatv.YDGtnkysas....dsrsgdlgsiaar  (152)
FMDA12    ...ftRLaLPYtaphrvlatv.YNGtnkysas....gsgvrgdfgsLapr  (153)
FMDA22    ...ftRLaLPYtaphrvlatv.YNGtskysag....gtgrrgdlgpLaar  (153)
FMDAsia1  ..itRLaLPYtaphrvlatv.YNGkttygee....ptmrgdravLask   (151)
FMDC1     ..ltRLaLPYtaphrvlatg.YtGtttytas.....trgdlahLtat    (150)
FMDC3     ..ltRLaLPYtaphrvlatt.YtGttaytas.....arrgdlahLaaa   (151)
FMDO1Bfs  ..ltRLaLPYtaphrvlatv.YNGecrysrn...avpnlrgdlqvLaqk  (154)
FMDO1K    ..ltRLaLPYtaphrvlatv.YNGecrysrn...avpnlrgdlqvLaqk  (154)
FMDSat1   ..ttRFaLPYtaphrvlatv.YNGdckykptgtaprenirgdlatLaar  (159)
FMDSat2   ..vtRFaIPFtaphrllstv.YNGeceytktvta...irgdrevLaqk   (153)
FMDSat3   ..vRFglPYtaphgvlatv.YNGnckysetqra.tsrrgdlavLaqr    (154)
EMCR      isnqisFvVPYnsplsvlsavwYNGhkrf.........dntgsLgia    (217)
EMCB      itnqisFvVPYnsplsvlpavwYNGhkrf.........dntgsLgia    (217)
EMCD      isnqisFvVPYnsplsvlpavwYNGhkrf.........dntgsLgia    (217)
Mengo     tsnqisFvVPYnsplsvlpavwYNGhkrf.........dntgdLgia    (217)
TMEBeAn   ns.qvsFvVPYnsplsvlpaawFNGwsdf.........gntkdFgva    (218)
TMEGd7    ns.qisFvVPYnsplsvlpaawFNGwsdf.........gntkdFgva    (218)
TMEDa     nt.qisFvVPYnsplsvlpaawFNGwsdf.........gntkdFlva    (216)
HALA      gn..iqlrLPwysylyavsga.LDGlgdk                      (221)
HACr326   gn..iqlrLPwysylyavsga.LDGlgdk                      (221)
HAhm175   gn..iqlrLPwysylyavsga.LDGlgdk                      (221)
HAMbb     gn..iqlrLPwysylyavsga.LDGlgdk                      (221)
Polio1M   gtapaRisVPYvgisnayshf.YDGfskvplk...dqsaalgdslYgaa  (232)
Polio1S   gtapaRisVPYvgisnayshf.YDGfskvplk...dqsaalgdslYgaa  (232)
Polio2La  gappaRisVPYvgianayshf.YDGfakvpla...gqastegdslYgaa  (232)
Polio2S   gappaRisVPYvgianayshf.YDGfakvpla...gqastegdslYgaa  (232)
Polio3Le  gaapaRisVPYvglanayshf.YDGfakvplk.tdandqigdslYsam   (231)
Polio3F   gaapaRisVPYvglanayshf.YDGfakvplk.sdandqvgdslYsam   (231)
Polio3S   gaapaRisVPYvglanayshf.YDGfakvplk.tdandqigdslYsam   (231)
CoxA21    gnappRMsIPYvgianayshf.YDGfarvple...gentdagdtfrglv  (227)
CoxB1     gnappRMsIPFisignayscf.YDGwtqf.........srngvtgin    (208)
CoxB3     gnappRMsIPFlsignaysnf.YDGwsef.........srngvYgin    (208)
CoxB4     gnappRMsIPFmsignaytmf.YDGwsnf.........srdgiYgyn    (211)
BEV       dgppaqFsVPFmssanaystv.YDGyarfm.........dtdpdrYgi1  (212)
Rhino1a   gqpfpRFsIPFlsiasayymf.YDGydg.........dntsskYgsv    (212)
Rhino1b   gqpfpRFsLPFlsiasayymf.YDGydg.........dnssskYgsi    (212)
Rhino2    gqaypRFsLPFlsvasayymf.YDGyde.......q..dqnYgta      (208)
Rhino14   g.dtsRFsVPYvglasayncf.YDGysh.........ddaetQygit    (216)
Rhino39   gqpypRFsLPFlsiasayymf.YDGydg.........dksssrYgvs    (210)
Rhino49   gqaypRFsLPFlsvasayymf.YDGyne.......q..ggnYgtv      (208)
Rhino89   gqpypRFtIPFmsiasayymf.YDGydg.........dsaaskYgsv    (214)
Consensus ---RF-LPY-----------YDG-----------Y----
Structure  b.bbGlbbb          bG2b   lllll.......lllllllllll
```

Figure 2. *Continued.*

227

```
FMDA10     vatQlp....asFnygaIqa......qaiheLlvRmkRaelyCPRPlla.  (191)  ikvtsqdrykqkiiapakql............  (212)
FMDA12     varQlp....asFnygalka......etiheLlvRmkRaelyCPRPlla.  (192)  ievssqdrhkqkiiapgkql............  (213)
FMDA22     vaaQlp....asFnfgaIqa......ttiheLlvRmkRaelyCPRPlla.  (192)  vevssqdrhkqkiiapaqll............  (213)
FMDAsial   vnkQlp....tsFnygaVka......eniteMiRiKRaelyCPRPilp.  (190)  ldtt.qdrrkqeiiapekql............  (210)
FMDC1      raghlp....tsFnfgaVka......etiteLlvRmkRaelyCPRPilp.  (189)  iqptg.drhkqplvapakql............  (209)
FMDC3      harhlp....tsFnfgaVka......etiteLlvRmkRaelyCPRPvlp.  (190)  vqptg.drhkqpliapakql............  (210)
FMDO1Bfs   vartlp....tsFnygalka......trvteLlyRmkRaetyCPRPlla.  (193)  ihpte.arhkqkivapvkqtl...........  (213)
FMDO1K     vartlp....tsFnygalka......trvteLlyRmkRaetyCPRPlla.  (193)  ihpte.arhkqkivapvkqtl...........  (213)
FMDSat1    iasEth..ipttFnygmlyt......kaevdVYRmkRaelyCPRPvlth  (201)  ydhngrdryktt lvkpakqls..........  (222)
FMDSat2    yssakh.slpstFnfgfVta......dkpvdVYyRmkRaelyCPRallpa  (196)  ythaggrfdapi.gvakqll............  (216)
FMDSat3    lenEttrclprtFnfgrLl......ceegdaYRmkRaelyCPRPlrvr  (197)  ythtt.dryktplvkpdkqmc...........  (217)
EMCR       .pnsdF...GtLffagtkp......dikftVYlRykNkrvfCPRPtvf.  (255)  fpwptsg.dkidmt.pragvlmle.........  (277)
EMCB       .pnsdF...GtLffagtkp......dikftVYlRykNmrvfCPRPtvf.  (255)  fpwpssg.dkidmt.pragvlmle.........  (277)
EMCD       .pnsdF...GtLffagtkp......dikftVYlRykNmrvfCPRPtvf.  (255)  fpwpssg.dkidmt.pragvlmle.........  (277)
Mengo      .pnsdF...GtLffagtkp......dikftVYlRykNmrvfCPRPtvf.  (255)  fpwptsg.dkidmt.pragvlmle.........  (277)
TMEBeAn    .pnadF...GrLwiqgnts.......asVriRyKkmkvfCPRPtlf.  (253)  fpwptpttkinadnpvp.ilele.........  (276)
TMEGd7     .ptsdF...GrLwiqgnss.......asVriRyKkmkvfCPRPtlf.  (253)  fpwptpttkinadnpvp.ilele.........  (276)
TMEDa      .pnadF...GrLwiqgnts.......asVriRyKkmkvfCPRPtlf.  (251)  fpwpvstrskinadnpvp.ilele........  (274)
HALA       .tdstF...GlVsiqianynhsdeylsfscylsvteQsefyFRapln.  (265)  snamlstesmmsriaagdlessvddprseedrrfe  (300)
HACr-326   .tdstF...GlVsiqianynhsdeylsfscylsvtqQsefyFPRapln.  (265)  snamlstesmmsriaagdlessvddprseedrrfe  (300)
HAHml75    .tdstF...GlVsiqianynhsdeylsfscylsvteQsefyFPRapln.  (265)  snamlstesmmsriaagdlessvddprseedkrfe  (300)
HAMbb      .tdstF...GlVsiqianynhsdeylsfscylsvteQsefyFPRapln.  (265)  snamlstesmmsriaagdlessvddprseedkrfe  (300)
Polio1M    .slNdF...GiLavrvVndhnptkvtskirVYlkpkHirvwCPRPpra.  (276)  vayygpg.vdykdgtltplstk......dltty.  (302)
Polio1S    .slNdF...GiLavrvVndhnptkvtskirVYlkpkHirvwCPRPpra.  (276)  vayygpg.vdykdgtltplstk......dltty.  (302)
Polio2La   .slNdF...GsLavrvVndhnptkltskirVYmKpkHvrvwCPRPpra.  (276)  vpyygpg.vdykdg.laplpgk......gltty.  (301)
Polio2S    .slNdF...GsLavrvVndhnptkltskirVYmKpkHvrvwCPRPpra.  (276)  vpyfgpg.vdykdg.ltblpek......gltty.  (301)
Polio3Le   .tvDdF...GvLavrvVndhnptkvtskvrIYmKpkHvrvwCPRPpra.  (275)  vpyygpg.vdyknn.ldplsek......gltty.  (300)
Polio3F    .avDdF...GvLairvVndhnptkvtskvrIYmKpkHvrvwCPRPpra.  (275)  vpyygpg.vdykdg.laplsek......gltty.  (300)
Polio3S    .tvDdF...GvLavrvVndhnptkvtskvrIYmKpkHvrvwCPRPpra.  (275)  vpyygpg.vdyrnn.ldplsek......gltty.  (300)
CoxA21     .siNdF...GvLavravnrsnphtihtsvrVYmKpkHircwCPRPpra.  (271)  vlyrgeg.vdmissaiqplkvd......sittf.  (298)
CoxB1      .tlNnM...GtLymrhVneagqgpikstvrIYfKpkHvkawvPRPpr1.  (252)  cqyekqknvnfnptgvtttrs......nittt.  (278)
CoxB3      .tlNnM...GtLyarhVnagstgpikstirIYfKpkHvkawiPRPpr1.  (256)  cqyekaknvnfqpsgvtttrqsit...tmtnt.  (281)
CoxB4      .slNnM...GtLyarhVndsspgglstirIYfKpkHvkayyPRPpr1.  (255)  cqykkaksvnfdveavtaeras......litt.  (281)
BEV        .psNFL...GfMyfrtLedaah..qvrfrIYaKikHtscwiRPraprq.  (253)  apykkrynlvfsgdsdricsnra....sltsy.  (281)
Rhino1a    .vtNdM...GtIcsriVtekqklsvvitthIYhKaKHtkawCPRPpra.  (256)  vpyihshvtnympetgdvttaivrrn..titta.  (287)
Rhino1b    .vtNdM...GtIcsriVtekqehpvvitthIYhKaKHtkawCPRPpra.  (256)  vpyihsrvtnyvpktgdvttaivpra...smktv.  (287)
Rhino2     .ntNnM...GsLcsriVtekhihkvhimtrIYhKaKHvkawCPRPpra.  (252)  leyrahrtnfkiedrsiqtaivtrp...iitta.  (283)
Rhino39    .vlNhM...GsMaFriVnehdehktlvkirVYhKaKHveawiRPapra.  (260)  lpytsigrtnypkntepvikrkg.....diksy.  (289)
Rhino49    .vtNdM...GtLctriVtnqqhlvevttrVYhKaKHvkawCPRapra.  (254)  vpyihsnvtnykvrdgeptlfikpre..nltta.  (285)
Rhino89    .stNnM...GsLcsriVtekhihsmhimtrIYhKaKHvkawCPRPpra.  (252)  leyrahrtnfkvedrdiktgitsra..iitta.  (283)
           .vtNdM...GtIcsriVtsnqkhdsnivcrIYhKaKHikawCPRPpra.  (258)  vayqhthstnyipsngeattqiktrpdvftvtnv.  (292)
Consensus  --N-M--G-L---V---------IY-K-KH----CRP----
Structure  .aaLal      bbbHbb          bbbbbbbIbbbbbbb
```

Figure 2. Continued.

228

extensive pairwise and profile fits, as portions of this region are crystal-lographically disordered, and in any case, there is only minimal structural similarity for this protein between rhino14 and mengovirus (32).

READING THE ALIGNMENTS

According to (arbitrary) formatting conventions, each aligned sequence in Fig. 2 is presented in its entirety, with dots (. . . .) inserted as necessary to maintain relative position. The aaa, bbb, lll, etc., characters in the "Structure" line show which rhino14 amino acids take part in α-helix (aaa), β-sheet (bbb), loop (lll), puff (ppp), or knob (kkk) structures according to the hydrogen bonding diagrams. The aligned residues for mengovirus are part of analogous elements with roughly equivalent lengths. Refer to Fig. 1 for orientation of these features and location of the mapped neutralization immunogenic sites of rhino14.

A "Consensus" sequence has been calculated to represent areas where a majority of the sequences conserve specific residues or a series of very similar residues. Based on a serotype election system equivalent to that of the comparative profiles (UW GCG, PRETTY program [15]), amino acids were polled pairwise at each position, and a vote was cast whenever residues matched or showed a high degree of similarity (defined here as Staden peptide comparison values of ≥1.2 [54]). Any amino acid garnering more than two-thirds of the potential votes at its position was designated the consensus residue and included on the appropriate line. The victorious electorate contributing to the win are highlighted with uppercase letters. Alignment positions without comparable winners are indicated with dashes. However, since the balloting parameters can be adjusted for what one specifically wants to see, neither the consensus line nor the case designations should be taken too seriously. They are used as a convenient way to rapidly identify the most highly conserved residues.

When picking through these compilations for selective comparisons, it is important to bear in mind that without firm crystallographic coordinates for every included sequence, all alignments must be cautiously decoded. These particular versions represent best estimates for how certain picornaviral sequences fit together. They are based on presently available data and are anticipated to have a certain degree of veracity and experimental value. However, biology is less predictable than mathematics and can't always be calculated by formula. The presented alignments might reasonably predict which amino acids form part of a homologous loop, for example, but similarities with rhino14 or mengovirus are certainly not sufficient to foresee the precise configuration of such a loop. Therefore, it is nearly impossible to determine which specific residues might identically substitute for one another (if indeed they do) in the random coil, surface regions of epitopes, or receptor binding sites. The correct phase of the β sheets (side chains in or out?) and even their exact start and stop points are also subject to some regional shifting in the absence of structural information. The carboxy- and amino-terminal extensions represent other regions where large errors might likewise be encountered. Some of these ends may weave tertiary and quaternary patterns characteristic only of individual strains. The extensions are highly variable in sequence, and their assigned alignment positions must be considered tenuous (at best).

COMMENTS ON THE VP4 ALIGNMENT

Protein VP4 (1A) is the smallest of picornaviral capsid proteins (17 to 85 amino acids [AA]). In virions it lies underneath the other proteins and winds an unpredictable course between the threefold and fivefold axes (23, 32).

For the HAVs, there is little direct experimental evidence for the existence of a VP4 (11, 12). Viral nucleotide sequences suggest that 21 to 23 codons may lie 5' to the start of the HAV VP2 region, but it is unclear how much of this fragment is actually translated and whether it encodes a VP4 analog. Nevertheless, alignments of this region strongly suggest that these short fragments share real similarities with other viral sequences. Starting with AA 7 of all HAV polyproteins is a characteristic consensus myristylation sequence (GxxxS/T), common to the amino termini of all other VP4s (10). The entero- and rhinoviruses maintain this site as the very beginning of the their open reading frames (except for the terminal methionine), while cardio- and aphthoviruses generate it through proteolytic scission of leader proteins (10, 41). Cleavage at this point in HAV polyproteins would produce a 6-AA leader (5 AA without the methionine) and a 17-AA VP4, the shortest for any virus. The putative VP4/VP2 junction (LA/D) is identical to that found in nearly all cardio- and aphthoviruses.

The cleavage junction between L and VP4 for FMDV has been repositioned in this alignment relative to original sequence references (7, 47). Conservation of the myristylation site (10) and strong amino acid identity among the amino termini of all cardio- and apthovirus sequences seemed appropriate justification. The shift would locate the new cleavage junction at the K/G amino acid pair, 201 codons into the polyprotein, and make the VP4s of all sequenced FMDVs 85 AA in length.

COMMENTS ON THE VP2 ALIGNMENT

Maturation cleavage between VP4 and VP2 is carried out by a unique autocatalytic mechanism. It has been proposed that interactions between RNA and regions of the VP0 protein may trigger proteolysis during virion assembly (39, 48). The β-A1 and β-A2 hairpin structure, contributed by the VP2 sequences, may play a role in this cleavage by orienting the protein/RNA and donating a putative catalytic serine to the process. Conservation of these elements among all picornaviruses (including the HAVs) has been described (39).

One unusual characteristic of the VP2 structure in rhino14 is the dominant surface protrusion, or puff, which contains most of the neutralization immunogenic site II (NIm-II) epitope. The β-E and β-F structures, which respectively begin and end this segment, share a high degree of sequence identity among all aligned viruses. Thus, the individual lengths of this loop can be accurately predicted for other viruses, too. In mengovirus the puff is shorter than that of rhino14 and takes a quite altered configuration. The TME viral sequences have an 11-AA insertion relative to mengovirus at the amino end of the puff and other dissimilar segments throughout its length. Dogmatically, all cardioviruses belong to the same serotype (50, 51), but if experimental antigenic differences can be detected between them (e.g., EMC virus and TMEBeAn), this region might be an interesting place to look for their origins. The puff is almost completely absent in FMDV and HAV. The remaining small fragments (bridging β-E and β-F) may possibly retain some analogous antigenic potential, though it is unlikely that, by themselves, these segments would form a dominant immunogen in the manner of the rhino14 NIm-II. More probably, they might combine with other loops (FMDV loop of VP1?) and terminal extensions to form a different, discontinuous epitope.

COMMENTS ON THE VP3 ALIGNMENT

Among the capsid proteins, the VP3s are most similar in overall size. The shortest (FMDC1, 219 AA) differs from the longest (HAV, 246 AA) by only 27 AA (12%). In contrast, the size variances in VP2 and VP1 proteins are 50 AA (25%) and 93 AA (47%), with average lengths of 247 and 256 AA, respectively.

The VP3 length differences occur primarily in a region between the beginning of β-B and the start of β-C. In mengovirus and rhino14, the β-B strand is interrupted (β-B1 and β-B2) by a 10- to 15-AA knoblike insertion with respectively different configurations. Recent epitope mapping with polio1S has located a cluster of neutralization escape mutants within the equivalent loop of this virus (38), although similar mutants have not (yet) been mapped within the knob for rhino14. The adjacent top corner bend (between β-B2 and β-C) has been determined to be antigenic in rhino14, however, forming part of the NIm-III epitope (48). Since this entire region is exposed on the surface of mengovirus too (49) (where escape mutant mapping is incomplete), it is reasonable to predict that the analogous segments in most picornaviruses may also contain neutralizing epitopes or contribute to epitopes. As a test for this hypothesis, it would be interesting to construct synthetic HAV or FMDV peptides containing this region. If the alignments are correct, antibodies to these peptides would be expected to have (neutralizing?) activity against native virus.

The VP3 knob protrusion lies on the south side of the deep canyon (rhino14) or pitlike (mengovirus) surface depression which may be involved in interactions with virion receptors. Portions of the β-C, α-A1, and α-A2 segments coat the walls of the rhino14 canyon and the mengovirus pit, and it is possible that alterations in these areas may significantly affect the putative receptor binding site (49). The loop connecting β-G and β-H lies nearby and is another potential region for hypervariable receptor reactions.

COMMENTS ON THE VP1 ALIGNMENT

Before resolution of any three-dimensional crystal structures, VP1 alone was considered as the major picornaviral surface protein. The most variable in length, sequence, and charge, it was believed to figure prominently in antigenicity and receptor recognition. But how are these properties affixed in the protein? Diversity makes VP1 sequences the hardest to align. On the other hand, they become the most valuable for predictions, if those properties mapped for one virus can be reasonably extrapolated to other strains.

The alignments show that VP1 differences among virus groups, though large, generally occur at discrete positions. Within every loop, whole segments are characteristically inserted, deleted, or substituted. The cardioviruses, for instance, have two large surface loops (I and II), totaling 36 AA, inserted at the carboxyl end of β-C in an arrangement unique to this genus. The insertion fills part of the deep groove that would be found in the analogous portion of other viruses, making the putative receptor binding site more pitlike than canyonlike (32). Sequence similarities suggest the TME viruses should look very much like mengovirus in this region.

Other smaller changes potentially alter the walls of the pits (or canyons) or affect the surface topography of epitopes. Deletion of 8 to 16 residues between β-D and β-E of the aphtho- and cardioviruses and HAV removes the potential NIm-1b epitope, for example. Relative insertions between β-B and β-C may correspondingly shift the range of NIm-1a or its analog. Unlike rhinovirus, then, these viruses may center a single epitope on the northern VP1 plateau between the canyon (or pit) and the fivefold axis.

The FMDV loop connecting the β-G and β-H strands is another region implicated in antigenic variation. Proteolytic reactions at trypsin-sensitive sites within this segment reduce the ability of FMDV to bind to cells and reduce its potential as a vaccine (8). On the basis of clear immunodominance, many determinations of FMDV strain sequences have focused exclusively on this loop without concurrent sequencing of the rest of the P1 region (2, 60). Synthetic peptides containing the loop fragment can elicit some protective antibodies in animals, but the response is never as strong or complete as with inoculation with intact virus (16). The sequence alignments reveal that the equivalent loop in rhino14 contributes residues towards the NIm-II epitope and suggest that perhaps, as for rhinovirus, the completed FMDV immunodominant site may actually be formed in part by segments donated from other proteins (VP2?) or other regions of VP1 (carboxy terminus?) (49). The alignments also indicate that the FMDV loop is almost completely missing from all HAV strains, and for these viruses, it probably does not participate in principal epitope formation in an equivalent manner.

Surprisingly, among these variable segments of (unrestricted?) rearrangement, the major structural elements are well conserved in size and sequence. The frequency of uppercase letters in α-A, α-Z, β-B, β-D, β-E, β-G1, β-H, and β-I in Fig. 2D shows the strength of these similarities among all viral strains. Homology (ancestry), analogy (restrictive convergence), and differential selective pressure may all be responsible for forming the alternating pattern of conserved and nonconserved VP1 segments. As the regions are mapped and correlated with functions, the biological limitations may become easier to understand.

PHYLOGENY BASED ON CAPSID ALIGNMENTS

Sequence alignments highlight individual differences among specific strains, but in total they also display a collective pattern of interviral relationships. Assuming all picornaviruses are descended from a common ancestor, the present genomic sequences serve as record of the cumulative mutations acquired by each strain during its evolutionary history. Within certain limits, then, the greater the number of mutations separating compared sequences, the greater the genetic distance between them.

Given a series of aligned homologous sequences, it is possible by pairwise comparisons to calculate a tree of maximum parsimony, representing the minimum mutational path required to convert one sequence into another. Parsimonious trees portraying these relationships are roughly equivalent to phylogenies if one presumes that mutational frequencies have remained relatively constant (for a review on the use of maximum parsimony, see reference 18).

To calculate trees for picornaviruses, the aligned capsid proteins were reconverted into their original coding sequences, forming equivalent nucleotide alignments. A maximum parsimony algorithm, designed by Walter Fitch and E. Margoldiash and modified by Mark Pallansch and co-workers (19, 46), was used to calculate the most probable relationships for picornaviral P1 sequences. The UW GCG DISTANCES program, using the same data, subsequently assigned quantitative distance values to each phylogenic node (connecting point). In this context, percent identity is defined as (total number of identical matches in a pairwise comparison-/average length of the compared sequences [without gaps]) ×100.

Figure 3 summarizes the relationships among 20 strains of viruses when compared throughout the complete length of the P1 region (3,063 nucleotides). The combined alignments (33 strains) were used to calculate this tree, but to simplify the displayed pattern, only one example of each serotype (20 viruses) is presented. The leftmost node connecting any pair of strains shows the minimum identity shared by those sequences. Progressively more distant nodes represent average values for the composite sequences. Maximum variance at any node for any pair of sequences is <1 to 2%.

Another, more complex tree showing relationships within serotypes (41 viruses) is illustrated in Fig. 4. This comparison was limited to VP1 sequences for which the most sequence information was available, but despite the shorter data base (1,158 nucleotides), the pattern of strain relationships is identical to that of the complete P1 comparison. For relative perspective, these distances are presented as amino acid identities.

As a basic pattern in either tree, the viruses divide neatly into four main groups or branches (I, II, III, and IV by arbitrary numbering). Group III, the HAV isolates, is the most distant branch, sharing less than 28% nucleotide identity with all other viruses. Among themselves, the HAV strains are very closely related, holding more than 97% of their amino acids in common.

Groups I and II include the aphtho- and cardioviruses, respectively. These groups share at about 40% nucleotide identity among respective members and are the most closely related of the major categories. The cardiovirus group (II) is further divided into TME virus and mengovirus

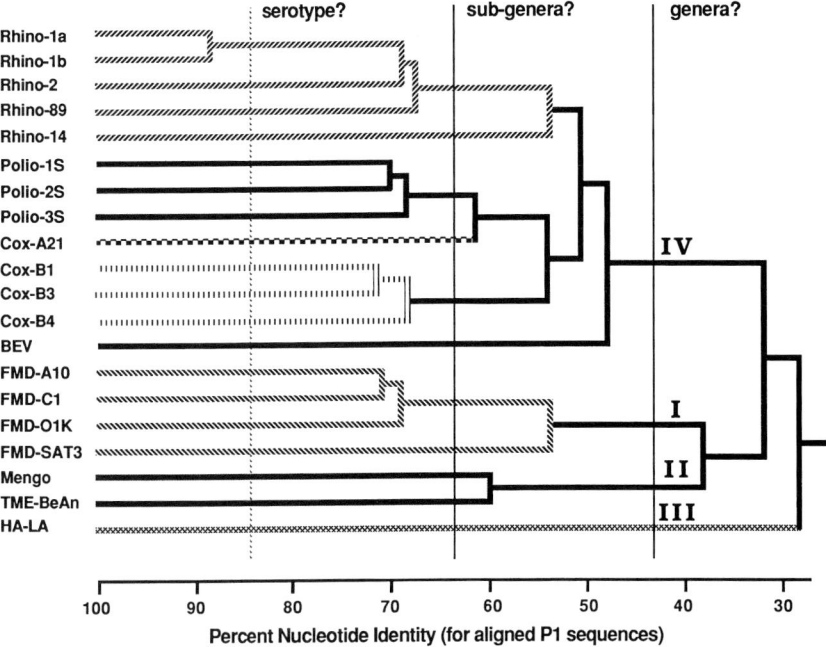

Figure 3. Relationship among picornaviruses according to complete P1 nucleotide alignments. Nucleotide alignments for 33 strains of picornaviruses were compared as described in the text. Relationships among 20 representative strains are presented graphically. The different line types allow easy identification of subgroups.

subgroup branches with about 60% shared identity. Among FMDVs (group I), the A, C, O, and Asia serotypes are more similar to each other than to the South African viruses (Sat), though A/C and Asia/O cluster independently into smaller, separate branches.

The group IV viruses (poliovirus, coxsackieviruses, rhinoviruses, and BEV) are interrelated by >50% nucleotide identity, with the human isolates sharing even higher values (>55%). The coxsackie- and polioviruses occupy branches separate from the rhinoviruses, but CoxA21 is more like the poliovirus strains (>62%) than it is like the CoxB strains (>55%). Within rhinoviruses, rhino14 is the most different, a characteristic already mentioned in other comparative studies (14). Interestingly, the remaining rhinovirus serotypes do not cluster according to receptor groups, indicating that the distinction between the major (M) and minor binding abilities must reside in a much smaller sampling of residues. All group IV members are related to groups I and II by about 33% nucleotide identity.

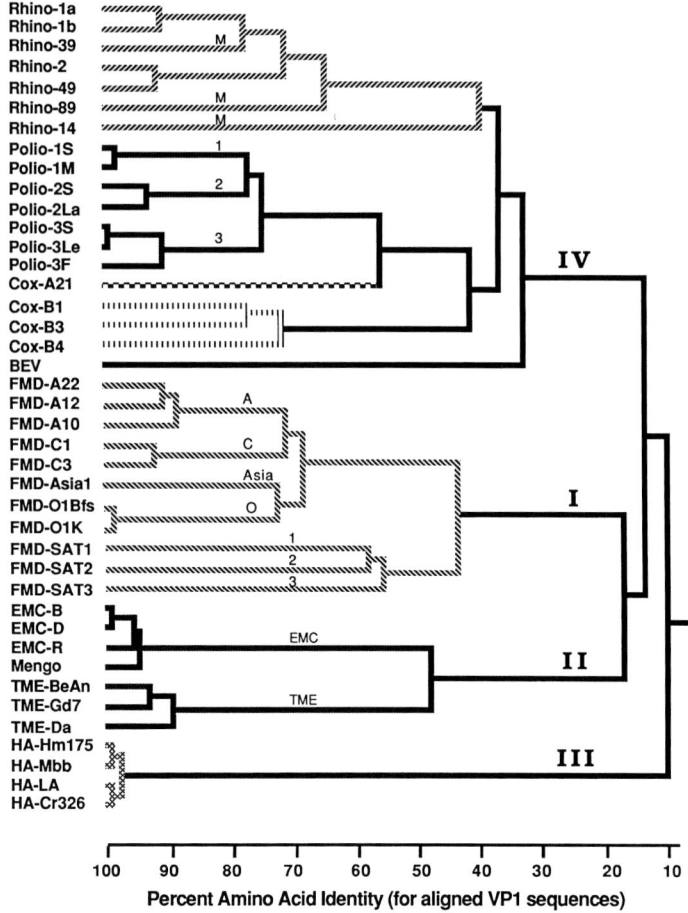

Figure 4. Relationship among picornaviral VP1 proteins. Nucleotide alignments for 41 viral strains were compared over the lengths of their VP1 coding sequences as described in the text. Data are presented as amino acid similarity. Some branches of particular serotypes (1, 2, 3, A, O, C, etc.) are marked for clarity. The "M" designates rhinovirus serotypes belonging to the major receptor binding group.

RECLASSIFICATION OF VIRAL SUBGROUPS

Traditional picornavirus classifications, based primarily on physical properties of the virions, are not entirely synonymous with the phylogenic groupings determined by sequence comparisons. Common taxonomy catalogs HAV as an enterovirus (like poliovirus) and assigns all rhinoviruses to a separate genus (50, 51). The present genera designations

(*Aphtho-*, *Cardio-*, *Entero-*, and *Rhinovirus*) and the measurable qualities that define them (buoyant density, pH lability, size, etc.) were established before the advent of genetic engineering and the ability to determine RNA sequences. The system emphasizes phenotypic rather than genotypic distinctions. In view of the overwhelming quantity of comparative data now available, it seems clear that some classification changes should be made if the true homologies among these viruses are to be accurately represented in taxonomy.

The pattern of branched relationships presented in Fig. 3 and 4 is characteristic for nearly any picornaviral sequence comparison. It recurs in trees based on P2, P3, or noncoding region alignments with minimum variation (unpublished data). Given significant sample size (200 to 300 nucleotides), even the values of nodes defining groups and subgroups remain quite constant throughout the genome (<5% maximum shift). Strains within serotypes can sometimes share slightly greater identity in their P2 and P3 regions then in their P1 proteins, but as a rule, subgroup differences within groups (e.g., BEV versus rhinovirus) and strain differences within serotypes (e.g., polio2S versus polio2La) are always progressively much smaller than the major differences among groups themselves (e.g., I versus II). Presentation of the data as amino acid comparisons instead of nucleotide identity does not alter the overall motif among strains or the locations of the major nodes.

The four phylogenic groups defined by the trees clearly represent the principal genotypic subdivisions within the family of picornaviruses and should reasonably be considered as primary genera. The requisite strain reassignments would merge the traditional entero- and rhinovirus groups while creating a separate division for the (obviously unique) HAVs. Cardio- and aphthovirus classifications would remain unchanged from the present system. Viruses within each new genus would share at least 45% aligned nucleotide identity with other common members, and within the P1 region they would also share >25% aligned amino acid identity. These values could easily serve as discriminating factors in the classification of new isolates. Comparison of a short fragment from anywhere in the genome would unambiguously identify the genotype. Alternatively, simple hybridization assays with prototype cDNAs could accurately catalog new isolates into the appropriate genus.

The phylogenic data also suggest that membership within any of 11 viral subgroups (subgenera?) can be reasonably predicted by sequence similarities. Isolates would belong to one of the following categories: poliovirus, coxsackievirus type B; coxsackievirus type A; BEV; rhino14-like; all other rhinoviruses; FMDV, Sat-like; FMDV, A/O/C/Asia-like; TME virus; mengoviruslike (EMC virus); or HAV, if they shared >66%

nucleotide identity with prototype member strains. These branches are clearly indicated from the comparison trees and, for the most part, follow classic picornaviral subdivisions.

It is not clear whether sequence similarities alone will be able to faithfully distinguish among new serotypes of the same subgroup. Within the present data base of viruses, separate serotypes generally share <85% nucleotide or amino acid identity throughout their P1 regions. Serotype distinctions are less obvious in the P2, P3, or noncoding regions, however, and comparison of fragments from these areas could be misleading. Therefore, antigenic cross-typing with defined serum still remains the safest way to determine serotypes accurately. Possibly, as future crystallography continues to refine our concepts of epitopes (and receptor binding classes?), the sequence units responsible for specific strain properties may become more identifiable and reliably predictable.

Acknowledgments. I thank everyone who kindly supplied sequence information to feed the voracious appetite of my computer, and I apologize that formatting constraints necessitated omission of some strains. Special thanks go to Pamela Hughes, Glyn Stanway, Hyone-Myong Eun, Yong-Soo Bae, Ji-Won Yoon, Berwyn Clarke, Gus Cordova, Ellen Doran, Bruce Korant, John Ticehurst, and Robert Ralston for allowing me to include their unpublished data, and to Michael Rossmann for gently correcting many of my errors.

This work is supported by a Public Health Service grant to A.C.P.

Literature Cited

1. **Baroudy, B. M., J. R. Ticehurst, T. A. Miele, J. V. Maizel, Jr., R. H. Purcell, and S. M. Feinstone.** 1985. Sequence analysis of hepatitis A virus cDNA coding for capsid proteins and RNA polymerase. *Proc. Natl. Acad. Sci. USA* **82:**2143–2147.

2. **Beck, E., G. Feil, and K. Strohmaier.** 1983. The molecular basis of the antigenic variation of foot-and-mouth disease virus. *EMBO J.* **2:**555–559.

3. **Beck, E., S. Forss, K. Strebel, R. Cattaneo, and G. Feil.** 1983. Structure of the FMDV translation initiation site and of the structural proteins. *Nucleic Acids Res.* **11:**7873–7885.

4. **Boothroyd, J. C., T. J. R. Harris, D. J. Rowlands, and P. A. Lowe.** 1982. The nucleotide sequence of cDNA coding for the structural proteins of foot-and-mouth disease virus. *Gene* **17:**153–161.

5. **Boothroyd, J. C., P. E. Highfield, G. A. M. Cross, D. J. Rowlands, P. A. Lowe, F. Brown, and T. J. R. Harris.** 1981. Molecular cloning of foot-and-mouth disease virus genome and nucleotide sequences in the structural protein genes. *Nature* (London) **290:**800–802.

6. **Callahan, P., S. Mizutani, and C. Colonno.** 1985. Molecular cloning and complete sequence determination of human rhinovirus type 14 genome RNA. *Proc. Natl. Acad. Sci. USA* **82:**732–736.

7. **Carroll, A. R., D. J. Rowlands, and B. E. Clarke.** 1984. The complete nucleotide sequence of the RNA coding for the primary translation products of foot-and-mouth disease virus. *Nucleic Acids Res.* **12:**2461–2472.

8. **Cavanagh, D., D. V. Sanger, D. J. Rowlands, and F. Brown.** 1977. Immunogenic and cell attachments sites of FMDV: further evidence for their location in a single capsid polypeptide. *J. Gen. Virol.* **35:**149–158.

9. **Cheung, A., J. Delamarter, S. Weiss, and H. Kuepper.** 1983. Comparison of the major antigenic determinants of different serotypes of foot-and-mouth disease virus. *J. Virol.* **48:**451–459.

10. **Chow, M., J. F. E. Newman, D. Filman, J. M. Hogle, D. J. Rowlands, and F. Brown.** 1987. Myristylation of picornavirus capsid protein VP4 and its structural significance. *Nature* (London) **327:**482–486.

11. **Cohen, J. I., J. R. Ticehurst, R. H. Purcell, A. Buckler-White, and B. M. Baroudy.** 1986. Complete nucleotide sequence of wild-type hepatitis A virus: comparison with different strains of hepatitis A virus and other picornaviruses. *J. Virol.* **61:**50–59.

12. **Coulepis, A. G., S. A. Locarnini, E. G. Westaway, G. A. Tannock, and I. D. Gust.** 1982. Biophysical and biochemical characterization of hepatitis A virus. *Intervirology* **18:**107–127.

13. **Dayhoff, M. O.** 1979. Matrices for detecting distant relationships, p. 353–358. *In* R. M. Schwartz and M. O. Dayhoff (ed.), *Atlas of Protein Sequence and Structure*, vol. 5, suppl. 3. National Biomedical Research Foundation, Silver Spring, Md.

14. **Deuchler, M., T. Skern, W. Sommergruber, C. Neubauer, P. Grundler, I. Fogy, D. Blass, and E. Kuechler.** 1987. Evolutionary relationships within the human rhinovirus genus: comparison of serotypes 89, 2 and 14. *Proc. Natl. Acad. Sci. USA* **84:**2605–2609.

15. **Devereux, J., P. Haeberli, and O. Smithies.** 1984. A comprehensive set of sequence analysis programs for the VAX. *Nucleic Acids Res.* **12:**378–395.

16. **DiMarchi, R., G. Brooke, C. Gale, V. Cracknell, D. Doel, and N. Mowat.** 1986. Protection of cattle against foot-and-mouth disease by a synthetic peptide. *Science* **232:**639–641.

17. **Earle, J. A., R. A. Skuce, E. M. Hoey, and S. J. Martin.** 1988. The complete nucleotide sequence of a bovine enterovirus. *J. Gen. Virol.* **69:**253–263.

18. **Fitch, W.** 1977. On the problem of discovering the most parsimonious tree. *Am. Nat.* **111:**223–257.

19. **Fitch, W. M., and E. Margoldiash.** 1967. Construction of phylogenic trees. *Science* **155:**279–284.

20. **Forss, S., K. Strebel, E. Beck, and H. Schaller.** 1984. Nucleotide sequence and genome organization of foot-and-mouth disease virus. *Nucleic Acids Res.* **12:**6587–6601.

21. **Gribskov, M., M. Homyak, J. Edenfield, and D. Eisenberg.** 1988. Profile scanning for three dimensional structure patterns in protein sequences. *Comp. Appl. Biol. Sci.* **4:**61–66.

22. **Gribskov, M., A. D. McLachlan, and D. Eisenberg.** 1987. Profile analysis: detection of distantly related proteins. *Proc. Natl. Acad. Sci. USA* **84:**4355–4358.

23. **Hogle, J. M., M. Chow, and D. J. Filman.** 1985. Three-dimensional structure of poliovirus at 2.9 Å resolution. *Science* **229:**1358–1365.

24. **Hughes, P. J., M. A. Evans, P. D. Minor, G. C. Schild, J. W. Almond, and G. Stanway.** 1986. The nucleotide sequence of a type 3 poliovirus isolated during a recent outbreak of poliomyelitis in Finland. *J. Gen. Virol.* **67:**2093–2102.

25. **Hughes, P. J., C. North, C. H. Jellis, P. D. Minor, and G. Stanway.** 1988. The nucleotide sequence of human rhinovirus 1B: molecular relationships within the rhinovirus genus. *J. Gen. Virol.* **69:**49–58.

26. **Iizuka, N., S. Kuge, and A. Nomoto.** 1987. Complete nucleotide sequence of the genome of coxsackievirus B1. *Virology* **156:**64–73.

27. **Jenkins, O., J. D. Booth, P. D. Minor, and J. W. Almond.** 1987. The complete nucleotide sequence of coxsackie B4 and its comparison to other members of the picornaviridae. *J. Gen. Virol.* **68:**1835–1848.

28. **Kitimura, N., B. Semler, P. G. Rothberg, G. R. Larsen, C. J. Adler, A. J. Dorner, E. A.**

Emini, R. Hanecak, J. J. Lee, S. van der Werf, C. W. Anderson, and E. Wimmer. 1981. Primary structure, gene organization and polypeptide expression of poliovirus RNA. *Nature* (London) **291**:547–553.

29. La Monica, N., C. Meriam, and V. R. Racaniello. 1986. Mapping of sequences required for mouse neurovirulence of poliovirus type 2 Lansing. *J. Virol.* **57**:515–525.

30. Lindberg, M. A., P. O. K. Stalhandske, and U. Pettersson. 1987. Genome of coxsackievirus B3. *Virology* **156**:50–63.

31. Linemeyer, D. L., J. G. Menke, A. Martin-Gallardo, J. V. Hughes, A. Young, and S. W. Mitra. 1985. Molecular cloning and partial sequencing of hepatitis A viral cDNA. *J. Virol.* **54**:247–255.

31a. Luo, M., M. Rossmann, and A. C. Palmenberg. 1988. Prediction of three-dimensional models for foot-and-mouth disease virus and hepatitis A virus. *Virology* **166**:503–514.

32. Luo, M., G. Vriend, G. Kamer, I. Minor, E. Arnold, M. G. Rossmann, U. Boege, D. G. Scraba, G. M. Duke, and A. C. Palmenberg. 1987. The atomic structure of mengo virus at 3.0 Å resolution. *Science* **235**:182–191.

33. Makoff, A. J., C. A. Paynter, D. J. Rowlands, and J. C. Boothroyd. 1982. Comparison of the amino acid sequence of the major immunogen from three serotypes of foot-and-mouth disease virus. *Nucleic Acids Res.* **10**:8285–8295.

34. Najarian, R., D. Caput, W. Gee, S. Potter, A. Renard, J. Merryweather, G. Van Nest, and D. Dina. 1985. Primary structure and gene organization of human hepatitis A virus. *Proc. Natl. Acad. Sci. USA* **82**:2627–2631.

35. Needleman, S. B., and C. D. Wunsch. 1970. A general method applicable to the search for similarities in the amino acid sequence of two proteins. *J. Mol. Biol.* **48**:443–453.

36. Nomoto, A., T. Omata, H. Toyoda, S. Kuge, H. Horie, Y. Kataoka, Y. Genba, Y. Nakano, and N. Imura. 1982. Complete nucleotide sequence of the attenuated poliovirus Sabin 1 strain genome. *Proc. Natl. Acad. Sci. USA* **79**:5793–5797.

37. Ohara, Y., S. Stein, J. Fu, L. Stillman, L. Klaman, and R. P. Roos. 1988. Molecular cloning and sequence determination of Theiler's murine encephalomyelitis viruses. *Virology* **164**:245–255.

38. Page, G. S., A. G. Mosser, J. M. Hogle, D. J. Filman, R. R. Rueckert, and M. Chow. 1988. Three-dimensional structure of poliovirus serotype 1 neutralizing determinants. *J. Virol.* **62**:1781–1794.

39. Palmenberg, A. 1987. Picornaviral processing: some new ideas. *J. Cell. Biochem.* **33**:191–198.

40. Palmenberg, A. C., E. M. Kirby, M. J. Janda, N. L. Drake, G. M. Duke, K. F. Potratz, and M. S. Collett. 1984. The nucleotide and deduced amino acid sequences of the encephalomyocarditis viral polyprotein coding region. *Nucleic Acids Res.* **12**:2969–2985.

41. Parks, G. D., G. M. Duke, and A. C. Palmenberg. Encephalomyocarditis virus 3C protease: efficient cell-free expression from clones which link viral 5' noncoding sequences to the P3 region. *J. Virol.* **60**:376–384.

42. Paul, A. V., H. Tada, K. von der Helm, T. Wissel, R. Kiehn, E. Wimmer, and F. Deinhardt. 1987. The entire nucleotide sequence of the genome of human hepatitis A virus (isolate MBB). *Virus Res.* **8**: 153–171.

43. Peaver, D. C., J. Borkowski, M. Calenoff, C. K. Oh, B. Ostrowski, and H. Lipton. 1988. Insights into Theiler's virus neurovirulence based on a genomic sequence comparison of the neurovirulent GD7 and less virulent BeAn strains. *Virology* **165**:1–12.

44. Peaver, D. C., M. Calenoff, E. Rozhon, and H. L. Lipton. 1987. Analysis of the complete nucleotide sequence of the picornavirus Theiler's murine encephalomyelitis virus indicates that it is closely related to cardioviruses. *J. Virol.* **61**:1507–1516.

45. Racaniello, V. R., and D. Baltimore. 1981. Molecular cloning of poliovirus cDNA and

determination of the complete nucleotide sequence of the viral genome. *Proc. Natl. Acad. Sci. USA* **78**:4887–4891.

46. Rico-Hesse, R., M. Pallansch, B. K. Nottay, and O. M. Kew. 1987. Geographic distribution of wild poliovirus type 1 genotypes. *Virology* **160**:311–322.

47. Robertson, B. H., M. J. Grubman, G. N. Wenddell, D. M. Moore, J. D. Welsh, T. Fischer, D. J. Dowbenko, D. G. Yanssura, B. Small, and D. G. Kleid. 1985. Nucleotide and amino acid sequence coding for polypeptides of foot-and-mouth disease virus type A12. *J. Virol.* **54**:651–660.

48. Rossmann, M. G., E. Arnold, J. W. Erickson, E. A. Frankenberger, J. P. Griffith, H.-J. Hecht, J. E. Johnson, G. Kamer, M. Luo, A. G. Mosser, R. R. Rueckert, B. Sherry, and G. Vriend. 1985. Structure of a human cold virus (rhinovirus 14) and functional relationship to other picornaviruses. *Nature* (London) **317**:145–153.

49. Rossmann, M. G., and A. C. Palmenberg. 1988. Conservation of the putative receptor attachment site in picornaviruses. *Virology* **164**:373–382.

50. Rueckert, R. R. 1976. On the structure and morphogenesis of picornaviruses, p. 131–213. *In* H. Fraenkel-Conrat and R. R. Wagner (ed.), *Comprehensive Virology*, vol. 6. Plenum Publishing Corp., New York.

51. Rueckert, R. R. 1985. Picornaviruses and their replication, p. 705–738. *In* B. Fields, D. M. Knipe, R. M. Chanock, J. L. Melnick, B. Roizman, and R. E. Shope (ed.), *Virology*. Raven Press, New York.

52. Sellers, P. H. 1974. On the theory and computation of evolutionary distances. *SIAM J. Appl. Math.* **26**:787–793.

53. Skern, T., W. Sommergruber, D. Blass, P. Gruendler, F. Fraundorfer, C. Pieler, I. Fogy, and E. Kuechler. 1985. Human rhinovirus 2: complete nucleotide sequence and proteolytic processing signals in the capsid protein region. *Nucleic Acids Res.* **13**:2111–2126.

54. Staden, R. 1982. An interactive graphics program for comparing and aligning nucleic acid and amino acid sequences. *Nucleic Acids Res.* **10**:2951–2961.

55. Stanway, G., P. Hughes, R. Mountford, P. Minor, and J. Almond. 1984. The complete nucleotide sequence of a common cold virus: human rhinovirus 14. *Nucleic Acids Res.* **12**:7859–7875.

56. Stanway, G., P. J. Hughes, R. C. Mountford, P. Reeve, G. C. Schield, and J. W. Almond. 1984. Comparison of the complete nucleotide sequences of the genomes of the neurovirulent poliovirus P3/Leon/37 and its attenuated Sabin vaccine derivative P3/Leon/12A1B. *Proc. Natl. Acad. Sci. USA* **81**:1539–1543.

57. Toyoda, H., M. Kohara, Y. Kataoka, T. Suganuma, T. Omata, N. Imura, and A. Nomoto. 1984. Complete nucleotide sequences of all three poliovirus serotype genomes: implication for genetic relationship, gene function and antigenic determinants. *J. Mol. Biol.* **174**:561–585.

58. Tracy, S., H.-L. Liu, and N. M. Chapmen. 1985. Coxsackievirus B3: primary structure of the 5′ non-coding and capsid coding regions of the genome. *Virus Res.* **3**:263–270.

59. Villanueva, N., M. Davila, J. Ortin, and E. Domingo. 1983. Molecular cloning of cDNA from foot-and-mouth disease virus C-1 Santa Pau (c-58). Sequence of protein-VP1-coding segment. *Gene* **23**:185–194.

60. Wenddell, G. N., D. G. Yansura, D. J. Dowbenkow, M. E. Hoatlin, M. J. Grubman, D. M. Moor, and D. G. Kleid. 1985. Sequence variation in the gene for the immunogenic capsid protein VP1 of foot-and-mouth disease virus type A. *Proc. Natl. Acad. Sci. USA* **82**:2618–2622.

Molecular Aspects of Picornavirus Infection and Detection
Edited by Bert L. Semler and Ellie Ehrenfeld
© 1989 American Society for Microbiology, Washington, DC 20006

Chapter 14

Human Enterovirus Infections: Molecular Approaches to Diagnosis and Pathogenesis

Harley A. Rotbart

The enteroviruses are a heterogeneous group of nearly 70 human pathogens which are responsible for a broad spectrum of clinical diseases. Like other picornaviruses, the enteroviruses are small (27-nm), single-stranded, nonenveloped RNA viruses of approximately 1.34-g/ml buoyant density (40). Enteroviruses distinguish themselves from rhinoviruses, fellow human picornaviruses, by their stability in acid, by their fecal-oral route of passage and transmission, and by their strict summer peak of disease activity. The prototypic enteroviruses, the polioviruses, remain the most clinically significant of the enteroviruses worldwide, causing paralytic disease in 4 of every 1,000 school-age children in developing countries (1). In this country, the polioviruses have been controlled with the introduction of vaccines in the late 1950s. The nonpolio enteroviruses, however, are responsible for 5 to 10 million symptomatic infections each year (all United States incidence figures from L. Anderson, Centers for Disease Control, Atlanta, Ga. [personal communication]). They are the most common etiologic agents of meningitis (75,000 cases per year) and of nonspecific febrile and exanthematous illnesses (5 million cases per year). They are also responsible for significant numbers of cases of myocarditis, hepatitis, pleurodynia, stomatitis, and neonatal sepsis (5, 30). Recently

Harley A. Rotbart • Department of Pediatrics and Department of Microbiology/Immunology, University of Colorado School of Medicine, 4200 East 9th Avenue, Box C227, Denver, Colorado 80262.

identified nonpolio enterovirus serotypes cause hemorrhagic conjunctivitis (4, 33) and poliomyelitis mimicking that due to the polioviruses (31). Several important diseases are suspected of having an enteroviral etiology without definitive proof; these include diabetes mellitus (2, 54), dermatomyositis (16, 48), congenital hydrocephalus (14), and amyotrophic lateral sclerosis (13). The enteroviruses cause infections which may persist for many years in immunocompromised individuals, often leading to death (50). Recently, a syndrome of late-onset muscular atrophy has been reported in individuals who suffered paralytic poliomyelitis 20 to 40 years previously (9, 24).

Beyond the obvious desire to determine the specific etiology of these diverse and important diseases, there are many reasons for seeking a rapid and accurate diagnostic test for the enteroviruses. It is often clinically impossible to distinguish enteroviral infections from those due to bacterial pathogens or other viruses, including herpes simplex, for which there are specific therapies (5). Although many enteroviral infections are self-limited and require no therapy, the fear that an illness may be bacterial or herpetic results in unnecessary hospitalization and antibiotic or antiviral treatment for thousands of enterovirus-infected patients each year. Certain enteroviral diseases are in fact severe enough to warrant specific therapy, were such available. Indeed, several experimental agents have been shown to be very effective against the enteroviruses in vitro (29, 34) and in animal models (B. A. Steinberg, A. A. Visosky, and M. A. McKinlay, *Program Abstr. 24th Intersci. Conf. Antimicrob. Agents Chemother.*, abstr. no. 432, 1984; B. A. Steinberg, A. A. Visosky, J. A. Frank, Jr., and M. A. McKinlay, *24th ICAAC*, abstr. no. 433). They have never been studied in humans, however, because the diagnosis of enterovirus infection is currently made too slowly to conduct an appropriate clinical drug trial. Finally, as alluded to earlier, a number of diseases are theorized to be due to the enteroviruses but have not yet been proven as such. A clear association with these viruses would facilitate the understanding and treatment of such conditions.

It is for these reasons, coupled with the inadequacies of current diagnostic techniques for the enteroviruses (see below), that my colleagues and I undertook the development of nucleic acid hybridization techniques for the detection of these pathogens. The summary of those efforts to date, as well as results of our most recent applications of these methods to the study of enteroviral pathogenesis, are presented in this chapter.

THE CURRENT STATE OF ENTEROVIRAL DIAGNOSIS

The Nobel Prize in medicine and physiology was awarded to J. F. Enders, F. C. Robbins, and T. H. Weller in 1954 for their success in cultivating poliovirus in tissue culture (12), an accomplishment which paved the way for vaccine development and provided a means for laboratory testing for the polio and nonpolio enteroviruses. Since then, tissue culture continues to be the mainstay of enteroviral diagnosis despite well-recognized limitations. Tissue culture is time-consuming and requires a high level of expertise. Of greater concern is the fact that certain of the enteroviruses will not grow in tissue culture, requiring inoculation into suckling mice for detection, a technique cumbersome enough to be omitted from almost all diagnostic laboratories. The sensitivity of routine tissue culture for the enteroviruses may be as low as 65 to 75% (6), and development of characteristic cytopathic effect may take too long to be of benefit to the patient. In the experience of my laboratory, cerebrospinal fluid (CSF) infections with the enteroviruses take a mean of 6.3 days for growth in culture (unpublished data), consistent with reported means to isolation from the CSF of 4.0 to 8.2 days (6, 51). Other sites may become positive sooner (51), but as meningitis is the most vexing of enteroviral infections for the clinician, CSF data are the most relevant. The use of additional cell lines improves the yield (8) at the cost of increasing the labor and resources required.

Immunodiagnostic techniques for the enteroviruses have been fraught with difficulties resulting from the extreme antigenic diversity among the serotypes (18, 52, 53). Although a common antigen may exist among the polioviruses and another among the coxsackievirus B types, checkerboard pools of antisera would be required to cover even the most common enteroviral serotypes responsible for human disease. The recent report of shared VP3-2C antigens among many of the serotypes (J. R. Romero, J. R. Putnak, and E. Wimmer, abstr. no. 967, *Pediatr. Res.* **20**:319A, 1986) is encouraging, and further testing will tell whether that observation will be of clinical utility. Serologic testing suffers from the same lack of a ubiquitous enteroviral antigen as immunoassays do, requiring, in this case, pools of antigens for testing. Coxsackievirus type B immunoglobulin M serology has the most proven clinical application (3, 10, 28), advantageous because of shared antigen and early appearance of the immunoglobulin M class of antibodies. Immunoglobulin G serology for the enteroviruses is useful for epidemiologic studies, but of little benefit to the individual patient.

RATIONALE FOR NUCLEIC ACID HYBRIDIZATION IN
ENTEROVIRAL DIAGNOSIS

Despite the serotypic heterogeneity among the enteroviruses, there has been reason to believe, well before the recent availability of specific genome sequence data, that enough conservation of sequences existed among these pathogens to make diagnosis by nucleic acid probes a possibility. First, the viruses are biologically similar. Although the most prevalent serotypes isolated from patients vary greatly from year to year, the clinical stigmata of enteroviral illnesses are very consistent over time; i.e., each disease syndrome associated with any enterovirus has been associated with many of them (5, 30). This commonality of pathogenesis extends even to poliomyelitis, which has been reported not only with the polioviruses, but with several other serotypes as well (30). Most recently, outbreaks of paralytic poliomyelitis due to enterovirus type 71 have been described (31). The epidemiologic predilection for all enteroviral disease to occur in the summer months, the physicochemical similarities which resulted in their initial classification as enteroviruses, and their common fecal-oral route of transmission provided further clues to their shared heritage. Finally, early and very limited hybridization (55) and sequence (19) data indicated that at least certain segments of the genomes of seemingly distantly related enteroviruses, such as poliovirus and coxsackievirus type B1, are tightly conserved.

TESTING THE HYPOTHESIS THAT NUCLEIC ACID PROBES
DERIVED FROM ONE OR A FEW ENTEROVIRAL SEROTYPES
COULD DETECT MANY OR ALL SEROTYPES

My colleagues and I prepared cDNA probes (37) directly from, or following restriction enzyme modification of, cloned sequences of poliovirus type 1, generously provided to us by B. Semler and E. Wimmer (42). The probes were labeled with ^{32}P by nick translation. Their location within the enteroviral genome is illustrated in Fig. 1. Lysates were prepared from tissue cultures (LLC-MK2 cells) infected with one of poliovirus type 1, 2, or 3, coxsackievirus type A9 or B1, or echovirus type 11. Cells infected with another RNA virus, respiratory syncytial virus, and uninfected cells served as control lysates. Dot-blot hybridization with the poliovirus-derived probes revealed cross-reactivity of these reagents with all six of the tested serotypes (37; Fig. 2), without nonspecific binding to the negative controls. Specificity was further verified by treatment of cells with antiviral agents prior to infection. Arildone and WIN 51711 inhibit enteroviral uncoating (29, 34); the former is most

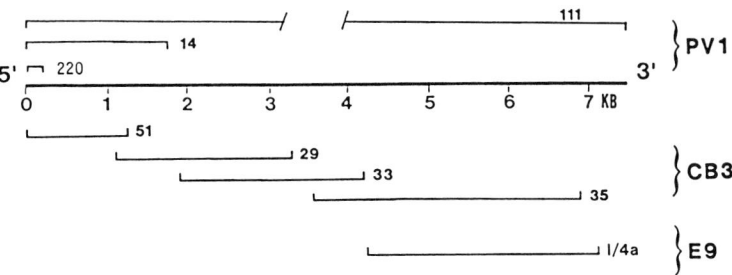

Figure 1. Orientation and designations of the enterovirus-derived probes. PV1, Poliovirus type 1; CB3, coxsackievirus type B3; E9, echovirus type 9.

active against the polioviruses, while the latter agent has a broader spectrum of activity against polio and nonpolio enteroviruses. Viral inhibition results in loss of hybridization signal (Fig. 3) (37). The detection of multiple serotypes, particularly by probe pDS14 (see Fig. 1 for all probe designations and relative locations), speaks to conserved sequences at the 5' end of those viruses tested.

Parallel work by Hyypia and colleagues showed that coxsackievirus type B3-derived cDNA probes detected coxsackie- and noncoxsackievirus enteroviruses in tissue culture infections (21). Interestingly, the probe used by these investigators included much of the 3' half of the coxsackievirus type B3 genome, indicating shared genomic sequences in that region as well.

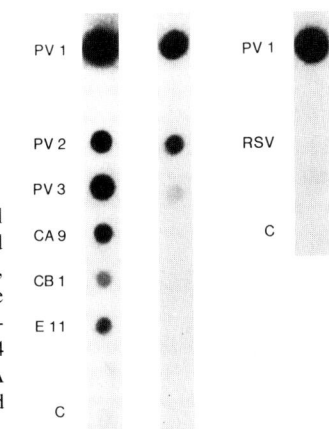

Figure 2. Dot-blot hybridization of LLC-MK2 cell lysates infected with enteroviruses (PV1, PV2, and PV3, poliovirus types 1, 2, and 3; CA9 and CB1, coxsackievirus types A9 and B1; E11, echovirus type 11). Controls were respiratory syncytial virus (RSV)-infected lysates or uninfected cell lysates (C). pDS14 and BAM220 are poliovirus type 1-derived cDNA probes (see Fig. 1). The autoradiograph was exposed for 18 h.

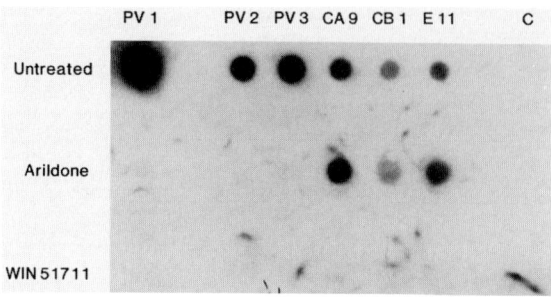

Figure 3. Dot-blot hybridization of enterovirus-infected cell lysates, pretreated with antiviral agents (Arildone or WIN 51711) or not pretreated at all. Uninfected cell lysates served as negative controls (C). The probe used was poliovirus type 1-derived pDS111 (see Fig. 1). The autoradiograph was exposed for 18 h.

CSF RECONSTRUCTION EXPERIMENTS

As noted, central nervous system infections (poliomyelitis, meningitis, and encephalitis) are among the most significant clinical manifestations of the enteroviruses. Any detection system, therefore, must work accurately in the presence of CSF; in fact, CSF will likely be the body fluid submitted most commonly for testing. My co-workers and I conducted reconstruction-type experiments (39) using normal CSF, left over in the hospital laboratories, from patients without signs of infection. To that fluid we added enteroviruses and various components of the inflammatory response observed in meningitis. Dot-blot hybridization conditions were optimized for these specimens (38, 39), and a broad variety of enteroviruses were tested with an expanded battery of cDNA probes (coxsackievirus type B3 cloned fragments were kindly provided by S. Tracy; 46). Viral detection was enhanced by the presence of both salt and formaldehyde added to the CSF sample (38). Heating the sample to 65°C further improved the sensitivity of the assay. Erythrocytes, mononuclear leukocytes, or polymorphonuclear leukocytes did not adversely affect the detection of viral RNA in the range of concentrations of those elements observed during clinical infection (Fig. 4). In contrast, immune serum globulin added in physiologic concentrations attenuated the hybridization signal, a phenomenon which was reversed by treatment of the CSF-virus specimen with proteinase K prior to the addition of salt and formaldehyde (39; Fig. 5). Albumin, also present in normal and infected CSF, did not similarly affect the hybridization assay. None of the cellular or protein components gave false-positive hybridization signals when tested alone, in the absence of virus (Fig. 4 and 5). Each of six probes used individually

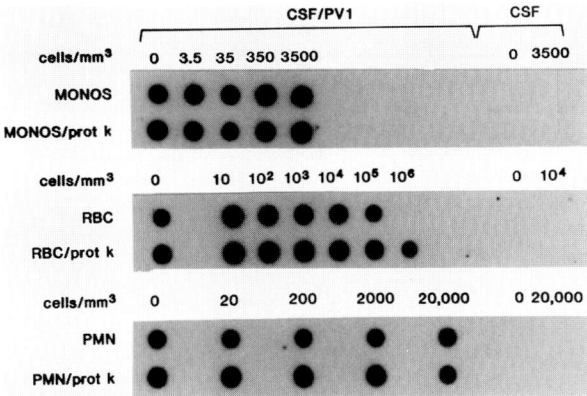

Figure 4. CSF specimens to which 10^5 TCID$_{50}$ of poliovirus type 1 (PV1) was added, followed by addition of increasing numbers of mononuclear cells (MONOS), erythrocytes (RBC), or polymorphonuclear cells (PMN). Negative controls were CSF to which cells, but not virus, had been added. Proteinase K (prot k) was included in parallel specimens, which were otherwise treated identically. The hybridization probe was pDS111 cDNA, and the autoradiograph was exposed for 48 h.

gave different patterns of hybridization with 12 representative enterovirus serotypes (Fig. 6) under conditions determined to be optimal for this type of assay. Combinations of probes, particularly 5′-end and 3′-end probe combinations, detected all serotypes tested with the exception of echovirus type 22 (Fig. 7). Herpes simplex virus or respiratory syncytial virus, added to CSF as negative controls, did not give false-positive signals, nor did CSF alone. The sensitivity of the assay ranged from 10^2 50% tissue

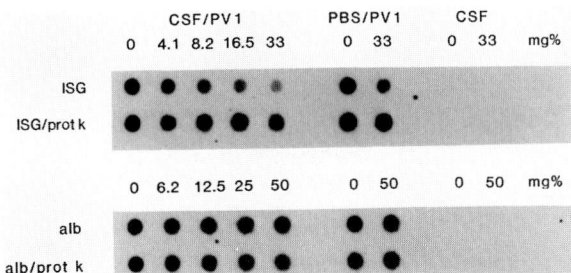

Figure 5. Specimens of CSF or phosphate-buffered saline (PBS), to which 10^5 TCID$_{50}$ of poliovirus type 1 (PV1) was added, followed by increasing concentrations of immune serum globulin (ISG) or albumin (alb). Proteinase K (prot k) was included in parallel specimens, which were otherwise treated identically. Negative controls were CSF to which protein, but no virus, had been added. Probe and autoradiography were as for Fig. 4.

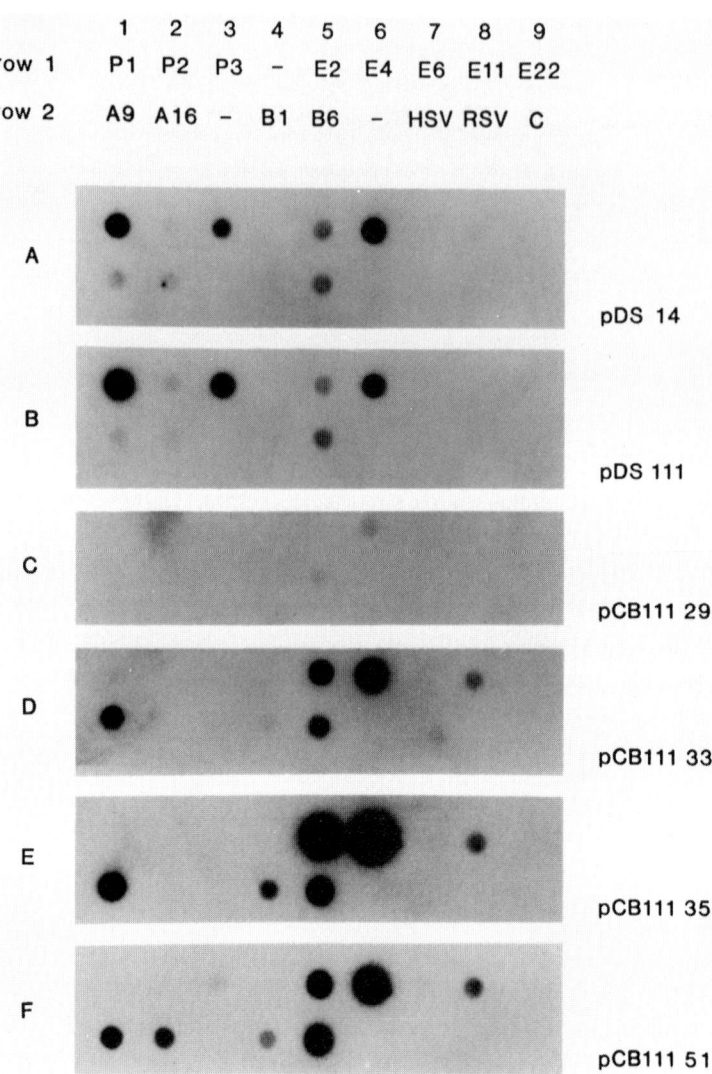

Figure 6. Hybridization of 12 enterovirus serotypes (P1, P2, and P3, poliovirus types 1, 2, and 3; E2, E4, E6, E11, and E22, echovirus types 2, 4, 6, 11, and 22; A9, A16, B1, and B6, coxsackievirus types A9, A16, B1, and B6) and two nonenteroviral control viruses (HSV, herpes simplex virus; RSV, respiratory syncytial virus) with probes derived from poliovirus type 1 or coxsackievirus type B3 (see Fig. 1). Each spot contained 7×10^5 TCID$_{50}$ of virus in CSF, except the negative control (C), which contained no virus. Autoradiography was for 24 h.

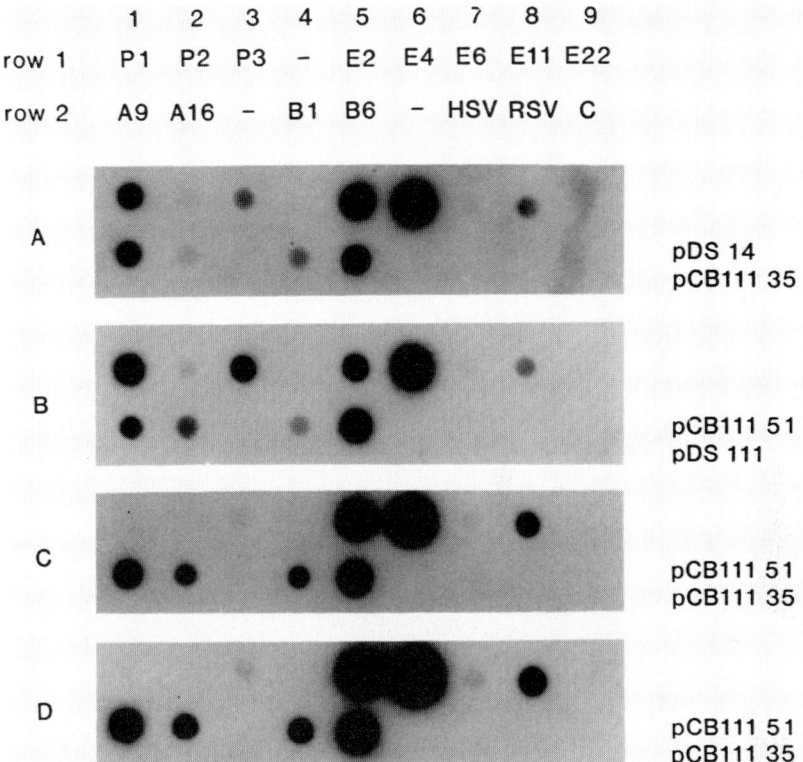

```
           1    2    3    4    5    6    7    8    9
row 1     P1   P2   P3    -   E2   E4   E6  E11  E22

row 2     A9  A16    -   B1   B6    -   HSV RSV  C
```

A   pDS 14
                                                           pCB111 35

B   pCB111 51
                                                           pDS 111

C   pCB111 51
                                                           pCB111 35

D   pCB111 51
                                                           pCB111 35

Figure 7. Combinations of poliovirus type 1 and coxsackievirus type B3 probes for detection of enteroviruses in CSF. For conditions and abbreviations, see legend of Fig. 6. Autoradiograph D is exactly the same as autoradiograph C except exposed for 48 h.

culture infective doses ($TCID_{50}$), for target serotypes which are closely related to the probe strains, to 10^5 $TCID_{50}$ for more distantly related viruses (39). Although few studies have addressed the issue, CSF from patients with enteroviral meningitis appears to contain between 10^0 and 10^4 $TCID_{50}$ of virions (51). This may underestimate the number of potentially available target RNA molecules by a factor of 100, the approximate particle-to-infectivity ratio for the enteroviruses.

THE FIRST CLINICAL TRIALS OF cDNA PROBES

Specimens of CSF collected from children with meningitis during a single summer were tested for the presence of enteroviruses by traditional

Table 1. Results of Dot-Blot Hybridization of Clinical CSF Specimens

Specimen[a]	n	Hybridization +
Culture +, clinical +	12	4
Culture −, clinical +	10	2
Culture −, clinical −	10	1
Bacterial meningitis[b]	3	1

[a] Specimens were obtained during a single summer season. All specimens were cultured for viruses; patients were also classified as to clinically suggestive (+) or not (−) of meningitis.
[b] Child with *H. influenzae* type b meningitis.

tissue culture techniques and by dot-blot hybridization (Table 1). A combination of the pDS14 and pCBIII35 cDNA probes resulted in a sensitivity, based on culture positivity, of only 33%, with a specificity of 85%. Additionally, two patients with positive hybridization results had clinical evidence of viral meningitis without growing virus from the CSF, likely false-negative culture results. Another child had a positive bacterial culture for *Haemophilus influenzae* type b with a false-positive hybridization assay for enterovirus.

We considered several possible strategies for overcoming the sensitivity and specificity shortcomings of this first trial: improvement of target retention, amplification of target sequences, and development of more sensitive probes.

IMPROVEMENT OF TARGET RETENTION IN CSF AND ON FILTER PAPER

To determine the extent of loss of target enteroviral RNA during the course of our solid-phase hybridization assay, we prepared ^{3}H-labeled enteroviral virions and ^{3}H-labeled enteroviral RNA. By adding these preparations separately to CSF, we were able to follow the retention of target molecules through the hybridization procedure by scintillography (36b). Free RNA, added to CSF, was sufficiently degraded by RNases within 15 s of contact with the body fluid to result in loss of >90% of counts during filtration through all commercially available membranes tested. Much of the remaining RNA was lost during subsequent hybridization incubations and washes. RNase inhibitors were protective, but only if added to the CSF before addition of the RNA. In contrast, radiolabeled virions survived exposure to CSF and the hybridization and wash procedures, with >65% retention of counts on membrane filters. We determined that formaldehyde added to the CSF-virion specimen (as

Figure 8. Serial dilutions of poliovirus type 1 (PV1) in CSF, hybridized after two different fixation protocols: SSC, a 1:1 mixture of the specimen and a salt solution, or SSC+FA, a 1:1 mixture of the specimen and a solution containing 3 parts salt and 2 parts formaldehyde. The undiluted specimen (10^0) contained 10^5 TCID$_{50}$ of virus; C is control CSF to which no virus had been added. Detecting probe was pDS111, and autoradiography was for 24 h.

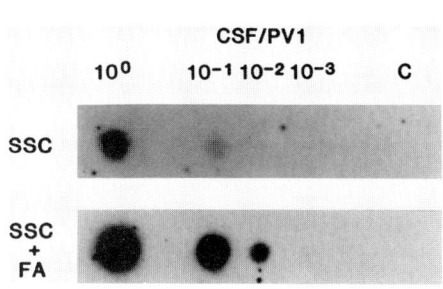

noted above) protects the RNA during the subsequent heating step in which viral capsids are presumably opened (allowing access of probes to target RNA during hybridization). Eliminating formaldehyde from the virion preparation, in fact, did result in markedly reduced hybridization signal (39; Fig. 8).

The lability of free RNA in CSF makes the preservation of intact virions of utmost importance. Addition of formaldehyde or another RNase inhibitor immediately after collection of the specimen from the patient will protect any RNA which becomes exposed by viral capsid opening during transport to the laboratory, as well as during the subsequent hybridization protocol. Unfortunately, free enteroviral RNA or partially exposed RNA (e.g., defective particles) in the specimen collected from the patient cannot be expected to remain viable for hybridization, regardless of technique.

AMPLIFICATION OF TARGET SEQUENCES

We have employed two strategies for increasing the number of target RNA molecules available for hybridization with enteroviral probes: biologic and enzymatic. In the biologic amplification approach, 10^2 TCID$_{50}$ of enterovirus (below the level of sensitivity of direct hybridization) was inoculated onto LLC-MK2 tissue culture cells and allowed to incubate for various lengths of time. The medium was decanted, and membrane filters were applied directly to the infected monolayer. The entire cell sheet was peeled off the culture dish onto the membrane, which was then fixed and hybridized with cDNA probes (H. A. Rotbart and M. J. Levin, *in* F. Tenover, ed., *DNA Probes in Infectious Diseases*, in press). Whereas cytopathic effect was not noted until after 60 h of incubation in tissue culture, the same inoculum of virus was detected with this "monolayer blot" technique in 36 h by scintillography (Table 2) or in

Table 2. Monolayer Blot Amplification of Virus in Tissue Culture[a]

Incubation in tissue culture (h)	Cytopathic effect	Autoradiography	Scintillography (^{3}H counts)
12	−	−	159
18	−	−	175
24	−	−	168
30	−	−	160
36	−	−	257[b]
42	−	+[b]	272
48	−	+	279
54	−	+	425
60	+[b]	+	880
66	+ +	+	2,147
72	+ + +	+	1,815
C[c]	−	−	146

[a] Monolayer blot amplification of $<10^2$ TCID$_{50}$ of coxsackievirus type A9 virions after incubation in tissue culture for the indicated lengths of time. Identical replicates were studied for cytopathic effect. Hybridization of monolayer blots was visualized by autoradiography (Fig. 9) and quantitated by scintillography.
[b] First positive blot by each technique.
[c] Control tissue culture, uninfected, hybridized at 72 h.

42 h by autoradiography (Fig. 9). Further enhancement of these methods may permit even briefer incubations in tissue culture before hybridization. We had hoped that enterovirus serotypes which do not produce cytopathic effect in tissue culture might still replicate their RNA, allowing

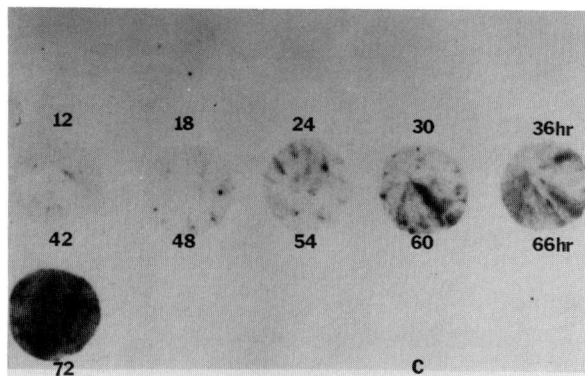

Figure 9. Monolayer blot hybridizations of coxsackievirus type A9-infected LLC-MK2 tissue culture cells after incubation for various times. Virus inoculum was $<10^2$ TCID$_{50}$ per well. Detecting probe was pDS14 cDNA; autoradiography was for 18 h.

detection with monolayer blot hybridization. Our attempts with several such coxsackievirus A serotypes, however, were unsuccessful, resulting in no detectable hybridization signals (unpublished data).

The enzymatic amplification of single genes or short sequences of DNA by repeated cycling of polymerase chain elongation reactions has recently been reported (33a). Applications of this technique, to date, have included the enhanced detection of the beta-globin gene and specific mutations therein (11, 41), diagnosis of factor VIII gene deficiency in hemophilia (25), and the detection of human immunodeficiency virus in infected cells (26). As of this writing, we are in the process of adapting polymerase chain amplification methods to the detection of enteroviruses in body fluids. By choosing flanking oligomeric primers based on specific sequence data reported for the enteroviruses that have been sequenced to date, the intervening sequences are replicated through 20 to 25 cycles of denaturation, annealing, and chain elongation with the aid of a temperature-stable polymerase. We expect as much as a 5-log increase in the number of target molecules available for detection by subsequent hybridization techniques (33a). The success of this technique for the enteroviruses will depend upon the ubiquitousness, among the many serotypes, of the oligomeric primer and probe sequences we have chosen.

DEVELOPMENT OF MORE SENSITIVE PROBES

There are several theoretical advantages of single-stranded RNA probes over the cDNA probes which we initially developed: RNA-RNA hybrids have greater affinity than do DNA-RNA hybrids; there are no vector sequences present to cause nonspecific hybridization; and there is no self-annealing among probe strands. A further advantage, unique to the probe preparation technique we chose, is the ability to separately produce labeled RNA transcripts of both senses. Cloned cDNA fragments of poliovirus type 1 (pDS14, provided by B. Semler and E. Wimmer [42]), coxsackievirus type B3 (pCBIII35, provided by S. Tracy [46]), and echovirus type 9 (pEC9I/4a, provided by B. Rosenwirth [49]) (Fig. 1) were excised from their original vectors by restriction enzyme digestion and recloned into transcription vectors (pDS14 into SP65, the others into pGEM2). Negative-sense, radioactive RNA transcripts were generated by linearizing the new recombinants with a restriction enzyme chosen to result in a positive-stranded DNA template, followed by reaction with promoter-specific polymerases in the presence of [^{3}H]- or [^{32}P]UTP (15, 32). We learned that the resultant RNA probe in the crude transcription mix performed as well as cleaner (DNase-treated or phenol-chloroform-extracted) preparations. Hybridization optimization experiments revealed that the ideal temperature of hybridization was 50°C; that longer reactions improved the signal obtained, with 24 h of hybridization ideal; and that formamide concentrations made little difference in the ultimate signal-to-noise ratio (36a).

These RNA probes were then tested in three types of experiments. First, in a comparative trial against its cDNA progenitor, the RNA probe derived from poliovirus type 1 was shown to be 10- to 100-fold more sensitive in detecting log dilutions of poliovirus type 1 and coxsackievirus

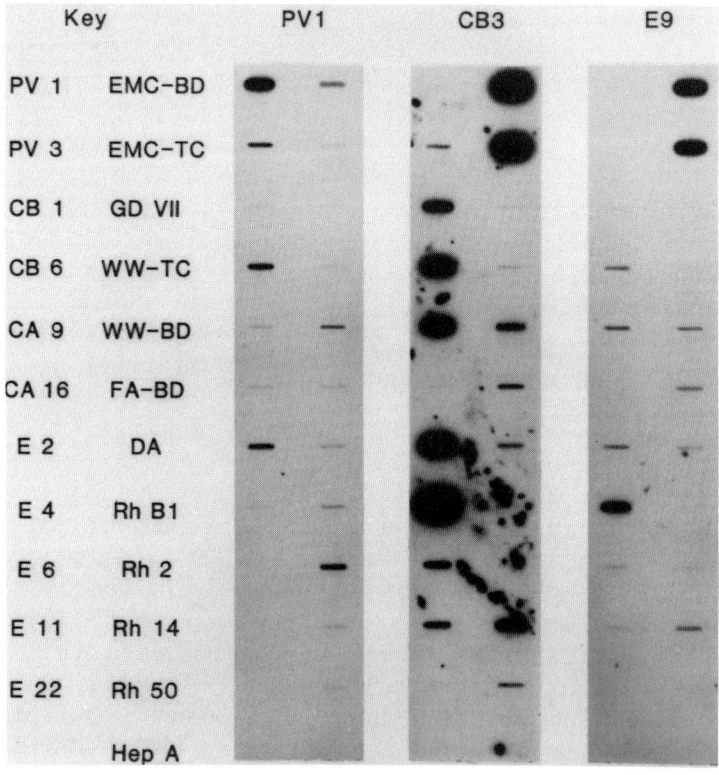

Figure 10. Slot-blot hybridization of human and animal picornaviruses with three single-stranded RNA probes derived from poliovirus type 1 (PV1), coxsackievirus type B3 (CB3), and echovirus type 9 (E9). All viruses were hybridized at an approximate titer of 10^5 TCID$_{50}$ (or 10^5 PFU) except for the following: Rh B1, 2, 14, and 50, 10^4 TCID$_{50}$; EMC 7 (BD) and WW-BD, 10^5 LD$_{50}$; FA-BD, 10^4 LD$_{50}$; Hep A, untitered. Abbreviations: PV, poliovirus; CB and CA, coxsackievirus types B and A; Rh, rhinovirus; E, echovirus; Hep A, hepatitis A virus; EMC, encephalomyocarditis virus. GD VII, WW, FA, and DA are all strains of Theiler's murine encephalomyelitis virus. TC, Tissue culture adapted; BD, brain derived. Autoradiography was for 18 h.

Table 3. Hybridization of a Variety of Enterovirus Serotypes with a
Combination of Three RNA Probes[a]

Serotype	^{3}H counts
Poliovirus	
1	4,820
2	2,762
3	4,829
Coxsackievirus	
B1	924
B6	1,914
A4[b]	5,845
A9	3,988
A14[c]	1,671
A16	3,590
A17[b]	1,639
A24[c]	710
Echovirus	
2	1,835
4	30,607
6	1,543
11	471
22	303
Control	89

[a] Probes derived from poliovirus type 1, coxsackievirus type B3, and echovirus type 9 (see
the text). Target virion titers were 10^5 TCID$_{50}$ unless noted otherwise.
[b] TCID$_{50}$ is unknown for this serotype.
[c] TCID$_{50}$ is 10^4 for this serotype.

type A9. Next, the three RNA probes were individually tested against 23
different serotypes of animal and human picornaviruses. Each probe
demonstrated significant hybridization with many of the target viruses
(Fig. 10), but no probe detected all serotypes. Finally, the three probes
were used in combination in a hybridization reaction to 16 human
enteroviruses; all 16 were detected (Table 3). These included two serotypes which do not grow in tissue culture (coxsackievirus types A4 and
A17), several that grow only sluggishly, and three that we have not
previously been able to detect with nick-translated cDNA probes (echovirus type 22 and coxsackievirus types A4 and A17).

To determine the sensitivity of this assay, we prepared and quantitated positive-sense RNA which was absolutely complementary to the
pCBIII35-derived negative-sense RNA probe. This positive-sense RNA
was used as a target to test the probe. Under our optimized solid-phase
hybridization conditions, 10 pg of target RNA was shown to be the lower
limit of sensitivity (36a).

Our first clinical trial with these RNA probes was recently completed

(H. A. Rotbart, E. A. F. Simoes, and M. J. Levin, *27th ICAAC*, abstr. no. 487, 1987). CSF samples were collected from patients in India with poliomyelitis at various stages of their disease. All fluids were negative for viral growth in tissue culture. Hybridization of these specimens with a combination of the poliovirus and coxsackievirus type B3-derived probes detected enteroviral (presumably polioviral) RNA in 22% of patients with poliomyelitis. Specificity, determined by concurrently testing CSF from Indian patients with other neurologic diseases, was 93%. Both figures may be underestimations, as many of the hybridization-negative patients may have cleared all polioviral RNA from their CSF by the time of sampling, and certain of the nonpoliomyelitis diseases may actually have been due to enteroviruses. Trials of the RNA probes on CSF specimens from patients in this country with meningitis and encephalitis are currently under way in our laboratory.

APPLICATIONS OF ENTEROVIRAL NUCLEIC ACID PROBES TO ISSUES OF PATHOGENESIS

We have recently developed in situ hybridization methods employing the RNA probes described above. The first application of these techniques was in the detection of both positive- and negative-sense target enteroviral RNA in the cytoplasms of infected cells (H. A. Rotbart, M. J. Abzug, M. J. Levin, R. S. Murray, and N. L. Murphy, *27th ICAAC*, abstr. no. 488). LLC-MK2 tissue culture cells were infected with coxsackievirus type B6 and incubated until cytopathic effect was observed. Cytospin preparations were then hybridized with one of two coxsackievirus type B3-derived RNA probes: either the negative-sense transcript of pCBIII35 or its exact complement, a positive-sense strand transcribed from the opposite cDNA template strand within the pGEM2 vector. This is made possible by the construction of the vector, which contains two promoters flanking a multiple cloning site sequence into which we cloned the fragment of coxsackievirus type B3 cDNA. The orientation of the insert within the vector and the linearization reaction determine which cDNA strand will serve as template. The appropriate polymerase was then selected for the active promoter. The resultant negative- and positive-sense RNA probes will detect, respectively, positive (message)- and negative (replicative intermediate)-sense target coxsackievirus RNA (Fig. 11). In a parallel experiment, we quantitated the relative amounts of hybridization to message versus replicative intermediate RNA. Lysates were prepared from infected cells and then fixed and hybridized on the solid phase with either the negative- or positive-sense, ^{3}H-labeled RNA

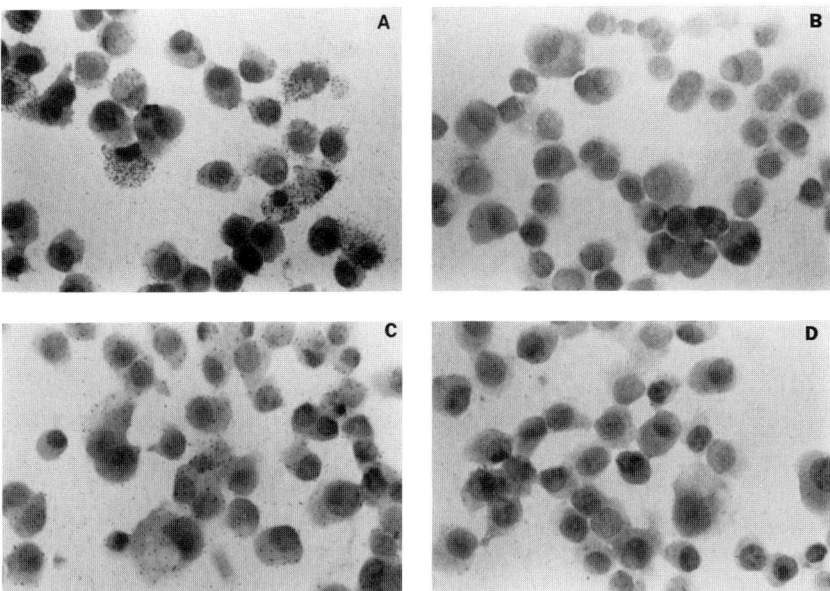

Figure 11. In situ hybridization of (A) coxsackievirus type B1-infected cells, with the negative-sense coxsackievirus type B3-derived RNA probe (grains indicate location of positive-sense viral RNA); (B) uninfected cells hybridized with the negative-sense probe; (C) coxsackievirus type B1-infected cells, with the positive-sense type B3-derived RNA probe (grains indicate location of negative-sense viral RNA); and (D) uninfected cells hybridized with the positive-sense probe.

probes. Table 4 confirms the detection of both senses of target RNA in approximately a 40:1 ratio of positive to negative sense.

This particular experimental design, and in situ hybridization for enteroviruses in general, has numerous applications to the study of enteroviral pathogenesis. It is well known, for example, that enteroviruses cause persistent infections in certain situations. Humans deficient in immunoglobulins develop persistent central nervous system infections lasting years, often with fatal outcome (50). The CSF may intermittently become culture negative for enterovirus, only to become positive again weeks, months, or years later with the same virus. The enteroviruses have been implicated in other chronic diseases, including myocarditis, diabetes mellitus, dermatomyositis/polymyositis, and postpoliomyelitis muscular atrophy syndrome. While virus may be recovered during the early presentation of disease in some of these patients, the disease symptoms persist for many years beyond evidence of active infection.

Table 4. Detection of Negative- and Postive-Sense RNA[a]

Target prepn	^{3}H counts hybridizing with:	
	Negative-sense probe[b]	Positive-sense probe[c]
Cell lysates		
Infected	45,336	2,449
Uninfected	1,193	1,333
(Net counts)	(44,143)	(1,116)
Purified virions[d]	3,245	163
Saline buffer	100	116
(Net counts)	(3,145)	(47)

[a] Slot-blot hybridization of coxsackievirus type B6-infected cell lysates was performed with RNA probes derived from coxsackievirus type B3. Positive- and negative-sense probes were generated from the identical cDNA insert in transcription vector GEM2. Uninfected lysates, hybridized with the same probes, served as negative controls. Purified virions were also used as a target to verify specificity of probes for the sense of the target (virions should contain exclusively positive-sense RNA). Quantitation was by cutting out the slots and subjecting them to scintillography.
[b] Detecting positive (message)-sense RNA.
[c] Detecting negative (replicative intermediate)-sense RNA.
[d] 10^5 TCID$_{50}$ of coxsackievirus type B6.

Documentation for persistent infection has been sought without success. Theiler's murine encephalomyelitis virus, a naturally occurring enterovirus infection of mice, causes a well-characterized chronic or persistent infection of the murine central nervous system, with virus recoverable many weeks to months after inoculation. The mechanisms for persistence of positive-sense single-stranded RNA viruses, without reverse transcriptase, are unknown. One possibility is via the negative-sense strand, either alone or complexed with the message sense. An analogous hypothesis has recently been proposed for herpes simplex virus persistence (44). The use of RNA probes of both senses on persistently infected tissues or cell lines may well delineate the relative importance of the two forms of viral RNA in establishing and maintaining persistence. In situ localization will define the extent and tropism of the virus.

In a more general way, in situ hybridization has been applied to the study of the pathogenesis of many viruses (17). The ability to identify specific organs, cell types, and intracellular distributions targeted by viruses has allowed for specific histopathologic correlations with clinical disease. My colleagues and I have begun to study enteroviral infections of brain, heart, liver, placenta, and fetus by in situ hybridization. Early results have encouraged us to believe this approach to be potentially of great utility in advancing our understanding of these viruses in a variety of clinical settings.

SUMMARY AND CONCLUSIONS

The enteroviruses are significant causes of human morbidity and mortality. Diagnostic testing, as it exists today, is inadequate for many reasons: tissue culturing is slow and too insensitive, animal inoculations are cumbersome and impractical, and immunoassays and further serological testing are complicated by the broad heterogeneity of enteroviral antigens. In the few years since my colleagues and I began experimenting with nucleic acid probes for diagnosis, much has been learned about these pathogens. A number of the enteroviral genomes have been fully sequenced (7, 22, 23, 27, 35, 45), revealing areas of homology corresponding to those predicted by the cross-hybridization studies done by ourselves and others (21, 37, 39, 46, 47). The atomic structures of poliovirus and rhinovirus have been resolved, illuminating clefts and crypts responsible for binding to cells and, possibly, susceptibility to antiviral therapy (20, 36). Drugs with broad in vitro and in vivo activity against the enteroviruses have been developed (29, 34) but have yet to be tested in humans. The surface has barely been scratched in addressing issues of enteroviral pathogenesis. Why, for example, are certain enteroviruses more virulent in adults (e.g., polioviruses, hepatitis A) while most are of much greater significance in infants and children? What determines the unique tissue tropisms of coxsackievirus B types for the heart, polioviruses for the central nervous system, other enteroviruses for muscle, liver, and pancreas—and how can these predilections clinically overlap to the great extent that they do? How do minor, point mutations in genome sequence attenuate virulent enteroviruses? To what extent do these viruses genetically recombine in nature to form hybrid strains such as those readily created in the laboratory (43), and what is the impact of such recombinations in the epidemiology of enteroviral diseases? Approaches which combine the power of molecular virology with the priorities of the clinician hold promise for resolving these questions and others in the years to come.

Acknowledgments. This work was supported in part by Public Health Service grants R23AI22945-02, DK34915, and P01AG07347-01 from the National Institutes of Health.

Literature Cited

1. **Assaad, F., and K. Ljungars-Esteves.** 1984. World overview of poliomyelitis: regional patterns and trends. *Rev. Infect. Dis.* 6:S302–S307.
2. **Barrett-Connor, E.** 1985. Is insulin-dependent diabetes mellitus caused by coxsackievirus B infection? A review of the epidemiologic evidence. *Rev. Infect. Dis.* 7:207–215.
3. **Bell, E. J., R. A. McCartney, D. Basquill, and A. K. R. Chaudhuri.** 1986. μ-Antibody

capture ELISA for the rapid diagnosis of enterovirus infections in patients with aseptic meningitis. *J. Med. Virol.* **19**:213–217.

4. **Centers for Disease Control.** 1987. Acute hemorrhagic conjunctivitis caused by coxsackievirus A24-Caribbean. *Morbid. Mortal. Weekly Rep.* **36**:245–251.

5. **Cherry, J. D.** 1981. Nonpolio enteroviruses: coxsackieviruses, echoviruses, and enteroviruses, p. 1316–1365. *In* R. D. Feigin and J. D. Cherry (ed.), *Textbook of Pediatric Infectious Diseases.* The W.B. Saunders Co., Philadelphia.

6. **Chonmaitree, T., M. A. Menegus, and K. R. Powell.** 1982. The clinical relevance of 'CSF viral culture': a two-year experience with aseptic meningitis in Rochester, N.Y. *J. Am. Med. Assoc.* **247**:1843–1847.

7. **Cohen, J. I., B. Rosenblum, J. R. Ticehurst, R. J. Daemer, S. M. Feinstone, and R. H. Purcell.** 1987. Complete nucleotide sequence of an attenuated hepatitis A virus: comparison with wild-type virus. *Proc. Natl. Acad. Sci. USA* **84**:2497–2501.

8. **Dagan, R., and M. A. Menegus.** 1986. A combination of four cell types for rapid detection of enteroviruses in clinical specimens. *J. Med. Virol.* **19**:219–228.

9. **Dalakas, M. C., J. L. Sever, D. L. Madden, N. M. Papadopoulos, I. C. Shekarchi, P. Albrecht, and A. Krezlewicz.** 1984. Late postpoliomyelitis muscular atrophy: clinical, virologic, and immunologic studies. *Rev. Infect. Dis.* **6**:S562–S567.

10. **Dorries, R., and V. Ter Meulen.** 1983. Specificity of IgM antibodies in acute human coxsackievirus B infections, analysed by indirect solid phase enzyme immunoassay and immunoblot technique. *J. Gen. Virol.* **64**:159–167.

11. **Embury, S. H., S. J. Scharf, R. K. Saiki, M. A. Gholson, M. Golbus, N. Arnheim, and H. A. Erlich.** 1987. Rapid prenatal diagnosis of sickle cell anemia by a new method of DNA analysis. *N. Engl. J. Med.* **316**:656–661.

12. **Enders, J. F., T. H. Weller, and F. C. Robbins.** 1949. Cultivation of the Lansing strain of poliomyelitis virus in culture of various human embryonic tissues. *Science* **109**:85–99.

13. **Fallis, R. J., and L. P. Weiner.** 1982. Further studies in search of a virus in amyotrophic lateral sclerosis, p. 355–360. *In* L. P. Rowland (ed.), *Human Motor Neuron Diseases.* Raven Press, New York.

14. **Gauntt, C. J., J. R. Gudvangen, Y. W. Brans, and A. E. Marlin.** 1985. Coxsackie group B antibodies in the ventricular fluid of infants with severe anatomic defects in the central nervous system. *Pediatrics* **76**:68–78.

15. **Green, M. R., T. Maniatis, and D. A. Melton.** 1983. Human beta-globin pre-mRNA synthesized in vitro is accurately spliced in Xenopus oocyte nuclei. *Cell* **32**:681–694.

16. **Gyorkey, R., G. A. Cabral, P. K. Gyorkey, G. Uribe-Botero, G. R. Dreesman, and J. L. Melnick.** 1978. Coxsackievirus aggregates in muscle cells of a polymyositis patient. *Intervirology* **10**:69–77.

17. **Haase, A., M. Brahic, L. Stowring, and H. Blum.** 1984. Detection of viral nucleic acids by in situ hybridization. *Methods Virol.* **7**:189–226.

18. **Herrmann, J. E., R. M. Hendry, and M. F. Collins.** 1979. Factors involved in enzyme-linked immunoassay of viruses and evaluation of the method for identification of enteroviruses. *J. Clin. Microbiol.* **10**:210–217.

19. **Hewlett, M. J., and R. Z. Florkiewicz.** 1980. Sequence of picornavirus RNA's containing a radioiodinated 5'-linked peptide reveals conserved 5' sequence. *Proc. Natl. Acad. Sci. USA* **77**:303–307.

20. **Hogle, J. M., M. Chow, and D. J. Filman.** 1985. Three-dimensional structure of poliovirus at 2.9 Å resolution. *Science* **229**:1358–1365.

21. **Hyypia, T. P., R. Stalhandske, R. Vainionpaa, and U. Pettersson.** 1984. Detection of enteroviruses by spot hybridization. *J. Clin. Microbiol.* **19**:436–438.

22. **Iizuka, N., S. Kuge, and A. Nomoto.** 1987. Complete nucleotide sequence of the genome of coxsackievirus B1. *Virology* **156**:64–73.

23. **Jenkins, O., J. D. Booth, P. D. Minor, and J. W. Almond.** 1987. The complete nucleotide sequence of coxsackievirus B4 and its comparison to other members of the picornaviridae. *J. Gen. Virol.* **68**:1835–1848.

24. **Johnson, R. T.** 1984. Late progression of poliomyelitis paralysis: discussion of pathogenesis. *Rev. Infect. Dis.* **6**:S568–S570.

25. **Kogan, S. C., M. Doherty, and J. Gitschier.** 1987. An improved method for prenatal diagnosis of genetic diseases by analysis of amplified DNA sequences. *N. Engl. J. Med.* **317**:985–990.

26. **Kwok, S., D. H. Mack, K. B. Mullis, B. Poiesz, G. Ehrlich, D. Blair, A. Friedman-Kien, and J. J. Sninsky.** 1987. Identification of human immunodeficiency virus sequences by using in vitro enzymatic amplification and oligomer cleavage detection. *J. Virol.* **61**:1690–1694.

27. **Lindberg, M. A., P. O. K. Stalhandske, and U. Pettersson.** 1987. Genome of coxsackievirus B3. *Virology* **156**:50–63.

28. **McCartney, R. A., J. E. Banatvala, and E. J. Bell.** 1986. Routine use of μ-antibody-capture ELISA for the serological diagnosis of coxsackie B virus infections. *J. Med. Virol.* **19**:205–212.

29. **McSharry, J. J., L. A. Caliguiri, and H. J. Eggers.** 1979. Inhibition of uncoating of poliovirus by Arildone, a new antiviral drug. *Virology* **97**:307–315.

30. **Melnick, J. L.** 1982. Enteroviruses, p. 187–251. *In* A. S. Evans (ed.), *Viral Infections of Humans: Epidemiology and Control*, 2nd ed. Plenum Publishing Corp., New York.

31. **Melnick, J. L.** 1984. Enterovirus type 71 infections: a varied clinical pattern sometimes mimicking paralytic poliomyelitis. *Rev. Infect. Dis.* **6**:S387–S390.

32. **Melton, D. A., P. A. Krieg, M. R. Rebagliati, T. Maniatis, K. Zinn, and M. R. Green.** 1984. Efficient in vitro synthesis of biologically active RNA and RNA hybridization probes from plasmids containing a bacteriophage SP6 promoter. *Nucleic Acids Res.* **12**:7035–7056.

33. **Mirkovic, R. R., R. Kono, and M. Yin-Murphy.** 1983. Enterovirus type 70: the etiologic agent of pandemic hemorrhagic conjunctivitis. *Bull. W.H.O.* **49**:341–346.

33a.**Mullis, K. B., and F. A. Faloona.** 1987. Specific synthesis of DNA in vitro via a polymerase catalyzed chain reaction. *Methods Enzymol.* **155**:335–350.

34. **Otto, M. J., M. P. Fox, and M. J. Fancher.** 1985. In vitro activity of WIN 51711, a new broad-spectrum antipicornavirus drug. *Antimicrob. Agents Chemother.* **27**:883–886.

35. **Racaniello, V. R., and D. Baltimore.** 1981. Molecular cloning of poliovirus cDNA and determination of the complete nucleotide sequence of viral genome. *Proc. Natl. Acad. Sci. USA* **78**:4887–4891.

36. **Rossmann, M. G., E. Arnold, J. W. Erickson, E. A. Frankenberger, J. P. Griffith, H. Hecht, J. E. Johnson, G. Kamer, M. Luo, A. G. Mosser, R. R. Rueckert, B. Sherry, and G. Vriend.** 1985. Structure of a human common cold virus and functional relationship to other picornaviruses. *Nature* (London) **317**:145–153.

36a.**Rotbart, H. A., M. J. Abzug, and M. J. Levin.** 1988. Development and application of RNA probes for the study of picornaviruses. *Mol. Cell. Probes* **2**:65–73.

36b.**Rotbart, H. A., M. J. Levin, N. L. Murphy, and M. J. Abzug.** 1987. RNA target loss during solid phase hybridization of body fluids—a quantitative study. *Mol. Cell. Probes* **1**:347–358.

37. **Rotbart, H. A., M. J. Levin, and L. P. Villarreal.** 1984. Use of subgenomic poliovirus DNA hybridization probes to detect the major subgroups of enteroviruses. *J. Clin. Microbiol.* **20**:1105–1108.

38. **Rotbart, H. A., M. J. Levin, and L. P. Villarreal.** 1985. Nucleic acid probes in the diagnosis of enteroviruses, p. 109–118. *In* L. de la Maza and E. M. Peterson (ed.),

Medical Virology IV: Proceedings of the 1984 International Symposium on Medical Virology. Lawrence Erlbaum Associates, London.

39. **Rotbart, H. A., M. J. Levin, L. P. Villarreal, S. M. Tracy, B. L. Semler, and E. Wimmer.** 1985. Factors affecting the detection of enteroviruses in cerebrospinal fluid with coxsackievirus B3 and poliovirus type 1 cDNA probes. *J. Clin. Microbiol.* **22**:220–224.

40. **Rueckert, R. R.** 1976. On the structure and morphogenesis of picornaviruses, p. 131–213. *In* H. Fraenkel-Conrat and R. R. Wagner (ed.), *Comprehensive Virology,* vol. 6. Plenum Publishing Corp., New York.

41. **Saiki, R. K., S. Scharf, F. Faloona, K. B. Mullis, G. T. Horn, H. A. Erlich, and N. Arnheim.** 1985. Enzymatic amplification of B-globin genomic sequences and restriction site analysis for diagnosis of sickle cell anemia. *Science* **330**:1350–1354.

42. **Semler, B. L., A. J. Dorner, and E. Wimmer.** 1984. Production of infectious poliovirus from cloned cDNA is dramatically increased by SV40 transcription and replication signals. *Nucleic Acids Res.* **12**:5123–5141.

43. **Semler, B. L., V. H. Johnson, and S. Tracy.** 1986. A chimeric plasmid from cDNA clones of poliovirus and coxsackievirus produces a recombinant virus that is temperature-sensitive. *Proc. Natl. Acad. Sci. USA* **83**:1777–1781.

44. **Stevens, J. G., E. K. Wagner, G. B. Devi-Rao, M. L. Cook, and L. T. Feldman.** 1987. RNA complementary to a herpesvirus α gene mRNA is prominent in latently infected neurons. *Science* **235**:1056–1059.

45. **Toyoda, H., M. Kohara, Y. Kataoka, T. Suganuma, T. Omata, N. Imura, and A. Nomoto.** 1984. Complete nucleotide sequences of all three poliovirus serotype genomes. *J. Mol. Biol.* **174**:561–585.

46. **Tracy, S.** 1984. A comparison of genomic homologies among the coxsackievirus B group: use of fragments of the cloned coxsackievirus B3 genome as probes. *J. Gen. Virol.* **65**:2167–2172.

47. **Tracy, S.** 1985. Comparison of genomic homologies in the coxsackievirus B group by use of cDNA dot-blot hybridization. *J. Clin. Microbiol.* **19**:371–374.

48. **Travers, R. L., G. R. V. Hughes, G. Cambridge, and J. R. Sewell.** 1977. Coxsackie B neutralisation titers in polymyositis/dermatomyositis. *Lancet* **ii**:1268.

49. **Werner, G., B. Rosenwirth, E. Bauer, J. Seifert, F. Werner, and J. Besemer.** 1986. Molecular cloning and sequence determination of the genomic regions encoding protease and genome-linked protein of three picornaviruses. *J. Virol.* **57**:1084–1093.

50. **Wilfert, C. M., R. H. Buckley, and T. Mohanakumar.** 1977. Persistent and fatal central nervous system echovirus infections in patients with agammaglobulinemia. *N. Engl. J. Med.* **296**:1485–1489.

51. **Wilfert, C. M., and J. Zeller.** 1985. Enterovirus diagnosis. *In* L. de la Maza and E. M. Peterson (ed.), *Medical Virology IV: Proceedings of the 1984 International Symposium on Medical Virology.* Lawrence Erlbaum Associates, London.

52. **Yolken, R. H., and V. M. Torsch.** 1980. Enzyme-linked immunosorbent assay for detection and identification of coxsackie B antigen in tissue culture and clinical specimens. *J. Med. Virol.* **6**:45–52.

53. **Yolken, R. H., and V. M. Torsch.** 1982. Enzyme-linked immunosorbent assay for detection and identification of coxsackieviruses A. *Infect. Immun.* **31**:742–750.

54. **Yoon, J., M. Ausatin, T. Onodera, and A. L. Notkins.** 1979. Virus-induced diabetes mellitus: isolation of a virus from the pancreas of a child with diabetic ketoacidosis. *N. Engl. J. Med.* **300**:1173–1179.

55. **Young, N. A.** 1973. Polioviruses, coxsackieviruses, and echoviruses: comparison of the genomes by RNA hybridization. *J. Virol.* **11**:832–839.

Molecular Aspects of Picornavirus Infection and Detection
Edited by Bert L. Semler and Ellie Ehrenfeld
© 1989 American Society for Microbiology, Washington, DC 20006

Chapter 15

Modification of Six Amino Acids in the VP1 Capsid Protein of Poliovirus Type 1, Mahoney Strain, Alters Its Host Range and Makes It Neurovirulent for Mice

Marc Girard, Annette Martin, Therese Couderc,
Radu Crainic, and Czeslaw Wychowski

Two recent experimental achievements have allowed considerable progress in the understanding of the poliovirus genome and its manipulation. First, it was demonstrated that genomic-length poliovirus cDNA molecules are infectious, i.e., that once transfected onto primate cell cultures, such molecules generate an infectious virus cycle (21). Increased infectivity of plasmids carrying a genomic-length poliovirus cDNA insert was obtained by using vectors carrying the simian virus 40 (SV40) origin of replication, enhancer sequence, and promoters to transfect COS-1 cells (26). Alternatively, increased infectivity was obtained by using vectors carrying the SV40 origin of replication and the whole SV40 early region, which encodes the SV40 small t and large T antigens, to transfect Vero or CV1 cells (10, 13). These observations have opened the way to the application of site-directed mutagenesis to the study of poliovirus (3, 6,

Marc Girard, Annette Martin, Therese Couderc, Radu Crainic, and Czeslaw Wychowski • Unité de Virologie Moléculaire, UA 545, Centre National de la Recherche Scientifique, and Unité de Virologie Médicale, Institut Pasteur, 75724 Paris Cedex 15, France.

12, 25, 30) and to the construction of hybrid viruses by in vitro DNA recombination between cDNA copies of different poliovirus genomes (1, 11, 20).

Second, recent studies by X-ray diffraction of poliovirus type 1 (PV-1) (8), rhinovirus type 14 (23), and mengovirus (17) crystals have led to fine resolution of the topography of the four structural polypeptides, VP1, VP2, VP3, and VP4, on the virus capsid and to the localization of the three major neutralization immunogenic sites NIm-I, NIm-II, and NIm-III (23). Location of the poliovirus neutralization sites was determined from the sequencing of neutralization escape mutants (4, 7, 18), from the study of the immunogenicity of synthetic oligopeptides with the amino acid sequence of type 1 VP1 (5, 28), and from the reactivity of a β-lactamase VP1 fusion protein with a type 1-specific neutralizing monoclonal antibody (MAb), C3 (27, 29). In the latter case, PV-1 VP1 was fused in phase to the first nine NH_2-terminal amino acids from pBR322 β-lactamase, after which deletions of various lengths were generated inside the VP1 sequence and the resulting truncated fusion proteins were reacted with C3. Deletions extending from the C-terminal end of VP1 up to VP1 amino acid 105 did not affect reactivity with C3, whereas deletions extending up to amino acid 92 or further upstream suppressed all reactivity. As a corollary, an oligopeptide with the sequence of VP1 amino acids 93 to 103 was found to react specifically with C3 and to competitively inhibit virus neutralization by C3 (29). The same peptide was able, once coupled to protein carriers, to induce poliovirus neutralizing antibodies in rabbits (9). The PV-1 neutralization epitope defined by VP1 amino acids 93 to 103, which we shall name epitope C3, is a continuous epitope (4) which lies in a loop of VP1 exposed at the surface of the capsid in close vicinity to the fivefold axis of the icosahedron (8).

The question was raised of whether the sequence of the C3 epitope could be changed without affecting the viability of the virus, although modifying its immunogenicity and, perhaps, inducing a modification of its host range. As a first attempt, we decided to substitute the sequence of the C3 epitope in the Mahoney strain of PV-1 with that of the corresponding region from the Lansing strain of PV-2. The latter has been adapted to mice (2) and induces fatal spinal paralysis in the animal upon intracerebral injection. In contrast, PV-1 is devoid of pathogenicity for mice. The study of viral recombinants constructed by in vitro DNA recombination between PV-1 Mahoney and PV-2 Lansing has shown that the mouse-adapted phenotype maps to the Lansing viral capsid (15).

More recently, it was shown that the region of the Lansing type 2 capsid corresponding to VP1 amino acids 93 to 105 was specifically implied in the determination of mouse neurovirulence (14, 16).

As will be shown below, a chimeric type 1/type 2 poliovirus in which a nine-amino-acid sequence from the PV-1 Mahoney capsid protein VP1 was replaced by the corresponding sequence from the Lansing type 2 capsid protein appears to be fully neurovirulent for mice.

CONSTRUCTION OF CHIMERIC TYPE 1/TYPE 2 POLIOVIRUSES

To be able to replace the sequence of the C3 epitope from PV-1 by the corresponding sequence from the Lansing strain of PV-2, two new restriction sites were first created by site-directed mutagenesis on either side of the C3 sequence in a Mahoney type 1 cDNA molecule. The rationale for this approach was that, once these sites were created, they would delimit a fragment of cDNA which would be easy to change for other appropriate cDNA fragments or synthetic oligodeoxynucleotides, following a cassette strategy.

Site-directed mutagenesis was performed using the technique of Morinaga et al. (19) on a fragment of the PV-1 cDNA corresponding to nucleotides 1 to 5601, which had been cleaved from plasmid pKK17 and was subcloned into a pBR327 derivative. Plasmid pKK17 contains a genomic-length Mahoney cDNA insert placed downstream from the SV40 late promoter. It also contains the SV40 origin of replication, the SV40 enhancer, and the whole SV40 early region (10). This plasmid generates infectious PV-1 when transfected onto Vero or CV1 cells with an efficacy of greater than 500 PFU/µg of DNA. An XhoI site is located in pKK17, 5′ from the poliovirus sequence, immediately downstream from the SV40 promoter.

The XhoI-BglII fragment from pKK17 was subcloned into a pBR327 derivative, the HindIII site and the EcoRV site of which had been destroyed and replaced by an XhoI site and a BglII site, respectively. The resulting recombinant plasmid, pAM3, was mutagenized by oligonucleotide site-directed mutagenesis (19).

We chose an HpaI site or an EcoRV site at position 2756 on the poliovirus cDNA insert, immediately upstream from the sequence coding for the C3 epitope (Fig. 1), and a HindIII site at position 2786, immediately downstream from the C3 sequence.

To carry out the site-directed mutagenesis, a preparation of plasmid pAM3 was digested by restriction enzyme XbaI, and another preparation was digested by restriction enzyme SalI. The four restriction fragments resulting from the first digestion were purified on a low-melting-point agarose gel. The smaller XbaI fragment (nucleotides 2546 to 2861), coding for the sequence of VP1 amino acids 28 to 128, was eliminated, and the three other fragments were mixed with the SalI-digested DNA and

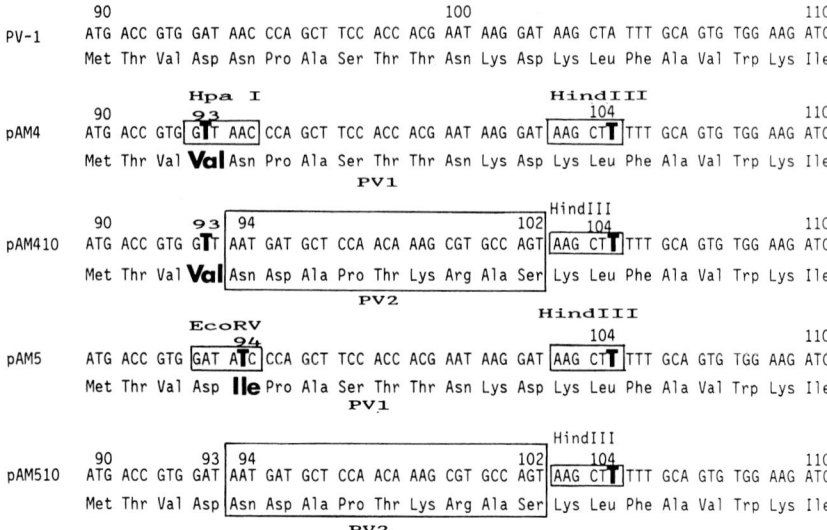

Figure 1. Construction of hybrid type 1/type 2 cDNA molecules by the exchange of nine codons in the VP1 sequence of PV-1. The restriction sites indicated (boxes) were generated in the PV-1 cDNA sequence by site-directed mutagenesis as diagrammed in Fig. 2 and described in the text. Boldface letters indicate the nucleotides and amino acids that were mutated as a consequence of the mutagenesis. The Mahoney type 1 sequence between the newly created sites was then replaced by the Lansing type 2 sequence (box) provided in the form of appropriate oligodeoxynucleotides. Numbering refers to amino acids in the VP1 of PV-1 and PV-2.

denatured. Reannealing provided a double-stranded DNA molecule with a single-stranded 315-base-pair gap covering the region to be mutagenized. Two synthetic oligonucleotides with a sequence complementary to that of the poliovirus cDNA, each of which carried the appropriate mismatch necessary to create the desired restriction site, were added simultaneously to the mix of pAM3 fragments (Fig. 2). In one case, the synthetic oligonucleotides were designed to create an *Hpa*I and a *Hind*III restriction site; in the other, they were designed to create an *Eco*RV and a *Hind*III site (Fig. 1).

The molecules were repaired by treatment with Klenow DNA polymerase and then ligated with T4 DNA ligase, and HB101 cells were transformed with the mixture. Resulting clones were screened by colony hybridization using as probes the ^{32}P-labeled oligonucleotides involved in the mutagenesis experiment. Double-mutant plasmids were selected on the basis of hybridization with both probes and checked for the presence of the new restriction sites. Two recombinant plasmids, pAM4 containing

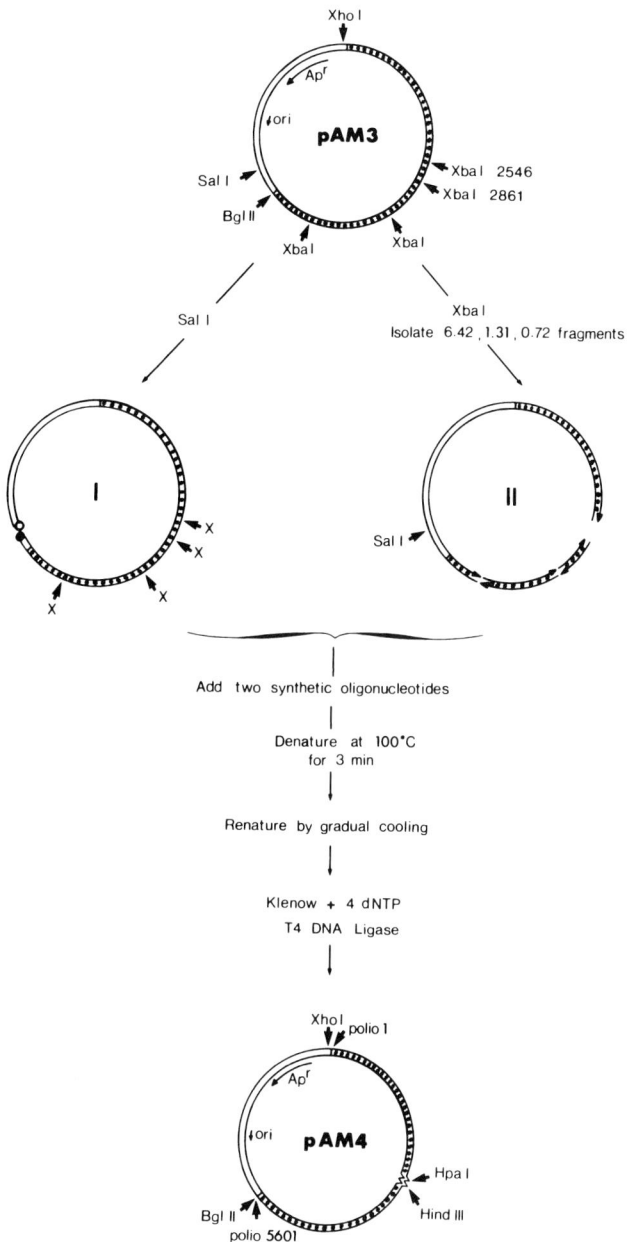

Figure 2. Steps in site-specific oligonucleotide-directed mutagenesis. Plasmid pAM3 contains the sequence of PV-1 nucleotides 1 to 5601 (dotted box) inserted between an *Xho*I and a *Bgl*II site (see text). The sequences of the oligodeoxynucleotides used for mutagenesis (GCTGGGTTAACCACGGTC and CCACACTGCAAAAAGCTTATCC, respectively) were complementary to that of the PV-1 cDNA with only one mismatch, creating an *Hpa*I and a *Hin*dIII site, respectively (see Fig. 1).

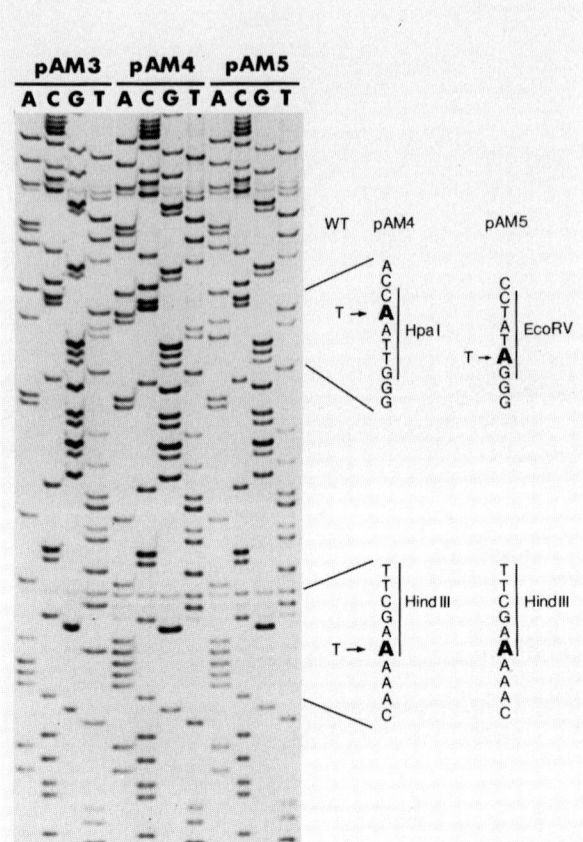

Figure 3. DNA sequence analysis of plasmids pAM3, pAM4, and pAM5. The DNA of the plasmids was sequenced by the method of Zagursky et al. (31), using as a precursor the oligodeoxynucleotide CCTCCGTAACTGGAC. Nucleotide sequences were obtained by the dideoxy chain termination method (24). The sequence of plasmids pAM4 and pAM5 differs from that of pAM3 as shown on the right-hand side of the figure.

an *Hpa*I and a *Hin*dIII site and pAM5 containing an *Eco*RV and a *Hin*dIII site, were thus selected. Their nucleotide sequence in the region coding for the C3 epitope was confirmed by sequencing by the method of Zagursky et al. (31) (Fig. 3).

The introduction of a *Hin*dIII site at position 2786 did not lead to any amino acid change (Fig. 1). In contrast, the creation of the *Hpa*I site at position 2756 led to an Asp-to-Val mutation at position 93 of VP1, and the creation of the *Eco*RV site resulted in an Asn-to-Ile mutation at position 94 (Fig. 1).

The unique restriction sites thus created allowed us to delete the sequence coding for VP1 amino acids 94 to 102 in either pAM4 or pAM5 and to replace it with the sequence of the corresponding antigenic region from the Lansing strain of PV-2. For that purpose, two complementary oligodeoxynucleotides, coding for the type 2 sequence (Fig. 1) and completed with the nucleotides required for the restoration of the *Hin*dIII site on the downstream side from that sequence, were inserted into each of the two vectors. The pAM4 vector was first treated with *Hpa*I and *Hin*dIII, and the pAM5 vector was treated with *Eco*RV and *Hin*dIII. HB101 cells were transformed with each of the ligation mixtures, and hybrid clones were screened by colony hybridization with the ^{32}P-labeled type 2 oligonucleotide as a probe. The presence of the *Hin*dIII site in the putative positive clones was then controlled. Resulting plasmids pAM41 and pAM51, containing a hybrid type 1/type 2 fragment of poliovirus cDNA, were checked by sequencing at the site of the insertion.

Deletion of the sequence encoding PV-1 VP1 amino acids 94 to 102 allows one to eliminate the mutation introduced at position 94 into plasmid pAM5 by the *Eco*RV restriction site, whereas the mutation introduced at VP1 position 93 into pAM4 by the *Hpa*I restriction site remains in pAM41 (Fig. 1).

To reconstitute a hybrid type 1/type 2 poliovirus cDNA of genomic length, both pAM41 and pAM51 were digested by *Xho*I and *Bgl*II restriction enzymes, and the *Xho*I-*Bgl*II fragments containing the first 5,601 base pairs of the poliovirus hybrid cDNAs were ligated with the *Xho*I-*Bgl*II restriction fragment from pKK17 containing the last 1,840 base pairs of the PV-1 Mahoney cDNA. Resulting plasmids pAM410 and pAM510, which contained a full-length PV-1 cDNA with a 27-nucleotide PV-2 insert in the region coding for capsid protein VP1, were transfected onto CV1 cells to regenerate the corresponding chimeric type 1/type 2 viruses v410 and v510. These two viruses should differ from each other in VP1 amino acid 93, which is mutated (Val) in the hybrid virus derived from pAM410 and is Mahoney-like (Asp) in that derived from pAM510 (Fig. 1). That this was indeed the case was determined by nucleotide sequencing of plasmids pAM410 and pAM510.

In parallel with these experiments, full-length double-mutant PV-1 cDNAs were reconstructed in a pKK17 vector from the DNAs of plasmids pAM4 and pAM5 (Fig. 1). Recombinant plasmids were transfected onto CV1 cells, generating PV-1 mutants vAM4 and vAM5 with an Asp-to-Val substitution at VP1 position 93 and an Asn-to-Ile substitution at VP1 position 94, respectively. Both mutant viruses grew with normal yields and formed large plaques on CV1, Vero, and HEp-2 cells. Their neutralization pattern has not been determined at this stage.

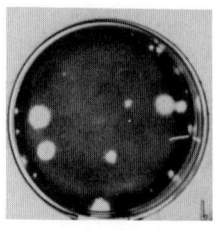

vKK17

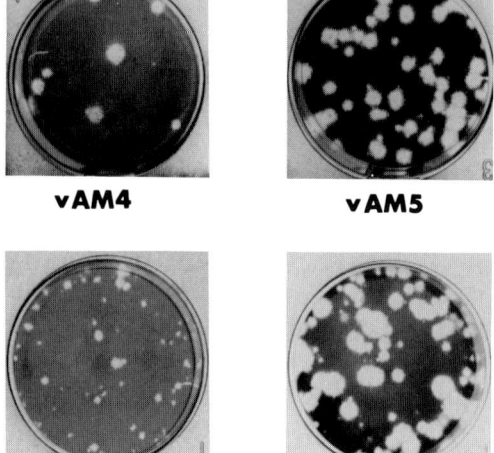

vAM4

vAM5

v410

v 510

Figure 4. Plaque size of v410, v510, and parent viruses. Stocks of wild-type virus vKK17, mutant viruses vAM4 and vAM5, and hybrid viruses v410 and v510 were prepared as described in the text and titrated by the plaque method on CV1 cells. The plates were stained after 2 days for all the viruses, except for v410, which was stained after 4 days at 37°C.

PROPERTIES OF THE CHIMERIC VIRUSES

Chimeric viruses v410 and v510 grew on CV1, Vero, HeLa, or HEp-2 cells, but whereas v510 produced normal-sized (large) plaques, v410 showed a small-plaque phenotype (Fig. 4). We do not know yet what is the molecular basis for this phenotype. The time course of v410 RNA synthesis was not different from that of v510 or of vKK17 (data not shown). Virus stocks raised on HEp-2 cells had titers of 2×10^8 PFU/ml for v410, 4×10^8 PFU/ml for v510, and 2×10^9 PFU/ml for vKK17.

Both v410 and v510 could be neutralized by a type 1-specific poliovirus-hyperimmune rabbit serum, using a plaque-reduction assay (Table 1). Immunoprecipitation experiments carried out using a type 1 VP1-specific immune rabbit serum showed that the hybrid VP1 proteins

Table 1. Neutralization by Neutralizing MAbs or Type-Specific Poliovirus Immune Sera[a]

| Virus | Neutralizing titer of MAb[b]: | | | | | |
| | Anti-PV-1 | | | Anti-PV-2 | | |
	Ic	C3	α-PV-1	IIo	HO2	α-PV-2
Mahoney	1,730	256	10,240	6	10	10
vKK17	1,905	83	21,379	9	10	19
Lansing	6	9	36	524	77,642	14,454
Sabin 2	4	4	42	16	12,407	10,852
vAM4	1,778	724	12,302	21	10	18
vAM5	1,318	467	12,589	21	10	16
v410	407	4	29,512	4	15,488	23
v510	1,318	5	8,709	3,162	9,549	54

[a] Neutralizing titers were determined by a plaque reduction test. An inoculum of about 50 PFU of virus was mixed with fourfold serial dilutions of each antibody (either MAb-containing mouse ascites fluids or type-specific poliovirus-neutralizing rabbit sera) in two wells of a 24-well Costar plate. After 2 h of incubation at 37°C, 3×10^5 Vero cells were added to each well, and the plates were further incubated for 3 h at 37°C, after which 0.6 ml of a 1.6% carboxymethyl cellulose solution in culture medium (Leibowitz L-15) was added. Cell cultures were stained with crystal violet after 3 days at 37°C in a 5% CO_2 atmosphere, and plaques were counted. Endpoints corresponding to 50% reduction in the number of plaques were calculated from the slope of the regression curve, using the least-squares method of the best-fitting straight line.

[b] Anti-PV-1 antibodies: Ic, Mab directed against site IIIB (VP3 amino acid 60); C3, MAb reacting with site I (VP1 amino acid 100); α-PV-1, anti-PV-1 rabbit serum. Anti-PV-2 antibodies: IIo, MAb directed against the type 2 MEF1 strain; HO2, a broadly reacting anti-PV-2 MAb (kindly supplied by A. Ostherhaus under the label 1-10C9E6); α-PV-2, anti-PV-2 rabbit serum.

from v410 and v510 could specifically be immunoprecipitated by type 1-specific antiserum (data not shown).

Neutralization studies using MAbs showed that none of the hybrid viruses was neutralized by the MAb specifically directed against the PV-1 C3 epitope (Table 1). This proved that both chimeric viruses had lost type 1 neutralization site NIm-I. Again, this was confirmed by an immunoprecipitation experiment which showed that C3 was no longer able to immunoprecipitate capsid protein VP1 from v410 or from v510 (data not shown). Both v410 and v510 were, on the other hand, still neutralized by MAb Ic, which is specifically directed against neutralization site NIm-III (4).

Further neutralization studies were carried out using a panel of type 2-specific MAbs, two of which (IIo and HO2) were able to neutralize v510 (Table 1). However, v410 could be neutralized by only one of these two MAbs (HO2). These results allowed us to map the type 2 epitopes reacting with these two MAbs to VP1 amino acids 93 to 103, which had remained yet undetermined.

These findings also prove that the type 1/type 2 chimeric polioviruses

Table 2. Mouse Neurovirulence of Chimeric Type 1/Type 2 Polioviruses and Parent Strains

Virus	LD_{50}^{a} (PFU)
vKK17	$>1.0 \times 10^7$
Lansing	1.0×10^4
v410	$>1.0 \times 10^7$
v510	2.2×10^4

[a] Calculated as described in the text.

have acquired a mosaic capsid with a type 2-specific neutralization epitope in a type 1 antigenic background. The heterogeneity found in the neutralization pattern of the two hybrid viruses must come from the presence of the Val instead of Asp residue at VP1 position 93 in v410. Thus, there might exist two different epitopes within the VP1 hybrid region of v510, of which one involves amino acid 93 but the other does not. The low level of neutralization of v410 and v510 by the anti-PV-2 rabbit antiserum could be due to the lack of response of rabbits to this antigenic site. This question will be answered by testing the immunogenicity of the chimeric viruses in rabbits, which is in progress.

NEUROVIRULENCE STUDIES

The Lansing strain of PV-2 causes fatal paralysis in mice after intracerebral injection, whereas the Mahoney strain of PV-1 does not (2, 14, 15). To study the possible neurovirulence of chimeric viruses v410 and v510, groups of six 30-day-old Swiss mice were inoculated intracerebrally with 0.03 ml of serial virus dilutions, and the animals were observed daily for 21 days for paralysis or death. Fifty percent lethal doses (LD_{50}) were calculated by the method of Reed and Muench (22). The LD_{50} of the PV-2 Lansing virus was 4 $\log_{10}$ PFU, whereas that of the PV-1 Mahoney virus was over 7 $\log_{10}$ PFU. These figures are similar to those reported by La Monica et al. (14, 15). The LD_{50} of v510 was very close to that of the Lansing strain (Table 2). Surprisingly, however, the LD_{50} of v410 was very high, in the same range as that of the Mahoney strain.

The experiment was repeated using groups of four mice which received from 4 $\log_{10}$ to 6.5 $\log_{10}$ PFU of virus intracerebrally and were scored as percent mice by days of survival. These figures were obtained by computing for each inoculated dilution the number of mice dead on a given day, multiplied by the number of days each mouse had survived. Thus, if the four mice survived the 21-day observation period, the score

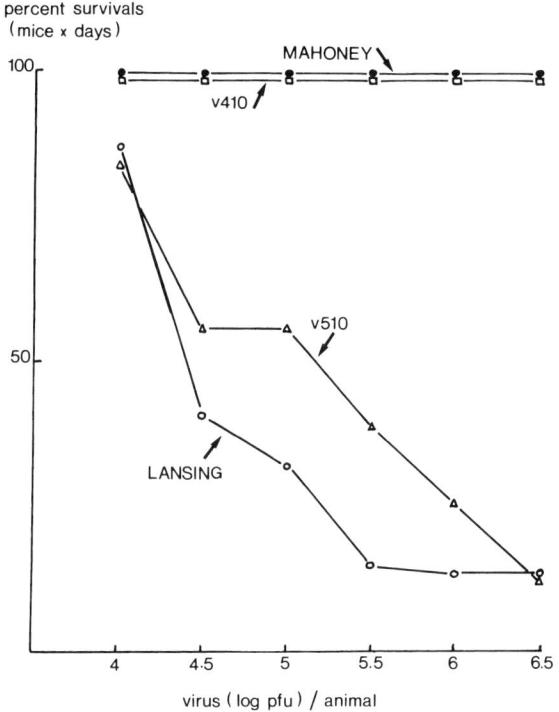

Figure 5. Neurovirulence of v410, v510, and parent viruses for mice. Groups of four mice were injected intracerebrally with the indicated amount of virus, and the animals were followed for 21 days. Results were plotted as percent mice by days of survival (see text) as a function of the dose of virus.

was $4 \times 21 = 84$, or 100%. If, on the other hand, three mice had been observed to die, for example, at day 5, and the fourth mouse had died at day 6, the score would have been $(3 \times 5) + (1 \times 6) = 21$, or 25%. This calculation allows one to get a quantitative representation of neurovirulence which takes into account not only the number of cases of paralysis or deaths observed in a given lot of animals but also the length of survival of each animal in that lot.

Chimeric virus v510 was slightly less neurovirulent than the prototype Lansing virus strain, as it took about 10 times more virus in the range between 4.5 $\log_{10}$ and 6.5 $\log_{10}$ PFU to achieve the same mouse-day score with v510 as with the Lansing strain (Fig. 5). In contrast, however, v410 scored as devoid of neurovirulence, like the Mahoney strain. Neural histopathology examination of the animals and determination of virus titers in the brain of mice infected with either chimeric virus will have to be performed to complete this study and to understand the basis for these widely different levels of pathogenicity.

CONCLUSION

Neurovirulence for mice has long been considered a unique but unexplained characteristic of the PV-2 Lansing strain. Recently, by studying the neurovirulence of Lansing-derived antigenic escape mutants resistant to neutralization with MAb, La Monica et al. (14) have been able to map the mouse-adapted phenotype of PV-2 to the region of VP1 amino acids 93 to 105. More specifically, these authors showed that amino acid substitutions at VP1 positions 100 and 101, and to a lesser extent at position 99, correlated with decreased neurovirulence for mice.

The mouse-adapted PV-2 Lansing strain differs from the nonadapted PV-1 Mahoney strain in the same region of the capsid protein by six single-amino-acid changes, namely, those at VP1 positions 95, 97, 99, 100, 101, and 102 (compare the sequences of PV-1 and of pAM510 in Fig. 3). It is remarkable that the mere substitution of these amino acid residues in a PV-1 Mahoney strain with those from the PV-2 Lansing strain conferred to the virus a level of neurovirulence for mice almost as high as that of the prototype PV-2 Lansing virus. It will be of interest to determine whether all six amino acid changes are required for such a host range modification, or only those at residue numbers 99, 100, and 101.

The second chimeric virus, v410, differed from v510 by only one amino acid substitution (Asp to Val) at VP1 position 93, yet it was devoid of neuropathogenicity for mice. This lack of neurovirulence cannot be ascribed to a slower rate of replication (v410 is not defective for replication in HeLa cells). It could perhaps be due to inefficient encapsidation of v410 or to a defect in capsid stability, as evidenced by the fact that v410 has a small-plaque phenotype and grows to lower yields than v510 in primate cells. A change at amino acid 93 from Asp to Gly was also observed by La Monica et al. in two of their Lansing MAb-resistant variants exhibiting reduced mouse neurovirulence (14). Interpretation of the role of the Asp residue at amino acid 93 was, however, made difficult in that study by the fact that both variants contained an additional mutation at position 100. Although unlikely, the possibility that the lack of neurovirulence of v410 for mice is due to an additional mutation outside antigenic site 1 cannot be ruled out at the present time.

These results confirm the conclusion that acquisition of poliovirus neurovirulence for mice depends on a host range mutation in the capsid protein loop formed by VP1 amino acids 93 to 103 on the surface of the virion. How can a modification of this loop, which extends outside of the capsid, affect neurovirulence in mice? The loop is located at quite a distance from the region of the canyon which is thought to interact with the host cell receptors (23). The conclusion that changing the amino acid sequence in the loop directly or indirectly alters the interaction of the virus with a mouse brain receptor is, however, difficult to escape. It will be of interest to determine the X-ray diffraction patterns of crystals of the v510 and v410 chimeras.

These results also show the feasibility of using poliovirus as a vector for the expression of foreign antigenic determinants, as independently demonstrated for type 1/type 3 chimeric viruses by Murray et al. (M. G. Murray, R. J. Kuhn, and E. Wimmer, *Modern Approaches to New Vaccines, Including Prevention of AIDS: Abstracts of Papers Presented at the 1988 Meeting*, p. 125) and by J. Almond (personal communication).

Acknowledgments. We thank J. Hogle for continuous encouragement and helpful discussion, A. Palmenberg for communication of picornavirus amino acid alignment, A. Osterhaus for the gift of MAb 1-10C9E6, and Maryse Tardi-Panit and Adina Candrea for skillful technical assistance.

Literature Cited

1. **Almond, J. W., G. D. Westrop, A. J. Cann, G. Stanway, D. M. A. Evans, P. D. Minor, and G. C. Schild.** 1985. Attenuation and reversion to neurovirulence of the Sabin poliovirus type 3 vaccine, p. 271–277. *In* R. A. Lerner et al. (ed.), *Vaccines 85: Molecular and Chemical Basis of Resistance to Parasitic, Bacterial and Viral Diseases.* Cold Spring Harbor Laboratory, Cold Spring Harbor, N.Y.

2. **Armstrong, C.** 1939. Successful transfer of the Lansing strain of poliomyelitis virus from the cotton rat to the white mouse. *Public Health Rep.* **54:**2302–2305.

3. **Bernstein, M. D., N. Sonenberg, and D. Baltimore.** 1985. Poliovirus mutant that does not selectively inhibit host cell protein synthesis. *Mol. Cell. Biol.* **5:**2913–2923.

4. **Blondel, B., R. Crainic, O. Fichot, G. Dufraisse, A. Candrea, D. Diamond, M. Girard, and F. Horaud.** 1986. Mutations conferring resistance to neutralization with monoclonal antibodies in type 1 poliovirus can be located outside or inside the antibody binding site. *J. Virol.* **57:**81–90.

5. **Chow, M., R. Yabrov, J. L. Bittle, J. Hogle, and D. Baltimore.** 1985. Synthetic peptides from four separate regions of the poliovirus type 1 capsid protein VP1 induce neutralizing antibodies. *Proc. Natl. Acad. Sci. USA* **82:**910–914.

6. **Dewalt, P. G., and B. L. Semler.** 1987. Site-directed mutagenesis of proteinase 3C results in a poliovirus deficient in synthesis of viral RNA polymerase. *J. Virol.* **61:**2162–2170.

7. **Diamond, D. C., B. A. Jameson, J. Bonin, P. I. Kohora, S. Abe, H. Itoh, T. Komatsu, M. Arita, S. Kuge, A. D. M. E. Ostrehaus, R. Crainic, A. Nomoto, and E. Wimmer.** 1985. Antigenic variation and resistance to neutralization in poliovirus type 1. *Science* **229:**1090–1093.

8. **Hogle, J. M., M. Chow, and D. J. Filman.** 1985. The three-dimensional structure of poliovirus at 2.9 Å resolution. *Science* **229:**1358–1365.

9. **Horaud, F., R. Crainic, S. van der Werf, B. Blondel, C. Wychowski, O. Akacem, P. Bruneau, P. Couillin, O. Siffert, and M. Girard.** 1987. Identification and characterization

of a continuous neutralisation epitope (C3) present on type 1 poliovirus. *Prog. Med. Virol.* **34**:129–155.

10. **Kean, K. M., C. Wychowski, H. Kopecka, and M. Girard.** 1986. Highly infectious plasmids carrying poliovirus cDNA are capable of replication in transfected simian cells. *J. Virol.* **59**:490–493.

11. **Kohara, M., T. Omata, A. Kameda, B. L. Semler, M. Itoh, E. Wimmer, and A. Nomoto.** 1985. In vitro phenotypic markers of poliovirus recombinant constructed from infectious cDNA clones of the neurovirulent Mahoney strain and the attenuated Sabin 1 strain. *J. Virol.* **53**:786–792.

12. **Kuge, S., and A. Nomoto.** 1987. Construction of viable deletion and insertion mutants of the Sabin strain of type 1 poliovirus: function of the 5′ noncoding sequence in viral replication. *J. Virol.* **61**:1478–1487.

13. **Kuhn, R. J., E. Wimmer, and B. L. Semler.** 1987. Expression of the poliovirus genome from infectious cDNA is dependent upon arrangements of eukaryotic and prokaryotic sequences in recombinant plasmids. *Virology* **157**:560–564.

14. **La Monica, N., W. J. Kupsky, and V. R. Racaniello.** 1987. Reduced mouse neurovirulence of poliovirus type 2 Lansing antigenic variants selected with monoclonal antibodies. *Virology* **161**:429–437.

15. **La Monica, N., C. Meriam, and V. R. Racaniello.** 1986. Mapping of sequences required for mouse neurovirulence of poliovirus type 2 Lansing. *J. Virol.* **57**:515–525.

16. **La Monica, N., and V. R. Racaniello.** 1986. Reduced mouse neurovirulence of poliovirus antigenic variants selected with monoclonal antibodies, p. 140. *In* F. Brown et al., *Vaccines 86: New Approaches to Immunization—Developing Vaccines against Parasitic, Bacterial and Viral Diseases.* Cold Spring Harbor Laboratory, Cold Spring Harbor, N.Y.

17. **Luo, M., G. Vriend, G. Kamer, E. Arnold, M. G. Rossman, U. Boege, D. G. Scraba, G. M. Duke, and A. C. Palmenberg.** 1987. The atomic structure of mengo virus at 3.0 Å resolution. *Science* **235**:182–191.

18. **Minor, P. D., G. C. Schild, J. Bootman, D. M. A. Evans, M. Ferguson, P. Reeve, M. Spitz, G. Stanway, A. J. Caan, R. Hauptmann, L. D. Clarke, R. C. Mountford, and J. W. Almond.** 1983. Location and primary structure of a major antigenic site for poliovirus neutralization. *Nature* (London) **301**:764–769.

19. **Morinaga, Y., T. Franceschini, B. Inouye, and M. Inouye.** 1984. Improvement of oligonucleotide-directed site-specific mutagenesis using double stranded plasmid DNA. *Bio/Technology* **2**:636–639.

20. **Omata, T., M. Kohara, S. Kuge, T. Komatsu, S. Abe, B. L. Semler, A. Kameda, H. Itoh, M. Arita, E. Wimmer, and A. Nomoto.** 1986. Genetic analysis of the attenuation phenotype of poliovirus type 1. *J. Virol.* **58**:348–358.

21. **Racaniello, V., and D. Baltimore.** 1981. Cloned poliovirus complementary DNA is infectious in mammalian cells. *Science* **214**:916–919.

22. **Reed, L. J., and M. Muench.** 1938. A simple method of estimating fifty percent endpoints. *Am. J. Hyg.* **27**:493–497.

23. **Rossman, M. G., E. Arnold, J. W. Crickson, E. A. Frankenberger, J. P. Griffith, H. J. Hecht, J. E. Johnson, G. Kamer, M. Luo, A. G. Mosser, R. R. Rueckert, B. Sherry, and G. Vriend.** 1985. Structure of a human common cold virus and functional relationship to other picornaviruses. *Nature* (London) **317**:145–153.

24. **Sanger, F., S. Nicklen, and A. Coulson.** 1977. DNA sequencing with chain-terminating inhibitors. *Proc. Natl. Acad. Sci. USA* **74**:5463–5467.

25. **Sarnow, P., H. D. Bernstein, and D. Baltimore.** 1986. A poliovirus temperature-sensitive

RNA synthesis mutant located in a non coding region of the genome. *Proc. Natl. Acad. Sci. USA* **83**:571–575.

26. **Semler, B. L., A. J. Dorner, and E. Wimmer.** 1984. Production of infectious poliovirus from cloned cDNA is dramatically increased by SV40 transcription and replication signals. *Nucleic Acids Res.* **12**:5123–5141.

27. **van der Werf, S., C. Wychowski, P. Bruneau, B. Blondel, R. Crainic, F. Horodniceanu, and M. Girard.** 1983. Localization of a poliovirus type 1 neutralization epitope in viral capsid polypeptide VP1. *Proc. Natl. Acad. Sci. USA* **80**:5080–5084.

28. **Wimmer, E., E. A. Emini, and B. A. Jameson.** 1984. Peptide priming of a poliovirus neutralization antibody response. *Rev. Infect. Dis.* **6**:505–509.

29. **Wychowski, C., S. van der Werf, O. Siffert, R. Crainic, P. Bruneau, and M. Girard.** 1983. A poliovirus type 1 neutralization epitope is located within amino acid residues 93 to 104 of viral capsid polypeptide VP1. *EMBO J.* **2**:2019–2024.

30. **Ypma-Wong, M. F., and B. L. Semler.** 1987. *In vitro* molecular genetics as a tool for determining the differential cleavage specificities of the poliovirus 3C proteinase. *Nucleic Acids Res.* **15**:2069–2088.

31. **Zagursky, R. J., K. Baumeister, N. Lomax, and M. L. Berman.** 1985. Rapid and easy sequencing of large linear double-stranded DNA and supercoiled plasmid DNA. *Gene Anal. Tech.* **2**:89–94.

Molecular Aspects of Picornavirus Infection and Detection
Edited by Bert L. Semler and Ellie Ehrenfeld
© 1989 American Society for Microbiology, Washington, DC 20006

Chapter 16

Genetic Analysis of Neurovirulence, Using a Mouse Model for Poliomyelitis

Vincent R. Racaniello, Nicola La Monica, Eric G. Moss, and Robert O'Neill

INTRODUCTION

Poliovirus is the causative agent of poliomyelitis, an acute disease of the central nervous system (CNS). This disease emerged in epidemic form in the United States at the beginning of the 20th century and increased in frequency and severity for 50 years. It is probably safe to say that public sentiment toward poliomyelitis was similar to that observed today for acquired immune deficiency syndrome. In the 1950s, over 40 years after poliovirus was first isolated, two excellent vaccines were developed which have effectively controlled paralytic poliomyelitis in many countries. Although poliovirus has fallen from public visibility in many countries, research on this virus has continued, making poliovirus one of the best-characterized animal viruses. Its complete chemical (12) and three-dimensional crystal structure (9) is known, and it is possible to construct viral recombinants and mutants by manipulating cloned infectious cDNA (22) and infectious RNA synthesized in vitro (29).

The continuing interest in poliovirus and the technical developments that have resulted from these studies have made it possible to approach previously intractable problems. One interesting, but complex, problem is to provide a molecular description of the pathogenesis of poliomyelitis.

Vincent R. Racaniello, Nicola La Monica, Eric G. Moss, and Robert O'Neill • Department of Microbiology, Columbia University College of Physicians and Surgeons, New York, New York 10032.

An approach to this problem is to identify viral functions that are essential for the production of disease. Such functions may be identified by studying attenuated poliovirus strains, those with reduced or abolished neurovirulence. Our laboratory has undertaken these studies with a mouse model for poliomyelitis, while other groups have used monkeys as experimental hosts. The purpose of this chapter is to summarize studies on the molecular basis of poliovirus attenuation obtained in both animal systems in an attempt to identify viral functions required for the production of paralytic disease.

POLIOVIRUS AND NEUROVIRULENCE

Poliovirus is a member of the *Picornaviridae*, a diverse virus family that contains a number of human pathogens. The polio virion is a naked icosahedron composed of four capsid proteins, VP1, VP2, VP3, and VP4. The virion contains a 7.5-kilobase RNA molecule of positive polarity that is translated in the cell to produce a polyprotein of approximately 250,000 M_r. Two virus-encoded proteinases process this polyprotein to produce functional viral gene products. The viral RNA contains a 5' noncoding region of approximately 743 nucleotides and a 3' noncoding region of approximately 73 nucleotides.

In a poliovirus infection, virus enters the body through the mouth and multiplies in the intestinal mucosa (see reference 6 for a review). Newly synthesized virus makes its way to the bloodstream and is then disseminated to many different tissues. Productive infection in certain tissues, whose identity in the human is unclear, results in a persisting viremia that is required for invasion of the CNS. The destruction of motor neurons by viral replication results in the characteristic flaccid paralysis associated with poliomyelitis.

Neurovirulence of poliovirus refers to the ability to replicate in and destroy cells of the CNS. This property is usually measured experimentally by inoculating virus into an animal host and quantitating the resulting paralysis and the pathological changes in the CNS. Both the route of inoculation and the animal host used must be considered since these factors may influence the outcome of neurovirulence tests (reviewed in). reference 21a).

Are there specific viral functions that enable poliovirus to be neurovirulent? This question has been addressed by studying attenuated viral mutants. The best-characterized attenuated mutants are the live oral polio vaccine strains, isolated by A. B. Sabin by multiple passage of neurovirulent viruses in various primate cells in vivo and in vitro (23, 24). These attenuated viruses are able to multiply well in cultured human and monkey cells and in the human intestine, but do not infect the CNS and do not cause paralytic disease. If we make the assumption that the Sabin vaccine strains contain mutations that reduce their neurovirulence but do not alter their replication in nonneural cell types, then it follows that these mutations must be in regions of the viral genome that are essential for the production of disease.

MUTATIONS THAT ATTENUATE THE SABIN VACCINE STRAINS

Two approaches have been taken to identify attenuating mutations in the genome of the Sabin vaccine strains. The first is to determine the nucleotide sequence of the vaccine strain and compare it with that of the neurovirulent parent. For example, the genomes of the Sabin strain of poliovirus type 1 (P1/Sabin) and its parent, P1/Mahoney, differ by 55 nucleotide substitutions out of a total genome length of 7,441 (19). The mutations, which are scattered throughout the viral genome, lead to 21 amino acid replacements: 7 in VP1, 2 each in VP2 and VP3, 1 in VP4, 3 in proteinase 2A, 2 in 2B, and 4 in RNA polymerase 3D. Five of the changes are located in the 5' noncoding region of the RNA, and two are located in the 3' noncoding region. A similar comparison of the nucleotide sequence of the genomes of P3/Sabin and its neurovirulent parent, P3/Leon, revealed that the two differ by only 10 point mutations (25, 26, 28). Three of the base changes result in amino acid substitutions: one in VP3, one in VP1, and one in proteinase 2A, while two changes are in the 5' noncoding region and two are in the 3' noncoding region. A similar analysis with P2/Sabin is not possible, since its parent, P2/P712, is an avirulent strain.

Although these studies show that a small number of mutations accompany the attenuation process, they do not demonstrate which changes are responsible for the attenuated phenotype. This question has been addressed by studying the neurovirulence of viral recombinants between the Sabin strains and their neurovirulent parents. Agol and his colleagues have constructed such recombinants by coinfecting cells with two different viruses, each containing a selectable genetic marker, and subjecting the viral progeny to the appropriate selection to obtain recombinant viruses (1–3). Neurovirulence was quantitated by intrathalamic inoculation in *Cercopithecus aethiops* monkeys followed by observation for paralysis and a determination of the histological lesion score. The results indicated that the 5' half of the viral genome is a major determinant of attenuation in the P3/Sabin strain. However, the 3' half of the P1/Sabin viral genome must also contain attenuating mutations.

Nomoto and his colleagues have used infectious cDNAs of

P1/Mahoney and P1/Sabin to construct a series of recombinant viruses that were used to determine the location in the viral RNA of mutations responsible for the attenuated phenotype (13, 18, 20). Neurovirulence was determined by intrathalamic inoculation into cynomolgus monkeys (*Macaca irius*) and the quantitation of paralysis, the lesion score, and the spread value. The last two values are calculated according to a standard procedure, after histopathological examination of spinal cord and brain tissue sections, and provide a gauge of the severity of lesion formation by the virus (31). The results can be summarized as follows. (i) Attenuating mutations are scattered throughout the P1/Sabin viral genome. (ii) A strong attenuating mutation is located in the 5' 1,122 nucleotides of the P1/Sabin genome. The strength of this attenuating mutation is approximately equal to the sum of all of the other attenuating mutations in the genome. (iii) The strong attenuating mutation within the 5' 1,122 bases of P1/Sabin is located at nucleotide 480 (18; A. Nomoto, personal communication). This nucleotide is an A in P1/Mahoney and a G in P1/Sabin viral RNA.

A similar approach has been used to identify attenuating mutations in the genome of the P3/Sabin vaccine strain (30). Recombinants between this strain and its neurovirulent parent, P3/Leon/37, were constructed, by manipulation of infectious cDNA, and inoculated intraspinally into cynomolgus monkeys. Neurovirulence was quantitated by determining the number of paralyzed monkeys and the lesion score. The results indicate the following. (i) A Lys-to-Arg mutation in VP1, a Thr-to-Ala mutation in 2A, four silent mutations in the coding region, and a mutation in the 3' noncoding region do not appear to play a role in the attenuation phenotype, since the introduction of these changes into a P3/Leon genome did not reduce the neurovirulence of this virus. (ii) The mutation in VP3 from Ser (P3/Leon) to Phe (P3/Sabin) is strongly attenuating. (iii) A base change from C (P3/Leon) to U (P3/Sabin) at position 472 is strongly attenuating. This nucleotide has been observed to change from U to C during replication of P3/Sabin in the gut of vaccinated infants and is accompanied by increased neurovirulence (8). In addition, a C has been found at position 472 of 14 viruses isolated from vaccine-associated cases of poliomyelitis (7). Therefore, two point mutations account for the attenuated phenotype of P3/Sabin.

MOUSE-ADAPTED PHENOTYPE OF P2/LANSING

Several poliovirus strains are able to cause paralytic disease in mice after intracerebral or intraspinal inoculation. For the past 5 years our laboratory has studied the infection of mice with P2/Lansing. The goal of these studies is to identify viral functions essential for the production of paralytic disease.

The P2/Lansing strain originally isolated from a fatal case of human poliomyelitis was unable to infect mice (4). However, intracerebral inoculation of a monkey brain preparation of P2/Lansing into a cotton rat resulted in paralytic disease (4). A cotton rat brain preparation made by continued intracerebral passage of the virus induced paralysis in white mice after intracerebral inoculation (5). In this way the P2/Lansing strain was adapted to cause disease in mice. When this strain is inoculated intracerebrally into mice, a fatal paralytic disease that clinically and pathologically resembles human poliomyelitis ensues (10, 11). In contrast to the human disease, the virus is not infectious by the oral route (10).

Adaptation of P2/Lansing to mice presumably occurred by the selection of viral variants able to replicate in the murine CNS. It was therefore of great interest to identify those regions of the P2/Lansing genome that contain changes which enable the virus to replicate in mice. This question could be addressed by constructing recombinants between mouse-adapted and mouse-avirulent P2/Lansing. Unfortunately, it was not possible to locate a strain of P2/Lansing that was unable to infect mice. However, it was known that P1/Mahoney cannot cause disease in mice, even after intracerebral inoculation of large amounts of virus (21). Therefore, it was possible to determine which sequences of P2/Lansing, when exchanged with the homologous sequences of P1/Mahoney, could confer the mouse-adapted phenotype to P1/Mahoney.

To address this question, La Monica et al. constructed viral recombinants between P2/Lansing and P1/Mahoney by manipulation of the cloned infectious cDNAs of both strains (16). The mouse neurovirulence of the viral recombinants was determined by intracerebral inoculation into 18- to 21-day-old Swiss Webster mice. Neurovirulence was quantitated by a determination of the 50% lethal dose (LD_{50}), the amount of virus that caused paralysis or death in 50% of inoculated mice. The results showed that a viral recombinant in which sequences encoding the capsid proteins of P1/Mahoney were exchanged with those of P2/Lansing was neurovirulent in mice. From these results we conclude that the capsid of P1/Mahoney is not compatible with replication in mice and that it cannot bind to receptors in the mouse brain or otherwise enter cells of the mouse CNS.

Which region of the P2/Lansing capsid can confer the mouse-adapted phenotype to P1/Mahoney? This question is of interest because when the functional block to P1/Mahoney replication in the mouse CNS is elucidated, it will be possible to identify amino acids involved with, for example, receptor binding. Since the three-dimensional structure of the

virion is known (9), it would be possible to map such functional regions on
the capsid. The capsid proteins of P1/Mahoney and P2/Lansing differ at
113 out of a total of 880 amino acids. These differences are scattered
throughout the three-dimensional structure and therefore do not provide
a clue to the function that is blocked in P1/Mahoney. An answer to this
question will require the construction of additional recombinants between
P1/Mahoney and P2/Lansing, with smaller fragments of the viral capsid
exchanged. These experiments are in progress.

REDUCED MOUSE NEUROVIRULENCE OF P2/LANSING
ANTIGENIC VARIANTS SELECTED WITH MAbs

Another approach to determining which regions of the P2/Lansing
capsid are important for the mouse-adapted phenotype is to introduce
mutations in this region and determine their effect on neurovirulence. One
way to isolate such viruses is to select for antigenic variants resistant to
neutralization with monoclonal antibodies (MAbs). It is known that such
variants of poliovirus contain amino acid changes in antigenic sites on the
surface of the virion. For a determination of whether such antigenic loops
are important for the mouse-adapted phenotype, antigenic variants of
P2/Lansing were selected and their mouse neurovirulence was studied
(15).

The MAbs used were obtained from mice inoculated with P2/Sabin
(17) and were directed against antigenic site 1, an immunodominant loop
of capsid polypeptide VP1 located on the virion surface and composed of
amino acids 91 to 102. This amino acid sequence is identical in P2/Lansing
and P2/Sabin; indeed, six of the eight MAbs tested neutralized P2/Lansing
(15). These six MAbs were used to screen for MAb-resistant (*mar*)
variants, and 22 such variants were obtained. The LD_{50} of the variants
was determined along with four P2/Lansing viruses that had been subjected to the variant isolation procedure in the absence of MAbs. The
unselected isolates had LD_{50}s ranging from 3.7 to 5.0 $\log_{10}$ PFU, while the
LD_{50} of the *mar* variants ranged from 3.9 to over 7.1. Variants with LD_{50}s
of approximately 6 $\log_{10}$ PFU or higher were considered to have reduced
neurovirulence, since the highest LD_{50} observed for unselected isolates
was 5.0. By these criteria, at least 10 of the 22 *mar* variants were less
neurovirulent than the parent virus.

It was important to determine whether the reduced neurovirulence of
the *mar* variants is due to a general defect in viral growth. To address this
question, the efficiency of plating of the *mar* variants in HeLa cells at 33
and 39.5°C was determined. The results indicated that the reduced
neurovirulence of four variants can be ascribed to their temperature-

Table 1. Amino Acid Sequence at Positions 99 to 101 and Mouse Neurovirulence
of P2/Lansing and *mar* Variants[a]

Virus	Amino acid at position:									LD_{50}
	93	94	95	96	97	98	99	100	101	
Lansing	Asp	Asn	Asp	Ala	Pro	Thr	Lys	Arg	Ala	3.7–5.0
433R7.1							Asn			6.9, >7.3
433R8.1	Gly							Pro		6.8, >7.3
433R16.1	Gly							Pro		6.7, >6.4
433R17.1			Glu	Ser			Glu			6.0, 6.1
433R19.1									Asp	6.5, 6.8
433R26.1							Glu			6.2, 6.4

[a] The amino acid sequence was predicted from the RNA sequence determined by direct sequencing by the dideoxy method, using reverse transcriptase and an oligonucleotide primer (14). The LD_{50} was determined by intracerebral inoculation and is shown as $\log_{10}$ PFU. Two independent determinations of LD_{50} were made. Only values from attenuated, non-temperature-sensitive *mar* variants are shown.

sensitive phenotype. Nucleotide sequence analysis of viral RNA indicated that the temperature-sensitive phenotype of three of these is due to a deletion of amino acid 105 of VP1. An examination of the atomic structure of the P1/Mahoney capsid (9) indicates that amino acid 105 is not part of antigenic site 1, but rather is part of a β sheet that anchors one end of the loop and forms the core of VP1. Therefore, the deletion of amino acid 105 probably leads to an alteration of the structure of the antigenic loop and consequent resistance to neutralization.

Six other *mar* variants were not temperature sensitive and had reduced neurovirulence in mice (Table 1). A determination of the viral RNA sequence of these *mar* variants indicated that amino acid substitutions at positions 100 and 101 of VP1 correlated with a reduction in mouse neurovirulence. Thus, in 433R8.1 and 433R16.1, the substitution of Arg with Pro at amino acid 100 resulted in decreased neurovirulence. However, viruses in which Arg was replaced with Ser or Leu were fully neurovirulent (15). A change from Ala to Asp at amino acid 101 of 433R19.1 also resulted in reduced neurovirulence. It also appears that a change at amino acid 99 to Glu correlates with reduced neurovirulence. Although 433R7.1 contained a change to Asn at this position and had reduced neurovirulence, it is not clear whether this change was responsible for the phenotype, since neurovirulent variant 433R3.1 contained the same change (15). Although in some cases it is possible to correlate amino acid substitutions in antigenic site 1 with a reduction in neurovirulence, proof that specific amino acid changes result in a reduction of neurovirulence will require the introduction of the mutation into P2/Lansing cDNA by site-directed mutagenesis and a determination of the neurovirulence of the resulting viruses.

If antigenic site 1 is important for the mouse-adapted phenotype of P2/Lansing, then a substitution of the antigenic loop 1 of P1/Mahoney with that of P2/Lansing might enable the resulting hybrid virus to replicate in mice. This prediction has recently been confirmed (16a, 17a).

Attenuated variants 433R16.1 and 433R19.1 showed growth rates in HeLa cells similar to that of P2/Lansing, but replicated to low levels in the mouse brain (15). This observation indicates that a reduction in neurovirulence of these variants is due to a host range mutation and not to a general defect in viral growth. Thus, the mutations define a viral function essential for the production of disease in mice. What viral functions might be affected by mutations in antigenic site 1 so as to reduce viral neurovirulence in mice? One possibility is that such changes alter the interaction of the virus with a mouse brain receptor and reduce the efficiency of infection. Indeed, any function mediated by the viral capsid, such as penetration, uncoating, or RNA packaging, might be affected. Alternatively, the mutations might affect the stability of the virion in the mouse brain. Answers to these questions await the identification of a cell culture model that mimics the host range phenotype of the variants.

REDUCTION OF POLIOVIRUS NEUROVIRULENCE IN MICE BY A SINGLE-BASE CHANGE IN THE 5′ NONCODING REGION OF THE GENOME

Studies on the neurovirulence of recombinants between P3/Sabin and P3/Leon have identified a major attenuating mutation at nucleotide 472 of the P3/Sabin genome. This region must therefore contain a function that is essential for the production of disease in humans and monkeys. It was important to determine whether the same mutation could attenuate P2/Lansing in mice.

To address this question, La Monica et al. constructed three viral recombinants in which the 5′ noncoding region of P2/Lansing was replaced with the 5′ noncoding region of either P3/Leon, P3/Sabin, or P3/119, a vaccine-associated isolate (14). The genomes of these three recombinants are predicted to differ only at two positions, 220 and 472, both in the 5′ noncoding region. Table 2 summarizes these sequence differences and the results of neurovirulence tests in mice. Clearly, the P3/Leon-P2/Lansing and the P3/119-P2/Lansing recombinants PRV6.1 and PRV8.4 were as neurovirulent in mice as the P2/Lansing parent. In contrast, PRV7.3, which derived the 5′ noncoding region from P3/Sabin, was markedly attenuated.

When 2×10^6 PFU of P2/Lansing or recombinant PRV6.1 or PRV8.4 was inoculated intracerebrally into mice, all animals succumbed. In

Table 2. Bases at Positions 220 and 472 and Neurovirulence of Recombinant
Poliovirus and Neural Isolates in Mice

| Virus | Base at position: | | LD_{50}[a] |
	220	472	
P2/Lansing	NA[b]	NA	3.7–5.0
PRV6.1	G	C	<2.8
PRV7.3	U	U	>7.3
PRV8.4	U	C	4.0
PRV7BE.1	U	C	4.5
PRV7SC.1	U	C	4.6
PRV6SC.1	G	C	ND[c]

[a] LD_{50} by day 21.
[b] NA, Not applicable; the sequence of the P2/Lansing/37 5' noncoding region does not align
perfectly with those of P3/Leon/37 and P3/Sabin.
[c] ND, Not determined.

contrast, the inoculation of 2×10^6 PFU of recombinant PRV7.3 did not
produce disease in mice, while 3 of 10 mice inoculated with 2×10^7 PFU
showed signs of disease. Virus recovered from the brain and spinal cord
of these diseased mice was as neurovirulent as P2/Lansing, PRV6.1, and
PRV8.4 (Table 2). Sequence analysis of viral RNA indicated that these
neural isolates contained a mutation at nucleotide 472 from U to C; no
change had occurred at position 220. Thus, it appears that disease in mice
induced by high doses of attenuated PRV7.3 is caused by a highly
neurovirulent revertant containing a C at nucleotide 472, which probably
constitutes a subpopulation of the virus inoculum.

To investigate the basis for the attenuated phenotype of PRV7.3, its
capacity to replicate in the CNS of mice was compared with that of
P2/Lansing, PRV6.1, and PRV8.4. It was found that while the neurovirulent viruses replicated to a high titer of 10^6 to 10^7 PFU/g by day 2,
PRV7.3 replicated poorly, if at all: the titer of PRV7.3 remained at the
input level (10^4 PFU/g) for several days and rapidly declined to less than
100 PFU/g. In contrast, the replication of all four viruses was indistinguishable in cultured HeLa cells. Therefore, the inability of PRV7.3 to
replicate in mice represents a host restriction and not a general defect in
viral growth.

The mechanism by which a mutation in the 5' noncoding region
results in a virus that cannot replicate in the CNS is unknown. Since the
mutation of the base at position 472 affects replication only in neural
tissues and not in cultured HeLa cells, the altered viral function most
likely involves a host cell component. Perhaps the secondary structure of

the region, which appears to be altered by the mutation at position 472, is important. Possible functions of the 5' noncoding region include the control of viral RNA synthesis, translation, packaging, and virion uncoating. There is some evidence to suggest that the efficiency of translation of viral RNA may be lower in attenuated strains (27). In any case, the ability of the mutation at position 472 to attenuate a mouse-adapted strain of poliovirus suggests that the viral function in question is common to neural cells of different animal species.

MAPPING REGIONS RESPONSIBLE FOR ATTENUATION OF POLIOVIRUS P2/P712 IN MICE

Our results with P2/Lansing type 3 recombinants indicated that a mutation known to attenuate poliovirus in humans and monkeys also attenuates poliovirus in mice. Therefore, the mouse model should be suitable for identifying attenuating mutations in the genome of the P2/Sabin strain. We chose to identify attenuating mutations in the genome of the P2/P712 strain, which is the parent of the P2/Sabin vaccine (24) and is avirulent in humans, monkeys, and mice. An infectious cDNA of P2/P712 was constructed, and different restriction fragments from P2/Lansing were exchanged with the homologous fragment from P2/P712. The resulting recombinant viruses were then tested in a mouse neurovirulence assay.

The recombinants and their neurovirulence are shown in Fig. 1. Apparently, the 5' noncoding region of P2/P712 contains the strongest determinant of attenuation, as shown by the attenuated phenotype of virus RO7, which failed to paralyze any mice at the levels of inoculum used. The replacement of a smaller portion of the 5' noncoding region of P2/Lansing with P2/P712 (bases 457 to 630) produced virus LS1, which had lower neurovirulence than P2/Lansing but was more neurovirulent than RO7. These results suggest that the 5' noncoding region of P2/P712 contains multiple attenuating mutations. The region of P2/P712 encoding part of capsid protein VP3, VP1, 2A, 2B, and part of 2C also appears to contain an attenuating mutation, since recombinant SRL was less neurovirulent than P2/Lansing. The region of P2/P712 encoding VP4, VP2, and the N terminus of VP3, as well as the region from within the 2C coding region to the end of the P2/P712 genome, does not contain sequences capable of conferring to P2/Lansing the attenuated phenotype. These studies show that, like the P2/Sabin and P3/Sabin vaccine strains, P2/Sabin also contains an attenuating mutation in the 5' noncoding region and, in addition, contains other attenuating mutations in the central portion of the genome.

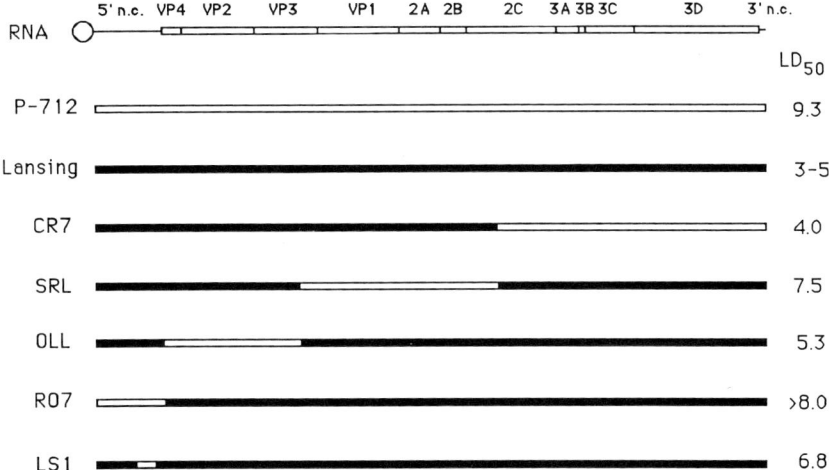

Figure 1. Maps of recombinants between P2/Lansing and P2/P712 and their neurovirulence in mice. The viral RNA is shown at the top, with noncoding (n.c.) and coding regions indicated. Symbols: □, P2/P712 sequences; ■, P2/Lansing sequences. The name of the virus is shown at the left; neurovirulence is shown at the right as the $\log_{10}$ PFU per LD_{50}.

As discussed previously, the P2/P712 strain is a naturally attenuated strain that is the progenitor of P2/Sabin. Thus, it is not possible to compare its sequence with that of P2/Sabin to identify specific base changes that participate in the attenuation process. One approach to this problem would be to identify bases that have changed in type 2 vaccine-associated isolates. Another approach would be to inoculate high levels of recombinant RO7 (Fig. 1) into mice; if paralysis results, an analysis of neural isolates from diseased mice might identify nucleotides which have changed and which contribute to the attenuation of P2/P712.

ATTENUATED AND NEUROVIRULENT POLIOVIRUS RECOMBINANTS DIFFER IN THEIR CAPACITY TO REPLICATE IN HUMAN NEUROBLASTOMA CELL LINE SKNSH

Our studies with mice, as well as those performed by others with monkeys, indicate that a function essential for the production of disease in the animal is present in the 5' noncoding region of the poliovirus genome. Since disruption of this function appears to be the basis of attenuation of all three vaccine strains, it is important to determine which viral function is involved. The possible functions for the poliovirus 5' noncoding region have been discussed above. A determination of which

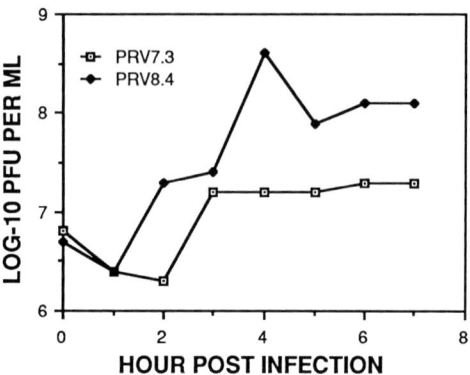

Figure 2. Replication of viral recombinants PRV7.3 and PRV8.4 in SKNSH cells. Cells were infected at a multiplicity of infection of 10, suspended in a Spinner bottle, and incubated at 37°C. At the indicated times postinfection, samples were taken and the total virus titer was determined by plaque assay on HeLa cell monolayers.

function is altered by an attenuating mutation requires a cell line which can distinguish between neurovirulent and attenuated viruses.

We began to search for a cell line in which viruses with a C or U at base 472 replicated differently. For these studies recombinant viruses PRV7.3 and PRV8.4 were used (Table 2), which differ only at base 472. Our previous results showed that, in one-step growth experiments, these two recombinants were indistinguishable in HeLa cells. Similar experiments in CHP100L cells, a neuroepithelioma line, and CALU cells, a lung carcinoma, produced the same result. However, the human neuroblastoma cell line SKNSH was able to distinguish between PRV7.3 and PRV8.4 (Fig. 2). In a one-step growth curve, PRV8.4 replicated with different kinetics and to higher titers than did the attenuated recombinant PRV7.3.

To determine the biochemical basis for the reduced replication of PRV7.3 in SKNSH cells, viral RNA and protein syntheses in infected cells were studied. The results will be reported elsewhere in detail and will be briefly summarized. In SKNSH cells at 1 h postinfection, many viral proteins can be observed in PRV8.4-infected cells and far lower amounts can be observed in cells infected with PRV7.3. By 2 h postinfection the levels of PRV7.3-specific proteins have risen to about 50% of that observed in PRV8.4-infected cells. In HeLa cells infected with either virus, there is no difference in the levels of viral proteins or the kinetics of their production. When viral RNA synthesis was examined by slot-blot analysis, the pattern was generally similar in both cell lines: at 1 h postinfection, input viral RNA was observed; at 2 h postinfection, levels of both viral RNAs increased, although PRV8.4 synthesized approximately twice as much RNA as did PRV7.3. The significance of this observation is not clear, especially since the pattern was the same in SKNSH and HeLa cells.

These results suggest that at early times after infection of SKNSH cells, when largely input levels of viral RNA are present, the genome of PRV8.4 is translated far better than that of PRV7.3. Although translation of PRV7.3 increases later in infection, levels of PRV7.3-specific proteins never reach those observed in PRV8.4-infected SKNSH cells, an observation which might explain the lower yield of the attenuated virus in this cell line. One of the effects of the base 472 mutation might, therefore, be to reduce the translation efficiency of the viral RNA in a way that is specific to cells of neuronal origin. This concept is consistent with the observation that in vitro translation of attenuated poliovirus strains is less efficient than translation of neurovirulent strains (27). Additional experiments are planned to determine whether other aspects of PRV7.3 viral replication are altered in SKNSH cells.

SUMMARY

The goal of our studies on poliovirus infection of mice is to identify viral functions that are essential for the production of disease. We have shown that such functions are present in both the capsid region and the 5' noncoding region of the poliovirus genome. For example, P1/Mahoney cannot infect mice because it does not carry the appropriate capsid; when provided with the P2/Lansing capsid, it can now infect mice. Thus, the P2/Lansing capsid is able to provide a function, perhaps binding to a receptor in the mouse CNS, which is not present in P1/Mahoney. An interesting problem will be to determine the nature of this function. The importance of the P2/Lansing capsid in infecting mice is underscored by the ability of mutations in the capsid to reduce neurovirulence. *mar* variants of reduced neurovirulence in mice were isolated, and the location of mutations in these variants defines a function essential for infecting mice, since these variants replicate quite well in cultured HeLa cells. Another challenge will be to identify the function that is disrupted in these variants.

Mutations in the 5' noncoding region known to attenuate poliovirus in humans and monkeys also attenuate P2/Lansing in mice. Thus, it appears that the function affected by these mutations is common to neural cells of several animal species. One target of the attenuating mutation at base 472 may be the process of translating viral RNA, although the precise mechanism for the effect of the mutation remains to be deciphered.

Our studies with a mouse model for poliomyelitis illustrate the value of studying the molecular biology of pathogenesis and replication at the same time. Studies in mice have clarified the effect of a mutation in the 5' upstream region and suggest that this region has a function essential for replication in animals. To identify this function and how it is altered by the mutation, we have begun to study the effects of the mutation on viral replication in cultured cells. The results of these studies should be an elucidation of the functional basis of viral attenuation and a better understanding of the normal processes of viral replication. Our long-range goal is to use a similar approach to identify and elucidate additional functions required for the production of disease in mice.

Literature Cited

1. **Agol, V. I., S. G. Drozdov, M. P. Frolova, V. P. Grachev, M. S. Kolesnikova, V. G. Kozlov, N. M. Ralph, L. I. Romanova, E. A. Tolskaya, and E. G. Viktorova.** 1985. Neurovirulence of the intertypic poliovirus recombinant v3/a1-25: characterization of strains isolated from the spinal cord of diseased monkeys and evaluation of the contribution of the 3'-half of the genome. *J. Gen. Virol.* **65:**309–316.

2. **Agol, V. I., S. G. Drozdov, V. P. Grachev, M. S. Kolesnikova, V. G. Kozlov, N. M. Ralph, L. I. Romanova, E. A. Tolskaya, A. V. Tyufanov, and E. G. Viktorova.** 1985. Recombinants between attenuated and virulent strains of poliovirus type 1: derivation and characterization of recombinants with centrally located crossover points. *Virology* **143:**467–477.

3. **Agol, V. I., V. P. Grachev, S. G. Drozdov, M. S. Kolesnikova, V. G. Kozlov, N. M. Ralph, L. I. Romanova, E. E. Tolskaya, A. V. Tyufanov, and E. G. Viktorova.** 1984. Construction and properties of intertypic poliovirus recombinants: first approximation mapping of the major determinants of neurovirulence. *Virology* **136:**41–55.

4. **Armstrong, C.** 1939. The experimental transmission of poliomyelitis to the Eastern cotton rat, *Sigmodon hispidus hispidus. Public Health Rep.* **54:**1719–1721.

5. **Armstrong, C.** 1939. Successful transfer of the Lansing strain of poliomyelitis virus from the cotton rat to the white mouse. *Public Health Rep.* **54:**2302–2305.

6. **Bodian, D., and D. H. Horstmann.** 1965. Polioviruses, p. 430–473. *In* F. L. Horsfall and I. Tamm (ed.), *Viral and Rickettsial Infections of Man.* J. B. Lippincott Co., Philadelphia.

7. **Cann, A. J., G. Stanway, P. J. Hughes, P. D. Minor, D. M. A. Evans, G. C. Schild, and J. W. Almond.** 1984. Reversion to neurovirulence of the live-attenuated Sabin type 3 oral poliovirus vaccine. *Nucleic Acids Res.* **12:**7787–7792.

8. **Evans, D. M. A., G. Dunn, P. D. Minor, G. C. Schild, A. J. Cann, G. Stanway, J. W. Almond, K. Currey, and J. V. Maizel.** 1985. Increased neurovirulence associated with a single nucleotide change in a noncoding region of the Sabin type 3 poliovaccine genome. *Nature* (London) **314:**548–550.

9. **Hogle, J. M., M. Chow, and D. J. Filman.** 1985. Three-dimensional structure of poliovirus at 2.9 Å resolution. *Science* **229:**1358–1365.

10. **Jubelt, B., B. Gallez-Hawkins, O. Narayan, and R. T. Johnson.** 1980. Pathogenesis of human poliovirus infection in mice. I. Clinical and pathological studies. *J. Neuropathol. Exp. Neurol.* **39:**138–148.

11. **Jubelt, B., O. Narayan, and R. T. Johnson.** 1980. Pathogenesis of human poliovirus infection in mice. II. Age-dependency of paralysis. *J. Neuropathol. Exp. Neurol.* **39:**149–158.

12. **Kitamura, N., B. L. Semler, P. G. Rothberg, G. R. Larsen, C. J. Adler, A. J. Dorner, E. A. Emini, R. Hanecak, J. J. Lee, S. van der Werf, C. W. Anderson, and E. Wimmer.**

1981. Primary structure, gene organization and polypeptide expression of poliovirus RNA. *Nature* (London) **291**:547–553.

13. **Kohara, M., T. Omata, A. Kameda, B. L. Semler, H. Itoh, E. Wimmer, and A. Nomoto.** 1985. In vitro phenotypic markers of a poliovirus recombinant constructed from infectious cDNA clones of the neurovirulent Mahoney strain and the attenuated Sabin 1 strain. *J. Virol.* **53**:786–792.

14. **La Monica, N., J. W. Almond, and V. R. Racaniello.** 1987. A mouse model for poliovirus neurovirulence identifies mutations that attenuate the virus for humans. *J. Virol.* **61**:2917–2920.

15. **La Monica, N., W. Kupsky, and V. R. Racaniello.** 1987. Reduced mouse neurovirulence of poliovirus type 2 Lansing antigenic variants selected with monoclonal antibodies. *Virology* **161**:429–437.

16. **La Monica, N., C. Meriam, and V. R. Racaniello.** 1986. Mapping of sequences required for mouse neurovirulence of poliovirus type 2 Lansing. *J. Virol.* **57**:515–525.

16a. **Martin, A., C. Wychowski, T. Couderc, R. Crainic, J. Hogle, and M. Girard.** 1988. Engineering a poliovirus type 2 antigenic site on a type 1 capsid results in a chimaeric virus which is neurovirulent for mice. *EMBO J.* **7**:2839–2847.

17. **Minor, P. D., M. Ferguson, D. M. A. Evans, J. W. Almond, and J. P. Icenogle.** 1986. Antigenic structure of polioviruses of serotypes 1, 2 and 3. *J. Gen. Virol.* **67**:1283–1291.

17a. **Murray, M. G., J. Bradley, X.-F. Yang, E. Wimmer, E. G. Moss, and V. R. Racaniello.** 1988. Poliovirus host range is determined by the eight amino acid sequence in neutralization antigenic site 1. *Science* **241**:213–215.

18. **Nomoto, A., M. Kohara, S. Kuge, N. Kawamura, M. Arita, T. Komatsu, S. Abe, B. L. Semler, E. Wimmer, and H. Itoh.** 1987. Study on virulence of poliovirus type 1 using in vitro modified viruses. *UCLA Symp. Mol. Cell. Biol.* **54**:437–452.

19. **Nomoto, A., T. Omata, H. Toyoda, S. Kuge, H. Horie, Y. Kataoka, U. Genba, and N. Imura.** 1982. Complete nucleotide sequence of the attenuated poliovirus Sabin 1 strain genome. *Proc. Natl. Acad. Sci. USA* **79**:5793–5797.

20. **Omata, T., M. Kohara, S. Kuge, T. Komatsu, S. Abe, B. Semler, A. Kameda, H. Itoh, M. Arita, E. Wimmer, and A. Nomoto.** 1986. Genetic analysis of the attenuation phenotype of poliovirus type 1. *J. Virol.* **58**:348–358.

21. **Racaniello, V. R.** 1984. Poliovirus type II produced from cloned cDNA is infectious in mice. *Virus Res.* **1**:669–675.

21a. **Racaniello, V. R.** 1988. Poliovirus neurovirulence. *Adv. Virus Res.* **34**:217–246.

22. **Racaniello, V. R., and D. Baltimore.** 1981. Cloned poliovirus complementary DNA is infectious in mammalian cells. *Science* **214**:916–919.

23. **Sabin, A. B.** 1985. Oral poliovirus vaccine: history of its development and use and current challenge to eliminate poliomyelitis from the world. *J. Infect. Dis.* **151**:420–436.

24. **Sabin, A. B., and L. R. Boulger.** 1973. History of Sabin attenuated poliovirus oral live vaccine strains. *J. Biol. Stand.* **1**:115–118.

25. **Stanway, G., A. J. Cann, R. Hauptmann, P. Hughes, R. C. Mountford, P. D. Minor, G. C. Schild, and J. W. Almond.** 1983. The nucleotide sequence of poliovirus type 3 Leon 12 a1b: comparison with poliovirus type 1. *Nucleic Acids Res.* **11**:5629–5643.

26. **Stanway, G., P. J. Hughes, R. C. Mountford, P. Reeve, P. D. Minor, G. C. Schild, and J. W. Almond.** 1984. Comparison of the complete nucleotide sequences of the genomes of the neurovirulent poliovirus P3/Leon/37 and its attenuated vaccine derivative P3/Leon 2a1b. *Proc. Natl. Acad. Sci. USA* **81**:1539–1543.

27. **Svitkin, Y. V., S. V. Maslova, and V. I. Agol.** 1985. The genomes of attenuated and virulent poliovirus strains differ in their in vitro translation efficiencies. *Virology* **147**:243–252.

28. Toyoda, H., M. Kohara, Y. Kataoka, T. Suganuma, T. Omata, N. Imura, and A. Nomoto. 1984. Complete nucleotide sequences of all three poliovirus serotype genomes. Implication for genetic relationship, gene function and antigenic determinants. *J. Mol. Biol.* **174**:561–585.

29. van der Werf, S., J. Bradley, E. Wimmer, F. W. Studier, and J. J. Dunn. 1986. Synthesis of infectious poliovirus RNA by purified T7 RNA polymerase. *Proc. Natl. Acad. Sci. USA* **83**:2330–2334.

30. Westrop, G. D., D. M. A. Evans, P. D. Minor, D. Magrath, G. C. Schild, and J. W. Almond. 1987. Investigation of the molecular basis of attenuation in the Sabin type 3 vaccine using novel recombinant polioviruses constructed from infectious cDNA. *FEMS Symp.* **32**:53–60.

31. World Health Organization. 1983. Requirements for poliomyelitis vaccine (oral). *Tech. Rep. Ser.* **687**:134.

Molecular Aspects of Picornavirus Infection and Detection
Edited by Bert L. Semler and Ellie Ehrenfeld
© 1989 American Society for Microbiology, Washington, DC 20006

Chapter 17

Expression of the Attenuation Phenotype of Poliovirus Type 1

Akio Nomoto, Noriyuki Kawamura, Michinori Kohara, and Mineo Arita

INTRODUCTION

Poliovirus, known to be the causative agent of poliomyelitis (16), is a human enterovirus that belongs to the *Picornaviridae*. This virus consists of a single-stranded RNA of plus-strand polarity and 60 copies each of four capsid proteins, VP1, VP2, VP3, and VP4, and occurs in three stable serologically distinct types, that is, type 1, type 2, and type 3.

To control severe paralytic disease caused by poliovirus, two kinds of vaccines were developed (16) and are currently being effectively used. One is inactivated virus (21); the other is attenuated virus strains (10, 20), of which those of A. B. Sabin were found to be most effective as oral live vaccines, that is, Sabin 1 (type 1), Sabin 2 (type 2), and Sabin 3 (type 3) strains. Although effective poliovirus vaccines have been available since the mid-1950s, paralytic poliomyelitis remains a serious threat in many countries of the world, especially in developing countries where effective vaccination against poliomyelitis is not sufficiently inclusive.

The virtue of the oral polio vaccines has been widely recognized. The Sabin vaccine strains, however, have an inherent problem of risk of

Akio Nomoto and Noriyuki Kawamura • Department of Microbiology, Tokyo Metropolitan Institute of Medical Science, Honkomagome, Bunkyo-ku, Tokyo 113, Japan.   **Michinori Kohara** • Japan Poliomyelitis Research Institute, Higashimurayama, Tokyo 189, Japan.   **Mineo Arita** • Department of Enteroviruses, National Institute of Health, Musashimurayama, Tokyo 190-12, Japan.

Table 1. Different Biological Characteristics between the Virulent
and Attenuated Strains of Poliovirus

Characteristic	Virulent	Attenuated
Monkey neurovirulence	Strong	Weak
Temperature sensitivity (*rct*)	Weak	Strong
Bicarbonate concn dependency (*d*)	Weak	Strong
Size of plaque	Large	Small

reversion from the attenuated to the neurovirulent phenotype upon repeated passages. Indeed, a very small number of cases of paralytic poliomyelitis continue to occur even in countries with extensive oral polio vaccine programs (10). Most of the cases are considered to be caused by the vaccines themselves, especially type 2 (Sabin 2) and type 3 (Sabin 3) vaccines. The molecular basis of the disease syndrome, however, is not known, although a wealth of data has recently been accumulated on poliovirus structure and function. Indeed, the chemical (6) and crystal (5) structures of poliovirus have already been elucidated.

The nucleotide sequences of the genomes of both the virulent (6, 9, 17, 24) and the attenuated (13, 23, 26) strains of all three poliovirus serotypes have been determined. Partial amino acid sequence analysis of the known viral proteins provided a precise genome organization of poliovirus. This study also revealed a fairly long untranslated region (approximately 750 nucleotides) at the 5′ terminus of the genome (6). The function of the 5′ noncoding sequence is not known. However, this genome region strongly influences the attenuation phenotype of poliovirus as mentioned below.

CHARACTERISTICS OF THE ATTENUATED SABIN 1 STRAIN OF POLIOVIRUS TYPE 1

The attenuated Sabin 1 strain was derived from the virulent Mahoney strain of poliovirus type 1 by multiple passages through host cells of nonhuman origin (19, 20). Although the molecular basis of pathogenesis of poliovirus is not clear at all, a number of biological characteristics of the attenuated poliovirus strains which are different from those of the virulent polioviruses have been detected.

Some of these biological characteristics are used as in vitro marker tests to analyze the properties of oral live vaccines (11) (Table 1). These include the sensitivity of viral proliferation to elevated temperatures (*rct* [reproductive capacity at different temperatures] marker), the sensitivity of viral plaque-forming ability to low concentrations of sodium bicarbonate under an agar overlay (*d* [delayed growth] marker), and the size of plaques produced in infected monolayers of primate cells (Table 1). Although humans are the only natural host of poliovirus, this virus can be transferred to monkeys, in which it also causes paralytic disease (16). Thus, the neurovirulent phenotype can be tested in experimental animals for the display of paralysis and the development of histological lesions spread in the central nervous system after intracerebral (intrathalamus) or intraspinal injection (Table 1).

Since all three Sabin vaccine strains, whose histories of isolation are different from strain to strain, share the characteristics described above, it is possible that these phenotypes of the attenuated Sabin vaccine strains have some correlation with certain steps in viral pathogenesis, a process which includes numerous biological steps.

STRUCTURAL DIFFERENCE IN THE GENOMES BETWEEN THE VIRULENT AND ATTENUATED STRAINS

The difference in the biological properties between the virulent Mahoney and attenuated Sabin 1 strains must be due to different genome structures that resulted from the attenuation process used to create the Sabin 1 virus. Elucidation of the total nucleotide sequence of the genomes of the virulent Mahoney (6, 17) and the attenuated Sabin 1 (13) strains made it possible to identify and map point mutations in the genome. These studies revealed 55 nucleotide substitutions within the 7,441 total nucleotides of the genome, not including poly(A) at the 3' terminus (Fig. 1) (13). These nucleotide changes were found to be scattered over the entire length of the genome and resulted in 21 amino acid replacements within the viral polyprotein (Fig. 1). A cluster of seven amino acid replacements was found in 1D (VP1); 1A (VP4), 1B (VP2), and 1C (VP3) contain one, two, and two different amino acids, respectively. The Tyr-Gly amino acid pair-specific proteinase 2A (27) contains three, 2B contains two, and RNA polymerase 3D contains four mutations. No amino acid replacement in 2C, 3AB, and the Gln-Gly amino acid pair-specific proteinase 3C (4) was observed. The 5'- and 3'-terminal noncoding regions contain five and two point mutations, respectively. Insertions or deletions have not been observed.

GENOME LOCI INFLUENCING THE ATTENUATION PHENOTYPE

Racaniello and Baltimore (18) have shown that a complete, cloned cDNA copy of the genome of the virulent Mahoney strain of poliovirus type 1 is infectious in mammalian cells. A similar clone of high specific

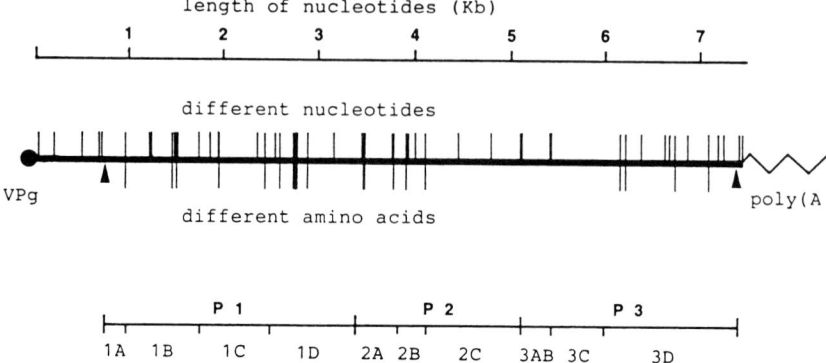

Figure 1. Location of nucleotide and amino acid differences between the Mahoney and Sabin 1 strains. The length of the entire genome of poliovirus type 1 is indicated at the top of the figure in kilobases (Kb) from the 5' terminus. Genome RNA and its gene organization are shown at the bottom of the figure. P1 represents a viral capsid protein region. 1A, 1B, 1C, and 1D are the recently adopted nomenclature of VP4, VP2, VP3, and VP1, respectively. P2 and P3 represent viral noncapsid protein regions. 2A and 3C are the Tyr-Gly amino acid pair-specific proteinase and Gln-Gly amino acid pair-specific proteinase, respectively. 3AB is a precursor protein for VPg which is attached to the 5' end of the genome. 3D is the virus-specific RNA polymerase. The positions of initiation and termination of viral polyprotein synthesis are indicated by closed triangles on the genome. The locations of nucleotide and amino acid differences between the Mahoney and Sabin 1 strains are indicated by lines over and under the genome RNA, respectively.

infectivity was constructed by Semler et al. (22). Omata et al. (15) isolated an infectious cDNA clone of the genome of the attenuated Sabin vaccine strain of poliovirus type 1. Furthermore, a highly infectious cDNA clone of the Sabin 1 strain was reported by Kohara et al. (7).

The availability of the total nucleotide sequences and infectious cDNA clones of both strains of poliovirus provided a molecular genetic approach for investigating the relationship between structure and function of the viral genome by using recombinant DNA technology. By this strategy, we have been constructing a number of recombinant viruses between the virulent Mahoney and the attenuated Sabin 1 viruses (8, 12, 14). The genome structures of recombinant viruses thus constructed are shown in Fig. 2. These recombinant viruses were tested for their biological properties, including monkey neurovirulence and in vitro phenotypic markers.

Monkey neurovirulence tests were performed as previously described (14). A total of 38 sections of the central nervous system were prepared to score the intensity of histological lesions, and the lesion scores were estimated by established procedures (Fig. 2) (1). For intra-

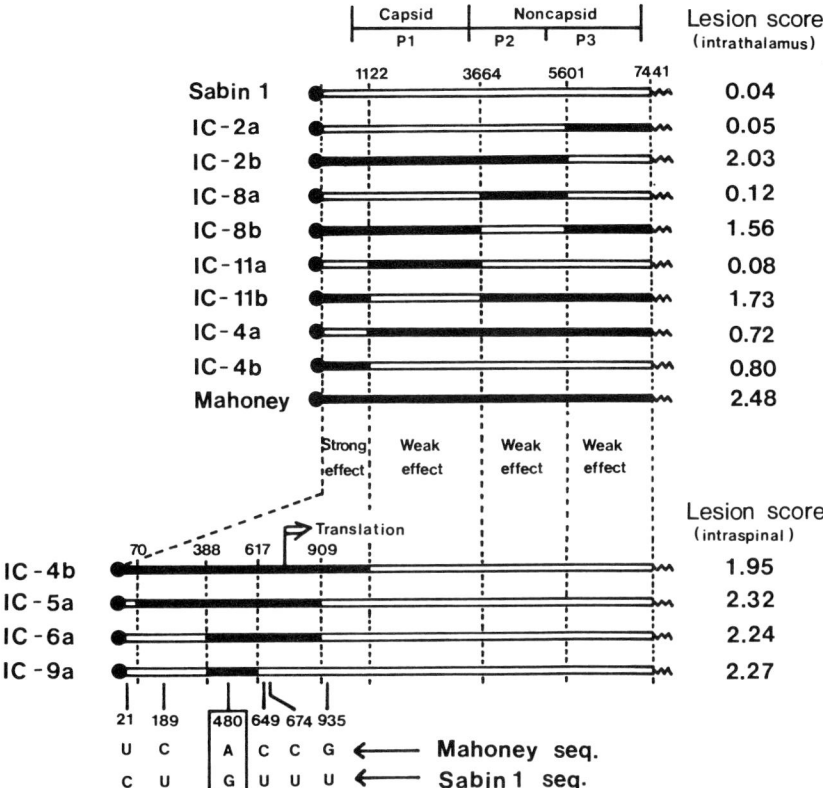

Figure 2. Genome structures of recombinant viruses between the virulent Mahoney and the attenuated Sabin 1 strains and lesion scores obtained from monkey neurovirulence tests on them. The expected genome structures of the recombinant viruses are shown by the combination of Mahoney (▬) and Sabin 1 (▭) sequences. Numbers over the genome are the corresponding restriction cleavage sites: 1122 for *Aat*II, 3664 for *Kpn*I, 5601 for *Bgl*II, 70 for *Kpn*I, 388 for *Nco*I, 617 for *Fok*I, and 909 for *Ban*II. Numbers under the genome are positions of different nucleotides between the genomes of the Mahoney and Sabin 1 strains. The different nucleotides upstream of the corresponding *Aat*II cleavage site are indicated at the bottom of the figure. The extent of influence to the neurovirulent phenotype is indicated by "strong" or "weak" under the corresponding genome region. PV1(SM) is omitted from the name of every recombinant virus. Lesion scores obtained from monkey neurovirulence tests by intrathalamus (intracerebral) or intraspinal injection of the viruses are shown on the right-hand side of the figure.

cerebral (intrathalamus) injection, 10^7 50% tissue culture infective doses of viruses were used, and for intraspinal injection, 10^6 50% tissue culture infective doses of viruses were used. Initial monkey neurovirulence tests were performed by using intracerebral inoculation to identify the genome region influencing neurovirulence or the attenuation phenotype of the virus (upper portion of Fig. 2). The lesion scores obtained are shown in Fig. 2. Many genome regions, and not simply one region, contribute to the phenotype of neurovirulence and attenuation (Fig. 2) (14). None of the recombinant viruses are either entirely neurovirulent or fully attenuated. It is of interest that the exchange of the genome region including nucleotide positions 1123 to 3664, encoding most of the viral capsid precursor protein P1, exerts only a weak influence on the phenotype. Thus, surface structures of the virion particle are not the main determinants of the attenuation phenotype of the Sabin 1 strain (14).

The genome region strongly influencing neurovirulence of poliovirus type 1 was identified to be the 5'-proximal 1,122 nucleotides, most of which represent the 5' untranslated region (14). Therefore, we further constructed recombinant viruses within the 5'-proximal 1,122 nucleotides of the genome (lower portion of Fig. 2). Monkey neurovirulence tests in which intraspinal injections performed on these recombinant viruses were used revealed that the adenine residue at nucleotide position 480 importantly influenced the neurovirulent phenotype expressed by this genome region. Thus, it appears that the point mutation at position 480 in the 5' noncoding region of the Sabin 1 strain significantly influences the attenuation phenotype. The mechanism for this phenomenon is unknown but could be a modulation in one or all of the following steps in viral proliferation: initiation of protein synthesis, RNA replication, and morphogenesis. Indeed, the genomes of the Mahoney and Sabin 1 strains differ in their efficiencies as mRNAs during in vitro translation (25). Moreover, a single-base change in the 5' untranslated region of poliovirus type 3 has been correlated with the attenuation phenotype of this strain (2).

In vitro marker tests were also carried out on many in vitro recombinant viruses. The data (not shown) indicated that determinants for the *rct* marker were spread over several areas of the entire viral genome, like those for neurovirulence (12, 14), and that the genome region between nucleotide positions 1123 and 3664 harbored strong determinants for both *d* and plaque size phenotypes (12, 14). Thus, the *rct* phenotype may have some correlation with the attenuation phenotype, but the *d* marker and plaque size were found to have little correlation with the phenotype. Two thermosensitive steps in the replication of the Sabin 1 virus have so far been identified. One is the step of the assembly of Sabin 1 virus capsid proteins into capsomeres (3); the other is the process of formation of VPg-pU(pU), suggested to play a role in the initiation of viral RNA synthesis (28). The latter experiment has recently been performed with extracts prepared from HeLa cells infected with the Sabin

1 virus, the Mahoney virus, and their recombinant viruses with respect to the genome region encoding virus-specific RNA polymerase 3D. These results suggest that the extent of viral multiplication in the central nervous system of monkeys might be one of the most important indicators of neurovirulence of poliovirus. Moreover, we conclude that the expression of the attenuation phenotype of the Sabin 1 strain of poliovirus is the result of a number of different biological characteristics (14).

ROLE OF NUCLEOTIDE POSITION 480 IN THE EXPRESSION OF THE ATTENUATION PHENOTYPE

As mentioned above, the existence of an adenine residue at nucleotide position 480 appeared to be sufficient to show a neurovirulent phenotype expressed by the genome region of the 5'-proximal 1,122 nucleotides. Therefore, it is possible that only one nucleotide exchange (A↔G) at nucleotide position 480 functions like a switch, determining the expression of these two phenotypes, neurovirulent or attenuated.

To test this possibility, we constructed a new recombinant virus, PV1(SM)IC-10b, whose genome structure is shown in Fig. 3. If the assumption that one nucleotide change at position 480 functions like a switch is correct, virus PV1(SM)IC-10b should show the attenuation phenotype of the Sabin 1 strain, since the genome of the recombinant virus has a guanine residue at position 480 and the sequence downstream of position 1123 is derived from the Sabin 1 virus. Monkey neurovirulence tests by intraspinal injection were performed on virus PV1(SM)IC-10b (Fig. 3). No monkeys died during the course of the test on viruses PV1(SM)IC-10b and Sabin 1 [PV1(Sab)IC-0], whereas two or three of five monkeys died when injected with PV1(SM)IC-4b and PV1(SM)IC-9a, respectively (Fig. 3). Thus, the new recombinant virus clearly showed reduced neurovirulence when compared with virus PV1(SM)IC-9a. Virus PV1(SM)IC-10b, however, did not show the attenuation phenotype of Sabin 1 when the incidence of paralysis and the lesion score were compared between the two viruses (Fig. 3). These parameters clearly indicated that PV1(SM)IC-10b had a neurovirulent phenotype stronger than that of the Sabin 1 virus. These results may suggest that only one nucleotide exchange (A→G) at nucleotide position 480 is not sufficient for the expression of the attenuation phenotype, probably because the functional structure formed in the 5' noncoding region that affects the expression of the attenuation phenotype is not attributed to the primary structure in the vicinity of position 480 but rather to the higher-ordered structure influenced by the nucleotide at position 480. Alternatively, the genome structure of PV1(SM)IC-10b may be unstable during replication

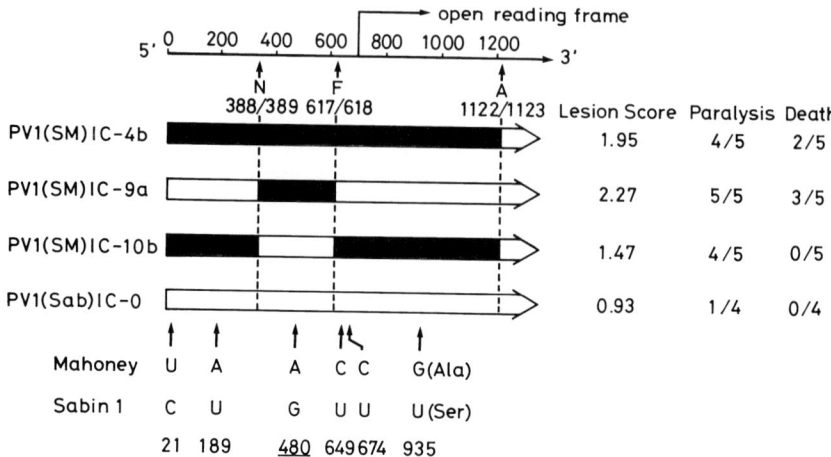

Figure 3. Genome structures of recombinant viruses between the virulent Mahoney and the attenuated Sabin 1 strains and the results of monkey neurovirulence tests on them. The expected genome structures of the recombinant viruses are shown by the combination of Mahoney (▬) and Sabin 1 (▭) sequences. The length of nucleotides is indicated at the top of the figure. N, F, and A represent the corresponding cleavage sites of *Nco*I, *Fok*I, and *Aat*II. Numbers under the restriction sites are nucleotide positions from the 5′ end of the viral genome. Sequences downstream of the *Aat*II cleavage site are derived from the Sabin 1 sequence. PV1(Sab)IC-0 is the virus recovered from primate cells transfected with an infectious cDNA clone of the Sabin 1 strain. Different nucleotides between the genomes of the Mahoney and Sabin 1 strains and their positions are indicated at the bottom of the figure. Lesion scores shown here were obtained by the test with intraspinal injection. The incidence of paralysis or rate of death shown here represents the number of monkeys paralyzed to the number of monkeys injected or the number of monkeys that died to the number of monkeys injected, respectively.

in the central nervous system of monkeys, and variants that have an adenine residue at position 480 in their genome may rapidly occur, resulting in the development of stronger lesions than expected.

GENETIC VARIATION OF POLIOVIRUS IN THE CENTRAL NERVOUS SYSTEM OF MONKEYS

For a determination of the stability of the genotype and phenotype of virus PV1(SM)IC-10b, the virus was injected into cynomolgus monkeys intraspinally. At 7 days after injection, viruses were recovered from lumbar cords of the monkeys. The recovered viruses were plaque purified and 38 isolates were obtained. Nucleotide sequence analysis of the 5′ noncoding region of the genomes and in vitro phenotypic marker tests were performed on the recovered viruses (data not shown).

Nucleotide sequence analysis revealed that the sequences of the 5' noncoding region, including nucleotide position 480, were unchanged in the genome of the recovered viruses tested. Thus, the 5' noncoding sequence of PV1(SM)IC-10b seems to be fairly stable during replication in the central nervous system. However, many variants were detected by in vitro biological tests, including *rct*, *d*, and plaque size marker tests. It is, therefore, possible that some variants derived from virus PV1(SM)IC-10b have relatively strong neurovirulence, resulting in unexpectedly high lesion scores and high incidences of paralysis displayed by monkeys injected with PV1(SM)IC-10b. Monkey neurovirulence tests on the variants derived from PV1(SM)IC-10b are under investigation.

Acknowledgments. This work was supported in part by a grant from the Ministry of Education, Science and Culture of Japan to Akio Nomoto and a grant from the Ministry of Health and Welfare to Mineo Arita.

Literature Cited

1. **Egashira, Y., N. Uchida, and H. Shimojo.** 1967. Evaluation of Sabin live poliovirus vaccine in Japan. V. Neurovirulence of virus strains derived from vaccinees and their contacts. *Jpn. J. Med. Sci. Biol.* **20:**281–302.

2. **Evans, D. M. A., G. Dunn, P. D. Minor, G. C. Schild, A. J. Cann, G. Stanway, J. W. Almond, K. Currey, and J. V. Maizel, Jr.** 1985. Increased neurovirulence associated with a single nucleotide change in a noncoding region of the Sabin type 3 poliovaccine genome. *Nature* (London) **314:**548–550.

3. **Fiszman, M., M. Reynier, D. Bucchini, and M. Girard.** 1972. Thermosensitive block of the Sabin strain of poliovirus type 1. *J. Virol.* **10:**1143–1151.

4. **Hanecak, R., B. L. Semler, C. W. Anderson, and E. Wimmer.** 1982. Proteolytic processing of poliovirus polypeptides: antibodies to polypeptide P3-7c inhibit cleavage at glutamine glycine pairs. *Proc. Natl. Acad. Sci. USA* **79:**3973–3977.

5. **Hogle, J. M., M. Chow, and D. J. Filman.** 1985. Three-dimensional structure of poliovirus at 2.9 angstrom resolution. *Science* **229:**1358–1363.

6. **Kitamura, N., B. L. Semler, P. G. Rothberg, G. R. Larsen, C. J. Adler, A. J. Dorner, E. A. Emini, R. Hanecak, J. J. Lee, S. van der Werf, C. W. Anderson, and E. Wimmer.** 1981. Primary structure, gene organization and polypeptide expression of poliovirus RNA. *Nature* (London) **291:**547–553.

7. **Kohara, M., S. Abe, S. Kuge, B. L. Semler, T. Komatsu, M. Arita, H. Itoh, and A. Nomoto.** 1986. An infectious cDNA clone of the poliovirus Sabin strain could be used as a stable repository and inoculum for the oral polio live vaccine. *Virology* **151:**21–30.

8. **Kohara, M., T. Omata, A. Kameda, B. L. Semler, H. Itoh, E. Wimmer, and A. Nomoto.** 1985. In vitro phenotypic markers of a poliovirus recombinant constructed from infectious cDNA clones of the neurovirulent Mahoney strain and the attenuated Sabin 1 strain. *J. Virol.* **53:**786–792.

9. **LaMonica, N., C. Meriam, and V. R. Racaniello.** 1986. Mapping and sequences required for mouse neurovirulence of poliovirus type 2 Lansing. *J. Virol.* **57:**515–525.

10. **Melnick, J. L.** 1984. Live attenuated oral poliovirus vaccine. *Rev. Infect. Dis.* **6**(Suppl.): 323–327.

11. **Nakano, J. H., M. H. Hatch, M. L. Thieme, and B. Nottay.** 1978. Parameters for

differentiating vaccine-derived and wild poliovirus strains. *Prog. Med. Virol.* **24**:178–206.

12. **Nomoto, A., M. Kohara, S. Kuge, N. Kawamura, M. Arita, T. Komatsu, S. Abe, B. L. Semler, E. Wimmer, and H. Itoh.** 1987. Study on virulence of poliovirus type 1 using in vitro modified viruses. *UCLA Symp. Mol. Cell. Biol. New Ser.* **54**:437–452.

13. **Nomoto, A., T. Omata, H. Toyoda, S. Kuge, H. Horie, Y. Kataoka, Y. Genba, Y. Nakano, and N. Imura.** 1982. Complete nucleotide sequence of the attenuated poliovirus Sabin 1 strain genome. *Proc. Natl. Acad. Sci. USA* **79**:5793–5797.

14. **Omata, T., M. Kohara, S. Kuge, T. Komatsu, S. Abe, B. L. Semler, A. Kameda, H. Itoh, M. Arita, E. Wimmer, and A. Nomoto.** 1986. Genetic analysis of the attenuation phenotype of poliovirus type 1. *J. Virol.* **58**:348–358.

15. **Omata, T., M. Kohara, Y. Sakai, A. Kameda, N. Imura, and A. Nomoto.** 1984. Cloned infectious complementary DNA of the poliovirus Sabin 1 genome: biochemical and biological properties of the recovered virus. *Gene* **32**:1–10.

16. **Paul, J. R.** 1971. A history of poliomyelitis. Yale University Press, New Haven, Conn.

17. **Racaniello, V. R., and D. Baltimore.** 1981. Molecular cloning of poliovirus cDNA and determination of the complete nucleotide sequence of the viral genome. *Proc. Natl. Acad. Sci. USA* **78**:4887–4891.

18. **Racaniello, V. R., and D. Baltimore.** 1981. Cloned poliovirus complementary DNA is infectious in mammalian cells. *Science* **214**:916–919.

19. **Sabin, A. B.** 1957. Properties of attenuated polioviruses and their behavior in human beings. *Spec. Publ. N.Y. Acad. Sci.* **5**:113–127.

20. **Sabin, A. B., and L. R. Boulger.** 1973. History of Sabin attenuated poliovirus oral live vaccine strains. *J. Biol. Stand.* **1**:115–118.

21. **Salk, J. E.** 1960. Persistence of immunity after administration of Formalin-treated poliovirus vaccine. *Lancet* **ii**:715–723.

22. **Semler, B. L., A. J. Dorner, and E. Wimmer.** 1984. Production of infectious poliovirus from cloned cDNA is dramatically increased by SV40 transcription and replication signals. *Nucleic Acids Res.* **12**:5123–5141.

23. **Stanway, G., A. J. Cann, R. Hauptmann, P. Hughes, L. D. Clarke, R. C. Mountford, P. D. Minor, G. C. Schild, and J. W. Almond.** 1983. The nucleotide sequence of poliovirus type 3 Leon 12a₁b: comparison with poliovirus type 1. *Nucleic Acids Res.* **11**:5629–5643.

24. **Stanway, G., P. J. Hughes, R. C. Mountford, P. Reeve, P. D. Minor, G. C. Schild, and J. W. Almond.** 1984. Comparison of the complete nucleotide sequences of the genomes of the neurovirulent poliovirus P3/Leon 37 and its attenuated Sabin vaccine derivative P3/Leon 12a₁b. *Proc. Natl. Acad. Sci. USA* **81**:1539–1543.

25. **Svitkin, Y. V., S. V. Maslova, and V. I. Agol.** 1985. The genomes of attenuated and virulent poliovirus strains differ in their in vitro translation efficiencies. *Virology* **147**:243–252.

26. **Toyoda, H., M. Kohara, Y. Kataoka, T. Suganuma, T. Omata, N. Imura, and A. Nomoto.** 1984. Complete nucleotide sequences of all three poliovirus serotype genomes: implication for genetic relationship, gene function and antigenic determinants. *J. Mol. Biol.* **174**:561–585.

27. **Toyoda, H., M. J. H. Nicklin, M. G. Murray, C. W. Anderson, J. J. Dunn, F. W. Studier, and E. Wimmer.** 1986. A second virus-encoded proteinase involved in proteolytic processing of poliovirus polyprotein. *Cell* **45**:761–770.

28. **Toyoda, H., C.-F. Yang, N. Takeda, A. Nomoto, and E. Wimmer.** 1987. Analysis of RNA synthesis of type 1 poliovirus by using an in vitro molecular genetic approach. *J. Virol.* **61**:2816–2822.

Molecular Aspects of Picornavirus Infection and Detection
Edited by Bert L. Semler and Ellie Ehrenfeld
© 1989 American Society for Microbiology, Washington, DC 20006

Chapter 18

Attenuation and Reversion of the Sabin Type 3 Vaccine Strain

P. D. Minor, G. Dunn, A. John, A. Phillips, G. D. Westrop, K. Wareham, and J. W. Almond

INTRODUCTION

The live attenuated vaccines described by Sabin and Boulger (9) have been used widely and have resulted in the elimination of poliomyelitis as a public health problem in many countries in the developed world. However, a small number of residual cases can be attributed to the use of the vaccine, and there is a continuing requirement to test every batch for safety in primates. The detailed understanding of the molecular biology of poliovirus from both a structural (3) and a genetic (1, 4, 7, 11, 13) point of view makes it feasible to study in molecular detail the basis of the attenuation to gain a fuller understanding of viral pathogenesis and the mechanisms by which vaccines protect from disease.

We have studied the type 3 strain of poliovirus as this presents the most severe problems of testing and has most often been implicated in disease. The strategy initially adopted was to compare the sequences of the genomic RNAs of three closely related viruses of differing neurovirulence, namely, Sabin vaccine strain P3/Leon 12a₁b, its virulent precursor isolated from a fatal case and designated P3/Leon/USA/1937, and a strain isolated from a fatal case of poliomyelitis after vaccination and

P. D. Minor, G. Dunn, A. John, and A. Phillips • National Institute for Biological Standards and Control, Blanche Lane, South Mimm, Potters Bar, Hertfordshire EN6 3QG, United Kingdom. **G. D. Westrop, K. Wareham, and J. W. Almond** • Department of Microbiology, Reading University, Reading RG1 5AQ, United Kingdom.

designated P3/119. Consideration of the three sequences showed that the process of attenuating Leon to produce P3/Leon 12a$_1$b had involved no more than 10 mutations, and in particular the changes most strongly implicated in attenuation were a mutation in the 5' noncoding region of the genome at residue 472, a mutation at the extreme 3' end of the genome immediately before the poly(A) tract, and at most 2 mutations leading to amino acid changes in the structural proteins (1). Isolates from vaccinees confirmed the significance of the mutation in the 5' noncoding region (2).

Complete cDNA copies of the poliovirus genome are infectious (8), raising the possibility of precise genetic manipulation to identify attenuating mutations. A number of recombinant viruses have thus been prepared between Leon and the Sabin vaccine strain and tested for neurovirulence by a modified version of the World Health Organization neurovirulence test (14). We believe that as a result of this procedure we have now identified the mutations responsible for the attenuated phenotype of the type 3 Sabin vaccine strain of poliovirus in the World Health Organization neurovirulence test.

VIRULENCE OF RECOMBINANT VIRUSES

The constructs examined are shown in Table 1. Viruses were recovered from cDNA by transfection of Hep2c cells at 35°C and usually subjected to plaque purification before testing. Experience has shown that in the course of a test of a batch of Sabin type 3 vaccine, about 2% of the animals may be expected to show paralysis and that histological lesion scores of between 0.2 and 0.9 will be recorded, indicating a low degree of invasion and neurological damage. In contrast, virulent viruses rapidly paralyze all animals and give lesion scores of greater than 2.2. The response to viruses of intermediate virulence is largely unknown.

Virus recovered from Leon cDNA, like the parent, was fully virulent, with all animals showing clinical signs and a lesion score of 2.71. Similarly, the virus recovered from Sabin cDNA was vaccinelike, with no animals showing signs and a lesion score of 0.41. Thus, the process of cloning and recovery did not have a significant effect on the virulence of these viruses.

The virulence of viruses consisting of the 5' portion of Sabin and the 3' portion of Leon (S5'/L) was significantly less than that of Leon, giving few animals with clinical signs and a low lesion score overall. In one test the score was in the range expected for the Sabin strain, and in another it was intermediate between that expected for Sabin and that expected for Leon. Similarly, viruses consisting of the VP3 of Sabin (residue 2034) in

Table 1. Virulence of Recombinant Polioviruses[a]

Virus	Residue										Proportion of animals showing clinical signs	Mean lesion score
	220	472	871	2034	3333	3464	4064	6127	7185	7432		
Leon											4/4	2.71
Sabin	s	s	s	s	s	s	s	s	s	s	0/8	0.41
S5'L	s	s	s								0/4	0.43
											1/3	1.14
SV3/L				s							0/4	1.37
											1/4	2.11
SV1/L					s						3/4	2.68
SP2/L						s					4/4	2.51
S3'/L								s	s	s	3/4	2.40
SLR1	s	s	s	s							0/3	0.28
SCC/L				s	s	s					0/4	1.32
											1/4	1.75
											2/4	2.10
											2/4	2.04
											3/4	2.47
SLR2				s	s		s	s	s	s	3/4	2.39
S5'3'/L	s	s	s					s	s	s	0/4	1.58
											0/4	1.51
LV3/S	s	s	s		s	s	s	s	s	s	1/4	1.32
L472/S	s		s	s	s	s	s	s	s	s	0/4	1.58
L472V3/S	s		s		s	s	s	s	s	s	3/4	2.07

[a] Constructs were made from cDNA clones by using convenient restriction sites. Viruses were recovered on Hep2c cells by transfection at 35°C essentially as described by others (7) and tested for neurovirulence according to standard procedures (14). s, Sabin type 3 nucleotide at the position indicated.

a genome otherwise derived from Leon (SV3/L) were of an intermediate character. While few animals showed clinical signs, the lesion scores were significantly higher than for Sabin and in one test approached those expected for a virulent virus. These findings are consistent with the view that mutations in the 5' portion of the genome and in VP3 both contribute to the attenuated phenotype and that neither fully attenuates the virus on its own. Constructs consisting of VP1 (residue 3333) or P2-3b (residue 3464) of Sabin in a genome otherwise derived from Leon were essentially Leon-like in virulence, resulting in clinical signs in most animals and high lesion scores. The recombinant containing the 3' portion of Sabin in a Leon genome (S3'/L) was also virulent. This implies that no mutation to the 3′ side of residue 2034 has a significant attenuating effect and that there are two mutations involved in the attenuation of Leon to give the Sabin vaccine, one in the 5′ noncoding region and the other in VP3. From previous studies (2) the mutation in the 5′ noncoding region is presumably at residue 472.

More complex constructs were examined. The genome of SLR1 was Sabin-like up to and including the mutation at residue 2034 in the region of the genome coding for VP3, while the remainder was derived from Leon. Virus recovered from this construct was highly attenuated and comparable to the Sabin strain itself, with no animals showing clinical signs and a very low lesion score. The reciprocal construct SLR2 was essentially fully virulent, consistent with the view that the only significant attenuating mutations were in the 5′ portion of the genome. If the mutation at residue 2034 is significant, the construct S5′3′/L, which contains the 5′ and 3′ portions of the Sabin genome in an otherwise Leon-like virus, should be only partially attenuated. The data in Table 1 show that this is the case, with no animals showing clinical signs, but an intermediate lesion score in two tests. The construct SCC/L contains Sabin-like sequences at residues 2034, 3333, and 3464 and is predicted to have a virulence similar to that of SV3/L. In fact, there was a wide variation in virulence, ranging from no animals showing signs and an intermediate lesion score to an essentially fully virulent virus. The variation in pathological effect may be explicable in terms of the observation that the mutation at residue 2034 renders growth of the virus sensitive to high temperatures, as outlined below. Neurovirulent mutants may thus have been selected in the recovery or growth of the viruses.

Three further constructs were prepared to test the conclusion that only the mutations at residue 2034 and in the 5′ noncoding region at residue 472 were involved in attenuation. The constructs were LV3/S, which differed from Sabin only at residue 2034, where it was Leon-like; L472/S, which similarly differed from Sabin only at residue 472, where it was Leon-like; and L472V3/S, which was Leon-like at both positions but otherwise identical to Sabin. Both LV3/S and L472/S were of intermediate virulence, as judged by the clinical effect and lesion score, consistent with their having a single attenuating mutation. The remaining virus, L472V3/S, was more virulent, with most animals affected and a lesion score of 2.07. It was thus virulent, but not to the same degree as Leon. This suggests that other mutations may make a small contribution to the attenuated phenotype, which might be difficult to detect in isolation. The major attenuating mutations thus reside in the 5′ noncoding region at residue 472 and in the region coding for VP3 at residue 2034.

ROLE OF THE 3' TERMINUS OF THE GENOME

It was possible that the 3' terminus of the genomic RNA immediately before the poly(A) tract was involved in generating the attenuated phenotype and might account for the slight residual attenuation of L472V3/S. This possibility arose initially from sequence comparisons of Leon, Sabin, and the virulent revertant P3/119, where cDNA from the virulent viruses was found to have lost one or two guanosine residues immediately before the poly(A) tract at base 7432 (1).

The structure of the 3' genomic sequences was examined by use of a phased primer, $5'$-dC · dC(dT)$_{10}$-$3'$. The sequence of the Sabin strain at the 3' terminus is ...GGAGG(A)$_n$, while that of Leon is ...GGAG(A)$_n$ and that of P3/119 previously reported is ...GG(A)$_n$ (1). The phased primer would thus be expected to anneal perfectly with the Sabin strain. Similarly, it should anneal perfectly to P3/119, but at a position which is effectively three bases closer to the 5' end of the genome. It would either fail to anneal to Leon or anneal at a position comparable to that expected for P3/119. An examination of the products formed in a dideoxy sequencing reaction should therefore reveal whether the 3'-terminal sequence is Sabin-like or not. The results for Sabin, Leon, and P3/119 are shown in Fig. 1. It can be seen that while the products of the reaction primed with Leon RNA (track B) were three residues smaller than for the reaction primed with Sabin RNA (track A), the Sabin and P3/119 products were identical (tracks A and C). This indicated that the differences at the 3' end of the genome which were reported previously did not reflect the virus population as a whole. They may have arisen from population heterogeneity or cloning artifacts. Similarly, heterogeneity has been reported for Mahoney (4, 7). As P3/119 is a highly virulent virus (1), the 3'-terminal sequences of the genome are presumably not involved in attenuation.

IN VITRO PROPERTIES OF THE SABIN TYPE 3 VACCINE STRAIN

The Sabin type 3 vaccine strain, like the Sabin type 1 and Sabin type 2 strains, is known to be temperature sensitive in its growth in vitro, in contrast to the parental Leon. The genetic basis of this temperature sensitivity was examined with the recombinant viruses by use of plaque formation at 40 and 35°C on Hep2c cells (Table 2). Leon showed a negligible drop in titer at 40 compared with 35°C, whether it was obtained from cDNA or from a conventional tissue culture pool. Similarly, the Sabin strain recovered from cDNA was indistinguishable from that from a tissue culture pool, showing a drop in titer of greater than 5 log$_{10}$. No plaques were formed at 40°C, although there was occasionally cell degeneration at high multiplicities.

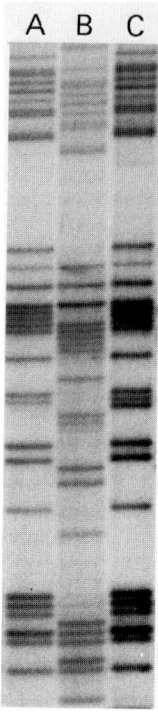

Figure 1. Structure of the 3′ end of the genome of Sabin (A), Leon (B), and P3/119 (C). Virion RNA was incubated in a standard dideoxy sequencing reaction mix with $(dC)_2(dT)_{10}$ as the primer, and the products were resolved on a standard gel system. The T track only is shown.

Of the recombinant viruses, SV1/L and S3′/L showed a negligible drop in titer at 40 compared with 35°C, similar to that observed with Leon, while S5′/L and SP2/L showed a larger drop of 1.4 and 1.7 $\log_{10}$. The largest effect was observed with SV3/L, which suffered a 3.4-$\log_{10}$ drop in titer at 40 compared with 35°C. In addition, it was possible to pick plaques at 40°C and the virus so selected, designated SV3/Lts+, was shown to be no longer temperature sensitive. In contrast, plaques picked from S5′/L and SP2/L at 40°C retained the parental phenotype. The frequency with which non-temperature-sensitive virus could be selected from SV3/L was consistent with the selection of a single point mutation (5).

The more complex constructs supported the view that the residue at base 2034 in the region coding for VP3 was the most important in generating the temperature-sensitive phenotype. Both SLR2 and S5′3′/L gave essentially the same titer at 40 and 35°C, whereas SLR1 and SCC/L showed marked reductions of 4.4 and 3.9 $\log_{10}$, respectively. Moreover, non-temperature-sensitive viruses could readily be selected from these

Table 2. Temperature Sensitivity of Virus Growth of Recombinant Viruses Generated from cDNA clones[a]

Virus	Residue										Log_{10} PFU at 35°C/PFU at 40°C
	220	472	871	2034	3333	3464	4064	6127	7165	7432	
Leon											0.6
rLeon											−0.1
Sabin	s	s	s	s	s	s	s	s	s	s	>6.0
rSabin	s	s	s	s	s	s	s	s	s	s	>5.0
S5'/L	s	s	s								1.4
SV3/L				s							3.4
SV1/L					s						0.5
SP2/L						s					1.7
S3'/L								s	s	s	0.4
SLR1	s	s	s	s							4.4
SLR2					s	s	s	s	s	s	0.5
SCC/L			s	s	s	s					3.9
S5'3'/L	s	s	s					s	s	s	0.6
SV3/Lts+				s							1.0
SLR1ts+	s	s	s	s							0.2
SCC/L1ts+		s	s		s	s					0.3
SCC/L2ts+		s	s	s	s	s					0.5
SCC/L3ts+		s	s	s	s	s					0.2

[a] The temperature sensitivity of virus growth was determined by plaque formation on Hep2c cells at 35 and 40°C under a 1% agar overlay. Cell sheets were incubated for 3 to 4 days. Plaques were picked with Pasteur pipettes before the cell sheets were stained with naphthalene black. s, Sabin type 3 nucleotide at the position indicated.

strains (Table 2), and the frequency was again consistent with the selection of a single point mutation.

The only base difference between Leon and Sabin associated with a significant temperature-sensitive phenotype was thus at residue 2034. However, it is likely that other mutations also contribute, as it proved possible to select temperature-resistant revertants from recombinants having the Sabin-like base at this position, but not from the Sabin strain itself.

Four independent isolates of temperature-resistant revertants were studied at the molecular level. They were isolated from separate plaque-purified populations of SLR1 and SCC/L, and their RNA sequences were determined through the region of the genome coding for the structural proteins. One, SCC/L1ts+, had a back mutation in residue 2034 such that amino acid 91 of VP3 was changed from a phenylalanine, as in the Sabin strain, to a serine, as in Leon. The other three all had mutations affecting residue 132 of VP1, which was a phenylalanine in the Sabin strain and a leucine in the mutant. Mutants SCC/L2ts+ and SCC/L3ts+ had different base changes with the same effect, and SLR1ts+ was shown to have other mutations characteristic of the parental SLR1 virus. The same mutation was thus independently selected three times, implying that it was a frequently occurring event in vitro. It was of interest that the suppressor mutation arose more frequently than the simple back mutation.

ISOLATES FROM VACCINE-ASSOCIATED CASES

Isolates of poliovirus type 3 from vaccine-associated cases have generally been found to be non-temperature sensitive. The genomic RNA of 10 such isolates was sequenced through the region coding for the structural proteins. In addition, two further isolates from vaccine-associated cases which were found to retain the temperature-sensitive phenotype were examined. The results are summarized in Table 3.

Three strains (116, 156, and 161) showed a back mutation at amino acid 91 of VP3. Several strains (115, 131, 156, and 161) possessed mutations in antigenic sites (5), and it can be demonstrated that the mutations found do not affect the temperature-sensitive phenotype when mutants of the Sabin type 3 strain are selected with monoclonal antibodies in vitro. The most common mutation site was in VP1 at residue 54, which was altered in 8 of the 12 viruses examined. However, two of the viruses, 131 and P3/382/UK/62, were still temperature sensitive, so that a mutation at this site appears to be insufficient alone to render the virus non-temperature sensitive. The remaining mutations listed in Table 3 are considered to be the most likely suppressors of the temperature-sensitive phenotype resulting from the presence of a phenylalanine at residue 91 of VP3. It is notable that residue 132 of VP1 was not found to be mutated in isolates from the vaccine-associated cases. A significant number of the mutated residues lie at the interface between adjacent protomers in the virion structure (J. M. Hogle, personal communication), suggesting that the temperature-sensitive phenotype may result from destabilization of the protomer-protomer interactions. The biological consequences of this are currently unclear. Residue 132 of VP1 lies away from the interface, lining a pocket in the beta barrel of VP1. An analogous pocket has been implicated in the uncoating of rhinoviruses (10), leading to the possibility that the temperature-sensitive phenotype of the Sabin strain may involve uncoating as well as virion stability.

Table 3. Differences between the Structural Proteins of the Sabin Type 3 Vaccine Strain and Related Viruses of Wild Type or Temperature-Sensitive Phenotype[a]

Virus strain	Temperature sensitivity	Amino acid difference from Sabin type 3 vaccine strain		
		Antigen site	Neutral	Possible supressor
106	wt			VP3 178Gln-Leu
115	wt	VP2 166Val-Ala		VP3 178Gln-Leu
116	wt		VP1 54Ala-Thr	VP3 91Phe-Ser
119	wt		VP1 54Ala-Val	VP2 200Arg-Lys
				VP2 215Leu-Met
122	wt			VP2 265Val-Ala
				VP3 108Thr-Ala
131	ts	VP3 77Asp-Asn	VP1 54Ala-Val	
132	wt			VP1 34Ala-Val
				VP3 175Thr-Ala
146	wt		VP1 54Ala-Val	VP2 215Leu-Met
156	wt	VP1 286Arg-Lys	VP1 54Ala-Thr	VP3 91Phe-Ser
		VP3 77Asp-Asn		
158	wt		VP1 54Ala-Val	VP2 215Leu-Met
				VP1 263Lys-Thr
161	wt	VP3 77Asp-Asn	VP1 54Ala-Thr	VP3 91Phe-Ser
P3/382/UK/62	ts		VP1 54Ala-Thr	VP1 263Lys-Ile

[a] The genomic RNA of the viruses was sequenced by the dideoxy method (2) through the region coding for the structural proteins. Predicted amino acid changes are shown. The temperature-sensitive phenotype was determined by plaque formation and was either Sabin-like (ts) or Leon-like (wt).

SEQUENCES IN THE 5′ NONCODING REGION AND REPLICATION OF POLIOVIRUSES IN THE HUMAN GUT

The functions of the 5′ noncoding region are not yet clear, but it has been reported that RNA from viruses possessing 5′ sequences characteristic of the Sabin type 3 vaccine strain is less efficiently translated than that from viruses possessing the analogous sequences from Leon (12). We have previously reported that the base at position 472 rapidly reverts to the wild type in the course of excretion of type 3 viruses by vaccinees. The data presented in Table 4 extend these findings to an American child, J, who excreted type 3 virus for at least 25 days after receiving a primary vaccination in the United States. In contrast to the findings reported by others for recipients of American vaccine, there was no difference in the timing of the reversion to the wild-type base in this child compared with four children from the United Kingdom.

The consensus sequence between the three Sabin strains in the region corresponding to bases 471 to 490 in the numbering system of Toyoda et al. (13) is shown in Fig. 2. In this numbering system, which

Table 4. Days Postvaccination When Reversion in 5′ Noncoding Region Was First Detected

Vaccinee	Days postvaccination[a]		
	Type 1 (G→A, base 486)	Type 2 (A→G, base 487)	Type 3 (U→C, base 475)
DM	6	2	2
EM	NR	4	3
KT1	2	4	3
KT2	NR	5	4
KT3	2	6	—
KT4	NR	5	—
Ju[b]	NR	—	—
DM[c]	8	—	—
EM[c]	NR	—	—
J[d]	—	—	3

[a] NR, No reversion; —, no isolate.
[b] Adult recipient, previously immunized in childhood.
[c] Second vaccination of child.
[d] First vaccination of child in the United States.

results in the maximal alignment of the sequences, residue 472 of type 3 corresponds to base 475. It can be seen that type 1 differs from the consensus sequence at three bases (477, 482, and 486), type 2 differs at one base (487), and type 3 differs at two bases (475 and 481). The conservation of these bases in virus excreted by vaccinees was examined and the results are presented in Table 4. In all six individuals where a type 2 isolate was made, the base at residue 487 reverted to the consensus within a week. This residue corresponds to base 481 of the type 2 sequence. In approximately half of the cases examined, the base at residue 486 reverted at a comparable time postvaccination. This residue corresponds to base 480, shown by Omata et al. to be a major determinant of attenuation in the Sabin type 1 strain (6). No other changes in the 5′

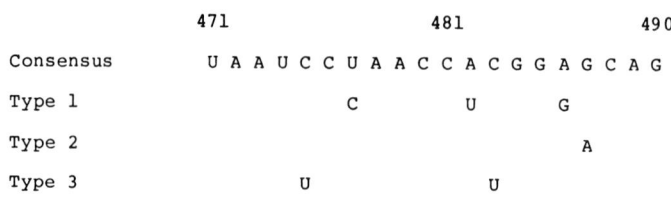

Figure 2. Sequence of type 1, type 2, and type 3 Sabin strains and the consensus sequence from residues 471 to 490 in the numbering system of Toyoda et al. (13).

noncoding region betwen residues 400 and 500 have been consistently detected.

These findings suggested that mutations in the 5' noncoding region might be generally significant in attenuating the neurovirulence of polioviruses. A number of vaccine strains have been obtained by workers other than Sabin, and many have been used in clinical trials, from which isolates were available. If such strains had mutations in the 5' noncoding region, it would be possible to map sequences necessary for replication in the human gut by identifying variant and invariant residues in viruses excreted by vaccinees. Sequences of the type 1 strains Cox and CHAT were determined through a region from bases 451 to 550, and both were shown to be identical to the parental virus Mahoney. The sequence of the type 3 strain USOL-D-BAC was determined through the same region. While there were a number of differences between USOL-D-BAC and the Sabin type 3 strain in this region, there were no detectable differences between the vaccine and virus excreted by vaccine recipients. This implies that the mutations in the 5' noncoding region may be a characteristic of the Sabin strains rather than of polio vaccines in general.

SUMMARY AND CONCLUSIONS

It has been shown that the major part of the attenuation of the Sabin type 3 vaccine strain can be explained by two base changes, one in the 5' noncoding region at residue 472 and the other at base 2034 in the region coding for the structural protein VP3. The change at position 2034 also accounts for the greater part of the temperature-sensitive phenotype of the Sabin 3 strain, possibly by affecting the stability of protomer-protomer interactions. The precise biological consequences of this mutation remain to be determined. All three Sabin strains may alter rapidly on replication in the human gut in sequences in the 5' noncoding region. The basis for this alteration remains to be explored.

Literature Cited

1. Cann, A. J., G. Stanway, P. J. Hughes, P. D. Minor, D. M. A. Evans, G. C. Schild, and J. W. Almond. 1984. Reversion to neurovirulence of the live-attenuated Sabin type 3 oral poliovirus vaccine. *Nucleic Acids Res.* 12:7787–7792.
2. Evans, D. M. A., G. Dunn, P. D. Minor, G. C. Schild, A. J. Cann, G. Stanway, J. W. Almond, K. Currey, and J. V. Maizel. 1985. A single nucleotide change in the 5' noncoding region of the genome of the Sabin type 3 poliovaccine is associated with increased neurovirulence. *Nature* (London) 314:548–550.
3. Hogle, J. M., M. Chow, and D. J. Filman. 1985. The three-dimensional structure of poliovirus at 2.9 A resolution. *Science* 229:1358–1365.
4. Kitamura, N., B. L. Semler, P. G. Rothberg, G. R. Larsen, C. J. Adler, A. J. Dorner, E. A. Emini, R. Hanecak, J. L. Lee, S. Vander Werf, C. W. Anderson, and E. Wimmer.

1981. Primary structure, gene organization and polypeptide expression of poliovirus RNA. *Nature* (London) **291**:547–553.

5. **Minor, P. D., M. Ferguson, D. M. A. Evans, J. W. Almond, and J. P. Icenogle.** 1986. Antigenic structure of polioviruses of serotypes 1, 2 and 3. *J. Gen. Virol.* **67**:1283–1291.

6. **Omata, T., M. Kohara, S. Abe, H. Itoh, T. Komatsu, M. Arita, B. L. Semler, E. Wimmer, S. Kuge, A. Kameda, and A. Nomoto.** 1985. Construction of recombinant viruses between Mahoney and Sabin strains of type 1 poliovirus and their biological characteristics, p. 279–283. *In* R. A. Lerner, R. M. Chanock, and F. Brown (ed.), *Vaccines 85: Molecular and Chemical Basis of Resistance to Parasitic, Bacterial, and Viral Diseases.* Cold Spring Harbor Laboratory, Cold Spring Harbor, N.Y.

7. **Racaniello, V. R., and D. Baltimore.** 1981. Molecular cloning of poliovirus cDNA and determination of the complete nucleotide sequence of the viral genome. *Proc. Natl. Acad. Sci. USA* **78**:4887–4891.

8. **Racaniello, V. R., and D. Baltimore.** 1981. Cloned poliovirus complementary DNA is infectious in mammalian cells. *Science* **214**:916–919.

9. **Sabin, A. B., and L. R. Boulger.** 1973. History of Sabin attenuated poliovirus oral live vaccine strains. *J. Biol. Stand.* **1**:115–118.

10. **Smith, T. J., M. J. Kremer, Ming Luo, G. Vriend, E. Arnold, G. Kamer, M. G. Rossman, M. A. McKinlay, G. D. Diana, and M. J. Otto.** 1986. The site of attachment in human rhinovirus 14 for antiviral agents that inhibit uncoating. *Science* **233**:1286–1293.

11. **Stanway, G., J. P. Hughes, R. C. Mountford, P. Reeve, and P. D. Minor.** 1984. Comparison of the complete nucleotide sequences of the genomes of the neurovirulent poliovirus P3/Leon/37 and its attenuated Sabin vaccine derivative P3/Leon12a₁b. *Proc. Natl. Acad. Sci. USA* **81**:1539–1543.

12. **Svitkin, Y. V., S. V. Maslova, and V. I. Agol.** 1985. The genomes of attenuated and virulent poliovirus strains differ in their *in vitro* translation efficiencies. *Virology* **147**:243–252.

13. **Toyoda, H., M. Kohara, Y. Kastaoka, T. Suganuma, T. Omata, N. Imura, and A. Nomoto.** 1984. Complete nucleotide sequence of all three poliovirus serotype genomes: implications for genetic relationship, gene function and antigenic determinants. *J. Mol. Biol.* **124**:561–585.

14. **World Health Organization.** 1983. Requirements for poliomyelitis vaccine (oral). *WHO Tech. Rep. Ser.* **687**:107–175.

INDEX